Understanding Boundary Value Problems, Weyl Functions and Differential Operators: Volume 3

Understanding Boundary Value Problems, Weyl Functions and Differential Operators: Volume 3

Williams Gerald

MURPHY & MOORE
www.murphy-moorepublishing.com

Published by Murphy & Moore Publishing,
1 Rockefeller Plaza,
New York City, NY 10020, USA

ISBN: 978-1-63987-549-8

Cataloging-in-Publication Data

Understanding boundary value problems, Weyl functions and differential operators. Volume 3 / Williams Gerald.
p. cm.
Includes bibliographical references and index.
ISBN 978-1-63987-549-8
1. Boundary value problems. 2. Boundary value problems--Weyl theory.
3. Differential operators. 4. Differential equations. I. Gerald, Williams.
QA379 .U53 2022
515.35--dc23

For information on all Murphy & Moore Publications
visit our website at www.murphy-moorepublishing.com

Contents

Preface

I am honored to present to you this unique book which encompasses the most up-to-date data in the field. I was extremely pleased to get this opportunity of editing the work of experts from across the globe. I have also written papers in this field and researched the various aspects revolving around the progress of the discipline. I have tried to unify my knowledge along with that of stalwarts from every corner of the world, to produce a text which not only benefits the readers but also facilitates the growth of the field.

Boundary value problems are differential equations with a set of additional constraints, called the boundary conditions. A solution which satisfies the boundary conditions is a solution to the boundary value problem and the differential equation. The Weyl distance function behaves in some ways like the distance function of a metric space. It takes values in a group of reflections, instead of taking values in the positive real numbers. This group of reflections is called the Weyl group. A differential operator is a function of the differentiation operator, also known as the derivative. Differentiation can be considered as an abstract operation that accepts a function and returns another function. This book provides significant information of this discipline to help develop a good understanding of boundary value problems, Weyl functions, differential operators and related fields. It presents this complex subject in the most comprehensible and easy to understand language. Constant effort has been made in this book, using case studies and examples, to make the understanding of these difficult concepts of mathematics as easy and informative as possible to the readers.

Finally, I would like to thank all the contributing authors for their valuable time and contributions. This book would not have been possible without their efforts. I would also like to thank my friends and family for their constant support.

Williams Gerald

7

Canonical Systems

Boundary value problems for regular and singular canonical systems of differential equations are investigated. After a brief introduction to Hilbert spaces of $\mathbb{C}^2$-valued vector functions which are square-integrable with respect to some 2×2 matrix-valued function in Section 7.1, the class of canonical systems to be studied here is introduced in Section 7.2. In Section 7.3 the notions of regular, quasiregular, and singular endpoints for canonical systems are explained. The number of square-integrable solutions at an endpoint of the interval is studied in Section 7.4. Together with a monotonicity principle from Chapter 5, this leads to a limit-circle/limit-point classification of singular endpoints in the same way as in Weyl's alternative in Chapter 6. The important concept of definiteness of canonical systems is defined and studied in Section 7.5, and a cut-off technique for solutions is provided. Afterwards, in Section 7.6, a symmetric minimal relation in the appropriate L^2-Hilbert space and its adjoint, the maximal relation, are associated with real definite canonical systems. The defect numbers of the minimal relation are specified for regular endpoints and for endpoints in the limit-circle or limit-point case. Boundary triplets and Weyl functions for canonical systems in the limit-circle case are constructed in Section 7.7, while the limit-point case is treated in Section 7.8. The connection between subordinate solutions and properties of the Weyl function, as well as the description of absolutely continuous and singular spectrum are studied in Section 7.9. Finally, in Section 7.10 some special classes of canonical systems of differential equations are discussed, among them weighted Sturm–Liouville equations.

7.1   Classes of integrable functions

The purpose of this section is to introduce classes of vector functions which are locally square-integrable with respect to a measurable nonnegative matrix function and to collect some useful properties of such functions.

Let $\imath \subset \mathbb{R}$ be an interval, not necessarily bounded, with endpoints $a < b$, not necessarily belonging to $\imath$. In the following an integral of a vector function or a matrix function over $\imath$ or over a subinterval is always understood in the componentwise sense. The linear space $\mathcal{L}^1_{\mathrm{loc}}(\imath)$ of locally integrable $\mathbb{C}^2$-valued vector functions consists of all measurable $\mathbb{C}^2$-valued vector functions f defined almost everywhere on $\imath$ such that for each compact subinterval $K \subset \imath$

$$\int_K |f(s)|\, ds < \infty.$$

Here $|x|$ denotes the Euclidean norm of x in $\mathbb{C}^2$. Note that for $f \in \mathcal{L}^1_{\mathrm{loc}}(\imath)$ and each compact subinterval $K \subset \imath$ the norm inequality

$$\left| \int_K f(s)\, ds \right| \leq \int_K |f(s)|\, ds \tag{7.1.1}$$

holds. A $\mathbb{C}^2$-valued vector function $f \in \mathcal{L}^1_{\mathrm{loc}}(\imath)$ is said to be integrable at the left endpoint a of the interval $\imath$ or integrable at the right endpoint b of the interval $\imath$ if for some, and hence for all $c \in \mathbb{R}$ with $a < c < b$

$$\int_a^c |f(s)|\, ds < \infty \quad \text{or} \quad \int_c^b |f(s)|\, ds < \infty, \tag{7.1.2}$$

respectively. Similarly, a measurable 2×2 matrix function Φ is locally integrable on $\imath$ if for each compact subinterval $K \subset \imath$

$$\int_K |\Phi(s)|\, ds < \infty;$$

here and in the following $|A|$ stands for the operator norm of a 2×2 matrix A. In particular,

$$\left| \int_K \Phi(s)\, ds \right| \leq \int_K |\Phi(s)|\, ds. \tag{7.1.3}$$

The linear space consisting of all locally integrable 2×2 matrix functions on $\imath$ will also be denoted by $\mathcal{L}^1_{\mathrm{loc}}(\imath)$; it will be clear from the context if the values of the functions in $\mathcal{L}^1_{\mathrm{loc}}(\imath)$ are vectors in $\mathbb{C}^2$ or 2×2 matrices. A 2×2 matrix function $\Phi \in \mathcal{L}^1_{\mathrm{loc}}(\imath)$ is said to be integrable at the left endpoint a of the interval $\imath$ or integrable at the right endpoint b of the interval $\imath$ if (7.1.2) holds for some, and hence for all $c \in \mathbb{R}$ with $a < c < b$ and f replaced by Φ.

Note that all norms on the linear space of 2×2 matrices are equivalent and that the operator norm $|A|$ of a 2×2 matrix A can be estimated by the Hilbert–Schmidt matrix norm as follows:

$$|A| \leq \|A\|_2 \leq \sqrt{2}|A|, \quad \text{where } \|A\|_2 := \sqrt{\sum_{i,j=1}^{2} |a_{ij}|^2}. \qquad (7.1.4)$$

For the product of 2×2 matrices A and B one has

$$|AB| \leq |A|\,|B| \quad \text{and} \quad \|AB\|_2 \leq \|A\|_2\,\|B\|_2. \qquad (7.1.5)$$

In the case where the 2×2 matrix A is nonnegative the trace norm of A can be estimated by the Hilbert–Schmidt matrix norm:

$$\|A\|_2 \leq \operatorname{tr} A \leq \sqrt{2}\|A\|_2, \quad \text{where } \operatorname{tr} A = a_{11} + a_{22},$$

which also gives

$$|A| \leq \operatorname{tr} A \leq 2|A|. \qquad (7.1.6)$$

The following definition introduces the semi-inner product space $\mathcal{L}_\Delta^2(\imath)$ of $\mathbb{C}^2$-valued functions which are square-integrable with respect to Δ; it is assumed that Δ is a 2×2 matrix function on $\imath$ and that $\Delta(s) \geq 0$ for almost every $s \in \imath$. In order to express the seminorm on $\mathcal{L}_\Delta^2(\imath)$ the notation

$$f(s)^* \Delta(s) f(s) = \big(\Delta(s) f(s), f(s)\big)$$

will be useful. Here $(\cdot, \cdot)$ denotes the standard scalar product in $\mathbb{C}^2$ and f is any $\mathbb{C}^2$-function defined on the interval $\imath$.

Definition 7.1.1. Let $\imath \subset \mathbb{R}$ be an interval and let Δ be a measurable 2×2 matrix function such that $\Delta(s) \geq 0$ for almost every $s \in \imath$. Then $\mathcal{L}_\Delta^2(\imath)$ denotes the linear space of all measurable functions f on $\imath$ with values in $\mathbb{C}^2$ which are *square-integrable with respect to* Δ, that is,

$$\|f\|_\Delta^2 = \int_\imath f(s)^* \Delta(s) f(s)\, ds = \int_\imath |\Delta(s)^{\frac{1}{2}} f(s)|^2\, ds < \infty. \qquad (7.1.7)$$

The semidefinite inner product $(\cdot, \cdot)_\Delta$ on $\mathcal{L}_\Delta^2(\imath)$ corresponding to the seminorm $\|\cdot\|_\Delta$ in (7.1.7) is given by

$$(f, g)_\Delta = \int_\imath g(s)^* \Delta(s) f(s)\, ds, \qquad f, g \in \mathcal{L}_\Delta^2(\imath). \qquad (7.1.8)$$

Theorem 7.1.2. *Let $\imath \subset \mathbb{R}$ be an interval and let Δ be a measurable 2×2 matrix function such that $\Delta(s) \geq 0$ for almost every $s \in \imath$. Then the linear space $\mathcal{L}_\Delta^2(\imath)$ equipped with the seminorm (7.1.8) is complete.*

Proof. Since the 2×2 matrix function Δ is measurable and nonnegative almost everywhere, there are measurable nonnegative functions e_1 and e_2, and a measurable 2×2 matrix function U with unitary values, such that

$$\Delta(s) = U(s)^* \Xi(s) U(s), \quad \text{where} \quad \Xi(s) = \begin{pmatrix} e_1(s) & 0 \\ 0 & e_2(s) \end{pmatrix},$$

for almost all $s \in \imath$. Hence, one has for all measurable functions f with values in $\mathbb{C}^2$ on $\imath$ that

$$\int_\imath f(s)^* \Delta(s) f(s) \, ds = \int_\imath (Uf)(s)^* \Xi(s) (Uf)(s) \, ds.$$

Written out in components this gives

$$\begin{aligned}
\int_\imath f(s)^* \Delta(s) f(s) \, ds &= \int_\imath |(Uf)_1(s)|^2 e_1(s) \, ds + \int_\imath |(Uf)_2(s)|^2 e_2(s) \, ds \\
&= \int_\imath |(Uf)_1(s)|^2 \, d\mu_1(s) + \int_\imath |(Uf)_2(s)|^2 \, d\mu_2(s),
\end{aligned} \tag{7.1.9}$$

where the measures μ_1 and μ_2 are absolutely continuous with respect to the Lebesgue measure m and their Radon–Nikodým derivatives are given by e_1 and e_2, respectively. Therefore, it is now clear that

$$f \in \mathcal{L}^2_\Delta(\imath) \quad \Leftrightarrow \quad (Uf)_1 \in \mathcal{L}^2_{d\mu_1}(\imath) \quad \text{and} \quad (Uf)_2 \in \mathcal{L}^2_{d\mu_2}(\imath).$$

This shows that the transformation U maps the space $\mathcal{L}^2_\Delta(\imath)$ bijectively onto $\mathcal{L}^2_{d\mu_1}(\imath) \times \mathcal{L}^2_{d\mu_2}(\imath)$ and from (7.1.9) one sees that the seminorms in $\mathcal{L}^2_\Delta(\imath)$ and $\mathcal{L}^2_{d\mu_1}(\imath) \times \mathcal{L}^2_{d\mu_2}(\imath)$ are preserved. Therefore, the completeness of $\mathcal{L}^2_\Delta(\imath)$ is a consequence of the completeness of $\mathcal{L}^2_{d\mu_1}(\imath)$ and $\mathcal{L}^2_{d\mu_2}(\imath)$. $\square$

The space $\mathcal{L}^2_\Delta(\imath)$ has the following approximation property.

Lemma 7.1.3. *Each element of the seminormed space $\mathcal{L}^2_\Delta(\imath)$ can be approximated by functions in $\mathcal{L}^2_\Delta(\imath)$ which have compact support.*

Proof. Let $(K_n)_{n \in \mathbb{N}}$ be a sequence of nondecreasing compact intervals such that $\imath = \bigcup_{n=1}^\infty K_n$. For $f \in \mathcal{L}^2_\Delta(\imath)$ put $f_n(s) = f(s)$ for $s \in K_n$ and $f_n(s) = 0$ elsewhere. Then $f_n \in \mathcal{L}^2_\Delta(\imath)$, f_n has support in K_n, and

$$\|f - f_n\|^2_\Delta = \int_\imath (f(s) - f_n(s))^* \Delta(s) (f(s) - f_n(s)) \, ds \to 0$$

as $n \to \infty$, by dominated convergence. $\square$

The space $\mathcal{L}^2_{\Delta,\mathrm{loc}}(\imath)$ consists of all $\mathbb{C}^2$-valued functions which are square-integrable with respect to Δ for each compact subinterval $K \subset \imath$, i.e.,

$$\int_K f(s)^* \Delta(s) f(s) \, ds < \infty.$$

A function $f \in \mathcal{L}^2_{\Delta,\mathrm{loc}}(\imath)$ is said to be *square-integrable with respect to* Δ *at the left endpoint* a of the interval $\imath$ or *square-integrable with respect to* Δ *at the right endpoint* b of the interval $\imath$ if for some, and hence for all $c \in \mathbb{R}$ with $a < c < b$,

$$\int_a^c f(s)^* \Delta(s) f(s)\, ds \; < \infty \quad \text{or} \quad \int_c^b f(s)^* \Delta(s) f(s)\, ds \; < \infty,$$

respectively. A function $f \in \mathcal{L}^2_{\Delta,\mathrm{loc}}(\imath)$ belongs to $\mathcal{L}^2_\Delta(\imath)$ if and only if f is square-integrable with respect to Δ at both endpoints a and b of $\imath$.

Clearly, if Δ is a nonnegative matrix function and f is a vector function, then

$$|\Delta(s) f(s)| = |\Delta(s)^{\frac{1}{2}} \Delta(s)^{\frac{1}{2}} f(s)| \leq |\Delta(s)^{\frac{1}{2}}| \, |\Delta(s)^{\frac{1}{2}} f(s)|.$$

Now the statement in the next lemma is a consequence of the Cauchy–Schwarz inequality and the fact that

$$|\Delta(s)^{\frac{1}{2}}|^2 = |\Delta(s)^{\frac{1}{2}} \Delta(s)^{\frac{1}{2}}| = |\Delta(s)|.$$

Lemma 7.1.4. *Let* Δ *be a locally integrable nonnegative* 2×2 *matrix function on* $\imath$ *and let* $K \subset \imath$ *be compact. If* $f \in \mathcal{L}^2_\Delta(K)$, *then* $\Delta f \in \mathcal{L}^1(K)$ *and*

$$\int_K |\Delta(s) f(s)|\, ds \leq \left(\int_K |\Delta(s)|\, ds \right)^{\frac{1}{2}} \left(\int_K f(s)^* \Delta(s) f(s)\, ds \right)^{\frac{1}{2}}.$$

In particular, if $f \in \mathcal{L}^2_\Delta(\imath)$, *then* $\Delta f \in \mathcal{L}^1_{\mathrm{loc}}(\imath)$ *and for all compact* $K \subset \imath$

$$\int_K |\Delta(s) f(s)|\, ds \leq \left(\int_K |\Delta(s)|\, ds \right)^{\frac{1}{2}} \|f\|_\Delta.$$

Let $\mathfrak{N} = \{ f \in \mathcal{L}^2_\Delta(\imath) : \|f\|_\Delta = 0 \}$, so that $\mathfrak{N}$ is a linear space, and consider the quotient space

$$L^2_\Delta(\imath) := \mathcal{L}^2_\Delta(\imath)/\mathfrak{N}$$

equipped with the scalar product induced by (7.1.8), that is, $(f,g)_\Delta = (\widetilde{f}, \widetilde{g})_\Delta$, where $\widetilde{f}, \widetilde{g} \in \mathcal{L}^2_\Delta(\imath)$ are representatives in the equivalence classes $f, g \in L^2_\Delta(\imath)$. From Theorem 7.1.2 it is clear that $L^2_\Delta(\imath)$ is a Hilbert space. When no confusion can arise, the equivalence classes in $L^2_\Delta(\imath)$ will also be referred to as functions that are square-integrable with respect to Δ. Note that the compactly supported functions in $L^2_\Delta(\imath)$ are dense in $L^2_\Delta(\imath)$ by Lemma 7.1.3.

Recall that a $\mathbb{C}^2$-valued vector function f on an open interval $\imath$ is *absolutely continuous* if there exists a $\mathbb{C}^2$-valued vector function $h \in \mathcal{L}^1_{\mathrm{loc}}(\imath)$ such that

$$f(t) - f(s) = \int_s^t h(u)\, du \tag{7.1.10}$$

for all $s, t \in \imath$. In this case, f is differentiable and $f' = h$ almost everywhere. The space of absolutely continuous $\mathbb{C}^2$-valued vector functions is denoted by $AC(\imath)$. When $a \in \mathbb{R}$, then $AC[a, b)$ stands for the subclass of $f \in AC(a, b)$ for which $h \in \mathcal{L}^1_{\text{loc}}(a, b)$ in (7.1.10) additionally belongs to $\mathcal{L}^1(a, a')$ for some, and hence for all $a < a' < b$, in which case

$$f(t) - f(a) = \int_a^t h(u)\, du$$

holds for all $t \in (a, b)$ and thus $f(a) = \lim_{t \to a} f(t)$. When $b \in \mathbb{R}$ there is a similar notation $AC(a, b]$ and for $f \in AC(a, b]$ one has $f(b) = \lim_{t \to b} f(t)$. The notation $AC[a, b]$ is analogous.

7.2 Canonical systems of differential equations

This section offers a brief review of so-called 2×2 canonical systems of differential equations. The existence and uniqueness result for linear systems of differential equations will be discussed and properties of the corresponding fundamental matrices will be derived.

Let $\imath = (a, b) \subset \mathbb{R}$ be an open, not necessarily bounded, interval and let H and Δ be 2×2 matrix functions defined almost everywhere on $\imath$ such that

$$H, \Delta \in \mathcal{L}^1_{\text{loc}}(\imath), \quad H(t) = H(t)^*, \quad \text{and} \quad \Delta(t) \geq 0 \qquad (7.2.1)$$

for almost every $t \in \imath$. Furthermore, let

$$J = \begin{pmatrix} 0 & -1 \\ 1 & 0 \end{pmatrix} \qquad (7.2.2)$$

and note that $J^* = -J = J^{-1}$. A *canonical system* is a system of differential equations of the form

$$Jf'(t) - H(t)f(t) = \lambda \Delta(t)f(t) + \Delta(t)g(t), \quad t \in \imath, \quad \lambda \in \mathbb{C}, \qquad (7.2.3)$$

where $g \in \mathcal{L}^2_{\Delta,\text{loc}}(\imath)$ is a function that is locally square-integrable with respect to Δ with values in $\mathbb{C}^2$. The condition $g \in \mathcal{L}^2_{\Delta,\text{loc}}(\imath)$ implies that Δg is locally integrable; cf. Lemma 7.1.4. In the general case of (7.2.3) one speaks of an *inhomogeneous system*, while if the term involving Δg is absent, that is,

$$Jf'(t) - H(t)f(t) = \lambda \Delta(t)f(t), \quad t \in \imath, \quad \lambda \in \mathbb{C}, \qquad (7.2.4)$$

one speaks of the corresponding *homogeneous system*.

A function f on $\imath$ with values in $\mathbb{C}^2$ is said to be a *solution* of the canonical system (7.2.3) if f belongs to $AC(\imath)$ and the equation (7.2.3) holds for almost every

$t \in \imath$. Observe that if f is a solution of (7.2.3), then f is also a solution of (7.2.3) when $g \in \mathcal{L}^2_{\Delta,\mathrm{loc}}(\imath)$ is replaced by $\widetilde{g} \in \mathcal{L}^2_{\Delta,\mathrm{loc}}(\imath)$ with $\Delta(g - \widetilde{g}) = 0$. Furthermore, if f is a solution of (7.2.3) and h is a solution of (7.2.4), then $f + h$ is a solution of (7.2.3). In fact, the collection of all solutions of the homogeneous system (7.2.4) forms a linear space. The following result on the existence and uniqueness of solutions of initial value problems for inhomogeneous canonical systems will be useful.

Theorem 7.2.1. *Let* $g \in \mathcal{L}^2_{\Delta,\mathrm{loc}}(\imath)$ *and* $\lambda \in \mathbb{C}$. *Fix some* $c_0 \in \imath = (a, b)$ *and* $\gamma \in \mathbb{C}^2$. *Then the initial value problem*

$$Jf'(t) - H(t)f(t) = \lambda\Delta(t)f(t) + \Delta(t)g(t), \qquad f(c_0) = \gamma, \qquad (7.2.5)$$

admits a unique solution $f \in AC(\imath)$. *Moreover, the mapping* $\lambda \mapsto f(t, \lambda)$ *is entire for every fixed* $t \in \imath$.

In order to prove this theorem one replaces the initial value problem (7.2.5) by an equivalent integral equation; recall that the functions H and Δ are locally integrable. The integral equation can be solved, for instance, by successive iterations, which also leads to the statement concerning the mapping $\lambda \mapsto f(t, \lambda)$ being entire, see, e.g., [754, Theorem 2.1].

In the next lemma a *Lagrange identity* for solutions of the inhomogeneous canonical system is obtained.

Lemma 7.2.2. *Assume that* $\lambda, \mu \in \mathbb{C}$ *and that* $g, k \in \mathcal{L}^2_{\Delta,\mathrm{loc}}(\imath)$. *Let* f, h *be solutions of the inhomogeneous equations*

$$Jf'(t) - H(t)f(t) = \lambda\Delta(t)f(t) + \Delta(t)g(t),$$
$$Jh'(t) - H(t)h(t) = \mu\Delta(t)h(t) + \Delta(t)k(t),$$

respectively. Then for every compact interval $[\alpha, \beta] \subset \imath$,

$$h(\beta)^* Jf(\beta) - h(\alpha)^* Jf(\alpha) = \int_\alpha^\beta \left(h(s)^*\Delta(s)g(s) - k(s)^*\Delta(s)f(s) \right) ds$$

$$+ (\lambda - \bar{\mu}) \int_\alpha^\beta h(s)^*\Delta(s)f(s)ds.$$

Proof. The assumptions that J is skew-adjoint and that $H(t)$ and $\Delta(t)$ are self-adjoint almost everywhere on $\imath$ lead to the identities

$$(h^* Jf)' = h^*(Jf') - (Jh')^* f$$
$$= h^*(\lambda\Delta f + \Delta g + Hf) - (\mu\Delta h + \Delta k + Hh)^* f$$
$$= h^*\Delta g - k^*\Delta f + (\lambda - \bar{\mu})h^*\Delta f,$$

which are valid almost everywhere on $\imath$. Integration over the interval $[\alpha, \beta]$ completes the argument. $\square$

Taking $\lambda = \mu = 0$ in Lemma 7.2.2, one obtains the following corollary. It provides the form of the Lagrange identity that will be studied in detail later in this chapter.

Corollary 7.2.3. *Assume that $g, k \in \mathcal{L}^2_{\Delta,\mathrm{loc}}(\imath)$. Let f, h be solutions of the inhomogeneous equations*

$$
\begin{aligned}
Jf'(t) - H(t)f(t) &= \Delta(t)g(t), \\
Jh'(t) - H(t)h(t) &= \Delta(t)k(t),
\end{aligned}
\tag{7.2.6}
$$

respectively. Then for every compact interval $[\alpha, \beta] \subset \imath$,

$$
h(\beta)^* Jf(\beta) - h(\alpha)^* Jf(\alpha) = \int_\alpha^\beta \left(h(s)^* \Delta(s) g(s) - k(s)^* \Delta(s) f(s) \right) ds.
$$

There is also a corollary of Lemma 7.2.2 involving solutions of the corresponding homogeneous system. Let $Y_1(\cdot, \lambda)$ and $Y_2(\cdot, \lambda)$ be solutions of (7.2.4) and define the *solution matrix*

$$
Y(\cdot, \lambda) = \begin{pmatrix} Y_1(\cdot, \lambda) & Y_2(\cdot, \lambda) \end{pmatrix}, \qquad \lambda \in \mathbb{C},
\tag{7.2.7}
$$

which is a 2×2 matrix function for each $\lambda \in \mathbb{C}$. Then the matrix function $Y(\cdot, \lambda)$ solves the equation (7.2.4) in the sense that it actually solves the matrix version of (7.2.4),

$$
JY'(t, \lambda) - H(t)Y(t, \lambda) = \lambda \Delta(t) Y(t, \lambda), \quad t \in \imath.
$$

Corollary 7.2.4. *Let $Y(\cdot, \lambda)$ be a solution matrix of the homogeneous canonical system (7.2.4). Then for every compact interval $[\alpha, \beta] \subset \imath$ and all $\lambda, \mu \in \mathbb{C}$,*

$$
Y(\beta, \mu)^* JY(\beta, \lambda) - Y(\alpha, \mu)^* JY(\alpha, \lambda) = (\lambda - \bar\mu) \int_\alpha^\beta Y(s, \mu)^* \Delta(s) Y(s, \lambda) ds.
$$

In particular, for all $[\alpha, \beta] \subset \imath$ and all $\lambda \in \mathbb{C}$,

$$
Y(\beta, \bar\lambda)^* JY(\beta, \lambda) = Y(\alpha, \bar\lambda)^* JY(\alpha, \lambda).
$$

It is a consequence of Corollary 7.2.4 that for every solution matrix $Y(\cdot, \lambda)$ the function

$$
t \mapsto Y(t, \bar\lambda)^* JY(t, \lambda)
$$

is constant on $\imath$. Hence, if for some $c_0 \in \imath$

$$
Y(c_0, \bar\lambda)^* JY(c_0, \lambda) = J,
\tag{7.2.8}
$$

then $Y(t, \bar\lambda)^* JY(t, \lambda) = J$ for all $t \in \imath$. This shows that

$$
Y(t, \lambda)^{-1} = -JY(t, \bar\lambda)^* J \quad \text{and} \quad Y(t, \bar\lambda)^{-*} = -JY(t, \lambda)J, \quad t \in \imath,
\tag{7.2.9}
$$

and thus it also follows that

$$Y(t, \lambda) J Y(t, \bar{\lambda})^* = J, \quad t \in \imath. \tag{7.2.10}$$

Let X be an invertible matrix which does not depend on λ and assume that $X^* J X = J$. Let $Y(\cdot, \lambda)$ be the solution matrix which is fixed by the initial condition

$$Y(c_0, \lambda) = X \tag{7.2.11}$$

for some $c_0 \in \imath$ and all $\lambda \in \mathbb{C}$. Then (7.2.8) is valid and hence (7.2.9) and (7.2.10) are satisfied; the matrix $Y(\cdot, \lambda)$ is a *fundamental matrix*, that is, its columns are linearly independent solutions of the homogeneous canonical system (7.2.4) on $\imath$. Frequently the fundamental matrix $Y(\cdot, \lambda)$ will be fixed by the initial condition

$$Y(c_0, \lambda) = I \tag{7.2.12}$$

for some $c_0 \in \imath$.

According to Theorem 7.2.1, there is a unique solution of the initial value problem (7.2.5). It is possible to express this unique solution in terms of the fundamental matrix $Y(\cdot, \lambda)$ determined by the initial condition (7.2.12) (and in a similar way with the initial condition (7.2.11)). In fact, for any $\lambda \in \mathbb{C}$, any $\gamma \in \mathbb{C}^2$, and any $g \in L^2_{\Delta, \mathrm{loc}}(\imath)$, the unique solution of the inhomogeneous initial value problem

$$J f' - H f = \lambda \Delta f + \Delta g, \quad f(c_0) = \gamma, \tag{7.2.13}$$

is provided by the *variation of constant formula*:

$$f(t) = Y(t, \lambda) \gamma + Y(t, \lambda) \int_{c_0}^t Y(s, \lambda)^{-1} J^{-1} \Delta(s) g(s) \, ds. \tag{7.2.14}$$

This can be seen by verifying that the second term on the right-hand side is a solution of the inhomogeneous equation that vanishes at c_0. Making use of (7.2.9), one recasts (7.2.14) as

$$f(t) = Y(t, \lambda) \gamma - Y(t, \lambda) \int_{c_0}^t J Y(s, \bar{\lambda})^* \Delta(s) g(s) \, ds. \tag{7.2.15}$$

In terms of the notation (7.2.7) for the fundamental matrix $Y(\cdot, \lambda)$ fixed by (7.2.12) the unique solution (7.2.15) of (7.2.13) can be written as

$$\begin{aligned}
f(t) = {} & Y_1(t, \lambda) \gamma_1 + Y_2(t, \lambda) \gamma_2 \\
& + Y_1(t, \lambda) \int_{c_0}^t Y_2(s, \bar{\lambda})^* \Delta(s) g(s) \, ds \\
& - Y_2(t, \lambda) \int_{c_0}^t Y_1(s, \bar{\lambda})^* \Delta(s) g(s) \, ds,
\end{aligned} \tag{7.2.16}$$

where $\gamma = (\gamma_1, \gamma_2)^\top$. This form of the solution will be used later.

The general form of the inhomogeneous equation (7.2.3) can be simplified by a transformation of the system. This transformation will be employed in Corollary 7.4.8.

Lemma 7.2.5. *Let $\lambda_0 \in \mathbb{R}$, $c_0 \in \imath$, and let $U(\cdot, \lambda_0)$ be a solution matrix which satisfies*

$$JU'(\cdot, \lambda_0) - HU(\cdot, \lambda_0) = \lambda_0 \Delta U(\cdot, \lambda_0), \quad U(c_0, \lambda_0)^* JU(c_0, \lambda_0) = J. \quad (7.2.17)$$

Assume that $g \in \mathcal{L}^2_{\Delta, \mathrm{loc}}(\imath)$ and let f be a solution of the inhomogeneous equation (7.2.3). Define the functions $\widetilde{f}$, $\widetilde{g}$, and $\widetilde{\Delta}$ by

$$\widetilde{f} = U(\cdot, \lambda_0)^{-1} f, \quad \widetilde{g} = U(\cdot, \lambda_0)^{-1} g, \quad \widetilde{\Delta}(\cdot) = U(\cdot, \lambda_0)^* \Delta(\cdot) U(\cdot, \lambda_0). \quad (7.2.18)$$

Then $\widetilde{\Delta}$ is a locally integrable nonnegative measurable matrix function,

$$\widetilde{f}^* \widetilde{\Delta} \widetilde{f} = f^* \Delta f \quad and \quad \widetilde{g}^* \widetilde{\Delta} \widetilde{g} = g^* \Delta g, \quad (7.2.19)$$

and, in particular, $\widetilde{g} \in \mathcal{L}^2_{\widetilde{\Delta}, \mathrm{loc}}(\imath)$. Moreover, the function $\widetilde{f}$ is a solution of the system of differential equations

$$J\widetilde{f}' = (\lambda - \lambda_0) \widetilde{\Delta} \widetilde{f} + \widetilde{\Delta} \widetilde{g}. \quad (7.2.20)$$

Conversely, if

$$\widetilde{f}, \widetilde{g} \in \mathcal{L}^2_{\widetilde{\Delta}, \mathrm{loc}}(\imath) \quad and \quad \widetilde{\Delta}(\cdot) = U(\cdot, \lambda_0)^* \Delta(\cdot) U(\cdot, \lambda_0)$$

satisfy the equation (7.2.20), then $f = U(\cdot, \lambda_0)\widetilde{f}$ and $g = U(\cdot, \lambda_0)\widetilde{g}$ satisfy the inhomogeneous equation (7.2.3).

Proof. First observe that it is a direct consequence of (7.2.17) that the function $U(\cdot, \lambda_0)$ satisfies

$$U(\cdot, \lambda_0)^* JU(\cdot, \lambda_0) = J;$$

cf. (7.2.8). In particular, this shows that $U(t, \lambda_0)$ is invertible for each $t \in \imath$.

Let $\widetilde{f}$, $\widetilde{g}$, and $\widetilde{\Delta}$ be defined by (7.2.18). Then it is clear that (7.2.19) holds. Since $g \in \mathcal{L}^2_{\Delta, \mathrm{loc}}(\imath)$ it also follows that $\widetilde{g} \in \mathcal{L}^2_{\widetilde{\Delta}, \mathrm{loc}}(\imath)$. Moreover,

$$Jf' - Hf = \lambda \Delta f + \Delta g \quad (7.2.21)$$

holds by assumption. Substituting $f = U(\cdot, \lambda_0)\widetilde{f}$ and $g = U(\cdot, \lambda_0)\widetilde{g}$ in (7.2.21), multiplying by $U(\cdot, \lambda_0)^*$ from the left, and using (7.2.17) a straightforward calculation leads to (7.2.20). Similarly, one verifies by a direct calculation that the converse statement holds. $\qquad \square$

It follows from (7.2.19) that the functions $\tilde{f}$ or $\tilde{g}$ in (7.2.18) are square-integrable with respect to $\tilde{\Delta}$ if and only if f or g are square-integrable with respect to Δ, respectively. The transformation in Lemma 7.2.5 implies that the boundary terms $h(x)^* J f(x)$ in the Lagrange formula of the original equation in Lemma 7.2.2 can be written in terms of the boundary terms of the corresponding solutions of the transformed equation (7.2.20).

Corollary 7.2.6. *Assume that* $\lambda, \mu \in \mathbb{C}$, $g, k \in \mathcal{L}^2_{\Delta,\mathrm{loc}}(\imath)$ *and that* f, h *are solutions of the inhomogeneous equations*

$$Jf'(t) - H(t)f(t) = \lambda\Delta(t)f(t) + \Delta(t)g(t),$$
$$Jh'(t) - H(t)h(t) = \mu\Delta(t)h(t) + \Delta(t)k(t).$$

Assume that $U(\cdot, \lambda_0)$ *with* $\lambda_0 \in \mathbb{R}$ *is a solution matrix which satisfies* (7.2.17) *and define the functions* $\tilde{f} = U(\cdot, \lambda_0)^{-1}f$ *and* $\tilde{h} = U(\cdot, \lambda_0)^{-1}h$ *as in* (7.2.18). *Then for each* $t \in \imath$

$$h(t)^* J f(t) = \tilde{h}(t)^* J \tilde{f}(t).$$

Recall that the functions H and Δ were assumed to be 2×2 matrix functions with complex entries. When these functions are real the solutions enjoy a certain symmetry property.

Definition 7.2.7. The canonical system $Jf' - Hf = \lambda\Delta f$ is said to be *real* if the entries of the 2×2 matrix functions H and Δ in (7.2.1) are real functions.

To deal with real canonical systems the notion of conjugate matrices is useful. For a matrix T the *conjugate matrix* $\overline{T}$ is the matrix whose entries are the complex conjugates of the entries of T. Let T and S be matrices, not necessarily of the same size, for which the matrix product TS is defined. Then clearly

$$\overline{TS} = \overline{T}\,\overline{S}. \qquad (7.2.22)$$

Lemma 7.2.8. *Assume that the canonical system* (7.2.4) *is real. Let* $Y(\cdot, \lambda)$ *be a solution matrix of* (7.2.4) *such that for all* $\lambda \in \mathbb{C}$

$$\overline{Y}(c_0, \overline{\lambda}) = Y(c_0, \lambda) \qquad (7.2.23)$$

for some point $c_0 \in \imath$. *Then*

$$\overline{Y}(\cdot, \overline{\lambda}) = Y(\cdot, \lambda) \qquad (7.2.24)$$

for all $\lambda \in \mathbb{C}$. *In particular,* (7.2.24) *holds when* $Y(\cdot, \lambda)$ *is a fundamental matrix fixed by* (7.2.11) *or* (7.2.12).

Proof. By definition, the solution matrix $Y(\cdot, \lambda)$ satisfies

$$JY'(\cdot, \lambda) - HY(\cdot, \lambda) = \lambda\Delta Y(\cdot, \lambda). \qquad (7.2.25)$$

By assumption, the entries of J, H, and Δ are real; hence taking complex conjugates and using (7.2.22) one sees that

$$J\overline{Y}'(\cdot, \lambda) - H\overline{Y}(\cdot, \lambda) = \bar{\lambda}\Delta\overline{Y}(\cdot, \lambda).$$

Therefore, the matrix function $\overline{Y}(\cdot, \bar{\lambda})$ satisfies the same equation (7.2.25) as $Y(\cdot, \lambda)$ and by (7.2.23) these matrix functions satisfy the same initial condition at c_0. Now the uniqueness in Theorem 7.2.1 leads to (7.2.24). □

The following observation is an easy consequence of Lemma 7.2.8.

Corollary 7.2.9. *Let the system $Jf' - Hf = \lambda\Delta f$ be real and let a fundamental matrix $Y(\cdot, \lambda_0)$ be fixed by the initial condition $Y(c, \lambda_0) = I$ for some $a < c < b$. Then for every $u \in \mathbb{C}^2$*

$$\int_\imath u^* Y(s, \lambda)^* \Delta(s) Y(s, \lambda) u \, ds = \int_\imath \bar{u}^* Y(s, \bar{\lambda})^* \Delta(s) Y(s, \bar{\lambda}) \bar{u} \, ds.$$

In particular,

$$Y(\cdot, \lambda)u \in \mathcal{L}^2_\Delta(\imath) \quad \Leftrightarrow \quad Y(\cdot, \bar{\lambda})\bar{u} \in \mathcal{L}^2_\Delta(\imath).$$

Proof. Clearly, for any $u \in \mathbb{C}^2$ and all $s \in \imath$ one has

$$u^* Y(s, \lambda)^* \Delta(s) Y(s, \lambda) u \geq 0.$$

Therefore,

$$u^* Y(s, \lambda)^* \Delta(s) Y(s, \lambda) u = \overline{u^* Y(s, \lambda)^* \Delta(s) Y(s, \lambda) u}$$
$$= \bar{u}^* Y(s, \bar{\lambda})^* \Delta(s) Y(s, \bar{\lambda}) \bar{u},$$

which gives the assertion. □

7.3 Regular and quasiregular endpoints

In this section the notions of regular and quasiregular for an endpoint of the interval $\imath$ are introduced; this makes it possible to extend Theorem 7.2.1, so that one may solve an initial value problem in an endpoint.

The following definition gives a classification for the endpoints of the canonical system (7.2.3).

Definition 7.3.1. An endpoint of the interval $\imath$ is said to be a *quasiregular endpoint* of the canonical system (7.2.3) if the locally integrable functions H and Δ in (7.2.1) are integrable up to that endpoint. A finite quasiregular endpoint is called *regular*. An endpoint is said to be *singular* when it is not regular. The canonical system (7.2.3) is called *regular* if both endpoints are regular; otherwise it is called *singular*.

The main result in this section implies that if the term $g \in \mathcal{L}^2_{\Delta,\mathrm{loc}}(\imath)$ in (7.2.3) is square-integrable with respect to Δ at an endpoint which is regular or quasiregular, then every solution of the inhomogeneous equation has a continuous extension to that endpoint, so that it is square-integrable with respect to Δ there.

Proposition 7.3.2. *Assume that the endpoint a or b of $\imath = (a,b)$ is regular or quasiregular and that $g \in \mathcal{L}^2_{\Delta,\mathrm{loc}}(\imath)$ is square-integrable with respect to Δ at a or b, respectively. Then each solution f of (7.2.3) is square-integrable with respect to Δ at a or at b and the limits*

$$f(a) := \lim_{t \to a} f(t) \quad or \quad f(b) := \lim_{t \to b} f(t) \tag{7.3.1}$$

exist, respectively. Moreover, for each $\gamma \in \mathbb{C}^2$ there exists a unique solution f of (7.2.3) such that $f(a) = \gamma$ or $f(b) = \gamma$, respectively, and the corresponding function $\lambda \mapsto f(t,\lambda)$ is entire for every $t \in \imath$ and $t = a$ or $t = b$, respectively.

Proof. It suffices to consider the case of the endpoint b. So let b be a regular or quasiregular endpoint, let $\lambda \in \mathbb{C}$, and fix $c \in (a,b)$. The proof is split in three separate steps.

Step 1. Any solution f of (7.2.3) with $f(c) = \eta$ satisfies

$$f(t) = \eta + \int_c^t J^{-1} \left(\lambda\Delta(s) + H(s)\right) f(s)\, ds + \int_c^t J^{-1}\Delta(s)g(s)\, ds \tag{7.3.2}$$

with $t \in \imath$. Recall that, since g is square-integrable with respect to Δ at b, it follows that Δg is integrable on $[c,b)$; cf. Lemma 7.1.4. By definition also $\lambda\Delta + H$ is integrable on $[c,b)$. Hence, Gronwall's lemma in Section 6.13 (see Lemma 6.13.2) shows that

$$|f(t)| \le \left(|\eta| + \int_c^t |\Delta(s)g(s)|\, ds\right) e^{\int_c^t |\lambda\Delta(s)+H(s)|\, ds}, \quad c \le t < b. \tag{7.3.3}$$

Thus, the solution f is bounded on $[c,b)$: $|f(s)| \le M$, $c < s < b$. In particular, this shows that

$$\int_c^b f(s)^* \Delta(s) f(s)\, ds \le M^2 \int_c^b |\Delta(s)|\, ds < \infty,$$

and hence f is square-integrable with respect to Δ at b. Moreover, it is clear from (7.3.2) that the limit $f(b) = \lim_{t \to b} f(t)$ in (7.3.1) exists.

Step 2. In the special case where h is a solution of the homogeneous system (7.2.4) with $h(c) = \eta$ it follows from (7.3.2) that

$$h(b) = \eta + \int_c^b J^{-1} \left(\lambda\Delta(s) + H(s)\right) h(s)\, ds. \tag{7.3.4}$$

The solution $h(\cdot, \lambda)$ actually depends on λ and according to Theorem 7.2.1 for each $c \leq t < b$ the function $\lambda \mapsto h(t, \lambda)$ is entire. It will be shown that also $\lambda \mapsto h(b, \lambda)$ is entire. In fact, from (7.3.4) it is clear that it suffices to prove that the mapping

$$\lambda \mapsto \int_c^b J^{-1} \left(\lambda \Delta(s) + H(s) \right) h(s, \lambda) \, ds \qquad (7.3.5)$$

is entire. To see this note that (7.3.3) and the equality $h(b) = \lim_{t \to b} h(t)$ imply that, for each compact set $K \subset \mathbb{C}$,

$$|h(t, \lambda)| \leq C_K \, e^{\int_c^b (|\Delta(s)| + |H(s)|) \, ds}$$

for all $c \leq t \leq b$ and for all $\lambda \in K$. Hence, by dominated convergence, the mapping in (7.3.5) is continuous, and an application of Morera's theorem implies that this mapping is holomorphic. Therefore, $\lambda \mapsto h(b, \lambda)$ is entire.

Step 3. Let $Z(\cdot, \lambda)$ be a fundamental matrix of the homogeneous equation (7.2.4) fixed by $Z(c, \lambda) = I$. Then, according to Step 1 and Step 2, one has

$$Z(t) = I + \int_c^t J^{-1} \left(\lambda \Delta(s) + H(s) \right) Z(s) \, ds$$

and Gronwall's lemma yields the estimate

$$|Z(t, \lambda)| \leq e^{\int_c^t |\lambda \Delta(s) + H(s)| \, ds}, \quad c \leq t < b. \qquad (7.3.6)$$

Thus, $Z(b, \lambda) = \lim_{t \to b} Z(t, \lambda)$ exists and it follows from Step 2 that the mapping $\lambda \mapsto Z(b, \lambda)$ is entire. Moreover, from $Z(t, \bar{\lambda})^* J Z(t, \lambda) = J$ for $c \leq t < b$ one concludes by taking the limit $t \to b$ that the matrix $Z(b, \lambda)$ is invertible for all $\lambda \in \mathbb{C}$. It is also clear that $Z(b, \bar{\lambda})^* J Z(b, \lambda) = J$. Thus, the function $U(\cdot, \lambda)$ defined by

$$U(t, \lambda) = Z(t, \lambda) Z(b, \lambda)^{-1}$$

is a fundamental matrix of the homogeneous equation which satisfies $U(b, \lambda) = I$ and $\lambda \mapsto U(t, \lambda)$ is entire for $c \leq t \leq b$. Therefore, if $\gamma \in \mathbb{C}^2$ is fixed one sees that

$$f(t) = U(t, \lambda)\gamma + U(t, \lambda) \int_t^b J U(s, \bar{\lambda})^* \Delta(s) g(s) \, ds$$

is the unique solution of the inhomogeneous equation with $f(b) = \gamma$; cf. (7.2.15). It remains to verify that $\lambda \mapsto f(t, \lambda)$ is entire for $c \leq t \leq b$. For this it suffices to check that

$$\lambda \mapsto U(t, \lambda) \int_t^b J U(s, \bar{\lambda})^* \Delta(s) g(s) \, ds = Z(t, \lambda) \int_t^b J Z(s, \bar{\lambda})^* \Delta(s) g(s) \, ds$$

is entire for $c \leq t \leq b$, which can be seen with the help of (7.3.6) in the same way as in Step 2. $\qquad \square$

Corollary 7.3.3. *Assume that the endpoints a and b of the canonical system (7.2.3) are regular or quasiregular and that $g \in \mathcal{L}^2_\Delta(\imath)$. Then each solution f of (7.2.3) belongs to $\mathcal{L}^2_\Delta(\imath)$ and both limits in (7.3.1) exist.*

The next statement follows from Corollary 7.2.3 and Corollary 7.3.3.

Corollary 7.3.4. *Assume that the endpoints a and b of the canonical system (7.2.3) are regular or quasiregular and that $g, k \in \mathcal{L}^2_\Delta(\imath)$. Let f, h be solutions of the inhomogeneous equations (7.2.6). Then*

$$h(b)^* Jf(b) - h(a)^* Jf(a) = \int_a^b \left(h(s)^* \Delta(s) g(s) - k(s)^* \Delta(s) f(s) \right) ds.$$

Finally, the next statement is a consequence of Proposition 7.3.2 and identity (7.2.10).

Corollary 7.3.5. *Assume that the endpoint a or b of the canonical system (7.2.3) is regular or quasiregular and let $Y(\cdot, \lambda)$ be a fundamental matrix of the canonical system (7.2.3). Then $Y(\cdot, \lambda)\phi$ is square-integrable with respect to Δ at a or b for every $\phi \in \mathbb{C}^2$ and $Y(\cdot, \lambda)$ admits a unique continuous extension to a or b such that $Y(a, \lambda)$ or $Y(b, \lambda)$ is invertible, respectively. In particular, the point c_0 in (7.2.12) can be chosen to be a or b, respectively.*

7.4 Square-integrability of solutions of real canonical systems

Let $\imath = (a, b)$ be an open interval and consider on this interval the homogeneous system $Jf' - Hf = \lambda \Delta f$. Recall that a solution f, depending on $\lambda \in \mathbb{C}$, is called square-integrable with respect to Δ at a or b if for some $c \in \imath$

$$\int_a^c f(s)^* \Delta(s) f(s) \, ds < \infty \quad \text{or} \quad \int_c^b f(s)^* \Delta(s) f(s) \, ds < \infty,$$

respectively. In this section the existence of such solutions is studied for *real* canonical systems; cf. Definition 7.2.7. The first main result asserts that if there are two linearly independent solutions which are square-integrable with respect to Δ at an endpoint for some $\lambda \in \mathbb{C}$, then for any $\lambda \in \mathbb{C}$ all solutions are square-integrable with respect to Δ at that endpoint. The second main result states that for any $\lambda \in \mathbb{C} \setminus \mathbb{R}$ there is at least one solution that is square-integrable with respect to Δ at an endpoint. A combination of these two results gives a general description of the existence of the solutions that are square-integrable with respect to Δ at an endpoint and leads to the limit-point and limit-circle classification.

In the rest of this section it will be assumed that the system (7.2.3) is real and the symmetry result in Corollary 7.2.9 will be used throughout.

Theorem 7.4.1. *Assume that for $\lambda_0 \in \mathbb{C}$ the equation $Jf' - Hf = \lambda_0 \Delta f$ has two linearly independent solutions which are square-integrable with respect to Δ at a or b. Then for any $\lambda \in \mathbb{C}$ each solution of $Jf' - Hf = \lambda \Delta f$ is square-integrable with respect to Δ at a or b, respectively.*

Proof. It is sufficient to show the result for one endpoint, say b. Assume without loss of generality that the endpoint a of the canonical system (7.2.3) is regular. Fix a fundamental solution $Y(\cdot, \lambda_0)$ by the initial condition $Y(a, \lambda_0) = I$. The columns $Y_1(\cdot, \lambda_0)$ and $Y_2(\cdot, \lambda_0)$ of $Y(\cdot, \lambda_0)$ belong to $\mathcal{L}_\Delta^2(\imath)$ by assumption. As the system is assumed to be real, one has

$$\int_a^b |\Delta(s)^{\frac{1}{2}} Y_i(s, \bar{\lambda}_0)|^2 \, ds = \int_a^b |\Delta(s)^{\frac{1}{2}} Y_i(s, \lambda_0)|^2 \, ds, \quad i = 1, 2; \tag{7.4.1}$$

cf. Corollary 7.2.9.

Let $\lambda \in \mathbb{C}$ and let $f(\cdot, \lambda)$ be any solution of $Jf' - Hf = \lambda \Delta f$. It will be shown that $f(\cdot, \lambda)$ is square-integrable with respect to Δ at b. Since the function $f(\cdot, \lambda)$ satisfies

$$Jf'(\cdot, \lambda) - Hf(\cdot, \lambda) = \lambda_0 \Delta f(\cdot, \lambda) + (\lambda - \lambda_0) \Delta f(\cdot, \lambda),$$

it follows from (7.2.16) (with $g = (\lambda - \lambda_0) f(\cdot, \lambda)$) that $f(\cdot, \lambda)$ can be written as

$$\begin{aligned} f(t, \lambda) = {} & Y_1(t, \lambda_0)\alpha_1 + Y_2(t, \lambda_0)\alpha_2 \\ & + (\lambda - \lambda_0)\big[Y_1(t, \lambda_0)y_2(t, \lambda) - Y_2(t, \lambda_0)y_1(t, \lambda)\big], \end{aligned} \tag{7.4.2}$$

where $f(a, \lambda) = (\alpha_1, \alpha_2)^\top$ and $y_i(\cdot, \lambda)$ is defined by

$$y_i(t, \lambda) = \int_a^t Y_i(s, \bar{\lambda}_0)^* \Delta(s) f(s, \lambda) \, ds, \quad i = 1, 2,$$

respectively. By applying the Cauchy–Schwarz inequality in the definition of $y_i(t, \lambda)$ and using (7.4.1) one obtains for $i = 1, 2$,

$$\begin{aligned} |y_i(t, \lambda)| &\le \sqrt{\int_a^t |\Delta(s)^{\frac{1}{2}} Y_i(s, \bar{\lambda}_0)|^2 \, ds} \sqrt{\int_a^t |\Delta(s)^{\frac{1}{2}} f(s, \lambda)|^2 \, ds} \\ &\le \sqrt{\int_a^b |\Delta(s)^{\frac{1}{2}} Y_i(s, \bar{\lambda}_0)|^2 \, ds} \sqrt{\int_a^t |\Delta(s)^{\frac{1}{2}} f(s, \lambda)|^2 \, ds} \\ &= \sqrt{\int_a^b |\Delta(s)^{\frac{1}{2}} Y_i(s, \lambda_0)|^2 \, ds} \sqrt{\int_a^t |\Delta(s)^{\frac{1}{2}} f(s, \lambda)|^2 \, ds}. \end{aligned}$$

Introduce the number $\alpha \ge 0$ and the nonnegative function φ by

$$\alpha = \max\{|\alpha_1|, |\alpha_2|\}, \quad \varphi(t) = \max\{|\Delta(t)^{\frac{1}{2}} Y_1(t, \lambda_0)|, |\Delta(t)^{\frac{1}{2}} Y_2(t, \lambda_0)|\},$$

so that $\varphi \in L^2(a,b)$. Multiply both sides in the identity in (7.4.2) from the left by $\Delta(t)^{\frac{1}{2}}$, then

$$|\Delta(t)^{\frac{1}{2}} f(t,\lambda)|$$

$$\leq 2\alpha\varphi(t) + 2|\lambda - \lambda_0|\varphi(t)\sqrt{\int_a^b \varphi(s)^2\, ds}\,\sqrt{\int_a^t |\Delta(s)^{\frac{1}{2}} f(s,\lambda)|^2\, ds}.$$

Therefore, one obtains that

$$|\Delta(t)^{\frac{1}{2}} f(t,\lambda)|^2 \leq \varphi(t)^2 \left(A + B \int_a^t |\Delta(s)^{\frac{1}{2}} f(s,\lambda)|^2\, ds \right), \tag{7.4.3}$$

where

$$A = 8\alpha^2, \quad B = 8|\lambda - \lambda_0|^2 \int_a^b \varphi(s)^2\, ds.$$

It follows from (7.4.3) by means of Lemma 6.1.4 with $u(t) = |\Delta(t)^{\frac{1}{2}} f(t,\lambda)|$, φ as above, and $r = 1$, that the function $f(\cdot,\lambda)$ is square-integrable with respect to Δ at b. $\qquad\square$

Next it will be shown that for each endpoint and any $\lambda \in \mathbb{C} \setminus \mathbb{R}$ there is at least one solution of the homogeneous canonical system (7.2.4) which is square-integrable with respect to Δ at that endpoint. The proof of this fact is based on the monotonicity principle in Section 5.2; cf. Corollary 5.2.14. To apply this result, let $Y(\cdot,\lambda)$ be a fundamental matrix of the canonical system (7.2.3) fixed as in (7.2.12) and consider the 2×2 matrix function $D(\cdot,\lambda)$ on $\imath$ defined by

$$D(t,\lambda) = Y(t,\lambda)^*(-iJ)Y(t,\lambda), \quad t \in \imath, \quad \lambda \in \mathbb{C}. \tag{7.4.4}$$

Observe that the function $t \mapsto D(t,\lambda)$, $t \in \imath$, is absolutely continuous for every $\lambda \in \mathbb{C}$ and that the matrices $D(t,\lambda)$ are self-adjoint and invertible for all $t \in \imath$ and $\lambda \in \mathbb{C}$.

According to the following theorem, the matrix function in (7.4.4) admits self-adjoint limits at a and b, which may be either self-adjoint matrices or self-adjoint relations with a one-dimensional domain and a one-dimensional multivalued part. Furthermore, the dimensions of the domains of the limit relations are directly connected with the number of linearly independent solutions of the homogeneous canonical system (7.2.4) that are square-integrable with respect to Δ.

Theorem 7.4.2. *For $\lambda \in \mathbb{C}^+$ or $\lambda \in \mathbb{C}^-$ the 2×2 matrix function $t \mapsto D(t,\lambda)$ is nondecreasing or nonincreasing on $\imath$, respectively. There exist self-adjoint relations $D(b,\lambda)$ and $D(a,\lambda)$ in $\mathbb{C}^2$ such that*

$$D(t,\lambda) \to D(a,\lambda) \quad \text{and} \quad D(t,\lambda) \to D(b,\lambda)$$

in the (strong) resolvent sense when $t \to a$ and $t \to b$, respectively, and

$$1 \leq \dim\big(\operatorname{dom} D(a,\lambda)\big) \leq 2 \quad \text{and} \quad 1 \leq \dim\big(\operatorname{dom} D(b,\lambda)\big) \leq 2.$$

Furthermore, $\phi \in \operatorname{dom} D(a,\lambda)$ or $\phi \in \operatorname{dom} D(b,\lambda)$ if and only if $Y(\cdot,\lambda)\phi$ is a solution of (7.2.4) that is square-integrable with respect to Δ at a or b, respectively.

Proof. It follows from Corollary 7.2.4 that

$$D(\beta,\lambda) - D(\alpha,\lambda) = 2\operatorname{Im}\lambda \int_\alpha^\beta Y(s,\lambda)^* \Delta(s) Y(s,\lambda)\, ds, \quad \lambda \in \mathbb{C}, \qquad (7.4.5)$$

holds for any compact interval $[\alpha,\beta] \subset \imath$. Hence, the matrix function $D(\cdot,\lambda)$ is nondecreasing for $\lambda \in \mathbb{C}^+$ and nonincreasing for $\lambda \in \mathbb{C}^-$. It follows from Corollary 5.2.14 that there exist self-adjoint relations $D(a,\lambda)$ and $D(b,\lambda)$ such that

$$\lim_{t\to a} \left(D(t,\lambda) - \mu\right)^{-1} = \left(D(a,\lambda) - \mu\right)^{-1}, \quad \mu \in \mathbb{C}\setminus\mathbb{R},$$

and

$$\lim_{t\to b} \left(D(t,\lambda) - \mu\right)^{-1} = \left(D(b,\lambda) - \mu\right)^{-1}, \quad \mu \in \mathbb{C}\setminus\mathbb{R}.$$

Next it will be shown that the dimension of the domains of the self-adjoint relations $D(a,\lambda)$ and $D(b,\lambda)$ is at least one. For this it is sufficient to prove that there exists at least one (finite) eigenvalue.

Note first that, by (7.4.4) and (7.2.12),

$$D(c_0,\lambda) = Y(c_0,\lambda)^*(-iJ)Y(c_0,\lambda) = -iJ$$

and hence the eigenvalues of $D(c_0,\lambda)$ are $\nu_-(c_0) = -1$ and $\nu_+(c_0) = 1$. As the function $D(\cdot,\lambda)$ is continuous on $\imath$, the same holds true for its eigenvalues $\nu_-(\cdot)$ and $\nu_+(\cdot)$. Since the matrices $D(t,\lambda)$ are self-adjoint and invertible for all $t \in \imath$ it follows that $\nu_-(t) < 0$ and $\nu_+(t) > 0$ for all $t \in \imath$. Recall that

$$\nu_-(t) = \inf_{|x|=1} \left(D(t,\lambda)x,x\right) \quad \text{and} \quad \nu_+(t) = \sup_{|x|=1} \left(D(t,\lambda)x,x\right),$$

and since $D(t_1,\lambda) \le D(t_2,\lambda)$, $t_1 \le t_2$, it follows that

$$\nu_-(t_1) \le \nu_-(t_2) \quad \text{and} \quad \nu_+(t_1) \le \nu_+(t_2), \quad t_1 \le t_2.$$

Therefore, it is clear that the limits of $\nu_-(t)$ and $\nu_+(t)$ exist and that

$$\nu_-(b) = \lim_{t\to b} \nu_-(t) \le 0 \quad \text{and} \quad 0 < \nu_+(b) = \lim_{t\to b} \nu_+(t) \le \infty.$$

In order to see the connection of these limits with the self-adjoint relation $D(b,\lambda)$ observe that for $\mu \in \mathbb{C}\setminus\mathbb{R}$

$$\frac{1}{\nu_-(t) - \mu} \quad \text{and} \quad \frac{1}{\nu_+(t) - \mu}$$

are the eigenvalues of the matrix $(D(t,\lambda) - \mu)^{-1}$. Therefore, again by continuity, one sees that

$$\frac{1}{\nu_-(b) - \mu} \quad \text{and} \quad \frac{1}{\nu_+(b) - \mu}$$

are the eigenvalues of the matrix $(D(b, \lambda) - \mu)^{-1}$. Hence, $\nu_-(b)$ is a nonpositive eigenvalue of the self-adjoint relation $D(b, \lambda)$, which implies $\dim(\operatorname{dom} D(b, \lambda)) \geq 1$. More precisely, if $\nu_+(b) < \infty$, then $\nu_+(b)$ is a positive eigenvalue of $D(b, \lambda)$, in which case $\dim(\operatorname{dom} D(b, \lambda)) = 2$, while if $\nu_+(b) = \infty$, then $D(b, \lambda)$ has a one-dimensional multivalued part and $\dim(\operatorname{dom} D(b, \lambda)) = 1$. Similar observations may be made for the self-adjoint relation $D(a, \lambda)$. In particular, it follows that $\dim(\operatorname{dom} D(a, \lambda)) \geq 1$.

Finally, it will be shown that $\phi \in \operatorname{dom} D(b, \lambda)$ if and only if the solution $Y(\cdot, \lambda)\phi$ of (7.2.4) is square-integrable with respect to Δ at b; the argument for the left endpoint a is the same. Suppose that $\lambda \in \mathbb{C}^+$, so that $D(\cdot, \lambda)$ is nondecreasing on $\imath$. In this case it follows from Corollary 5.2.13 and Corollary 5.2.14 that

$$\operatorname{dom} D(b, \lambda) = \{\phi \in \mathbb{C}^2 : \lim_{t \to b} \phi^* D(t, \lambda)\phi < \infty\}$$

and hence (7.4.5) implies that $\phi \in \operatorname{dom} D(b, \lambda)$ if and only if

$$\int_\alpha^b \phi^* Y(s, \lambda)^* \Delta(s) Y(s, \lambda)\phi \, ds < \infty,$$

that is, the solution $Y(\cdot, \lambda)\phi$ is square-integrable with respect to Δ at b. The case where $\lambda \in \mathbb{C}^-$ is dealt with in a similar way. $\square$

A combination of Theorems 7.4.1 and 7.4.2 leads to the following observation.

Corollary 7.4.3. *If for some $\lambda_0 \in \mathbb{C} \setminus \mathbb{R}$ the equation $Jf' - Hf = \lambda_0 \Delta f$ has, up to scalar multiples, only one nontrivial solution which is square-integrable with respect to Δ at a or b, then for any $\lambda \in \mathbb{C} \setminus \mathbb{R}$ the equation $Jf' - Hf = \lambda \Delta f$ has, up to scalar multiples, precisely one nontrivial solution which is square-integrable with respect to Δ at a or b, respectively.*

Proof. It is sufficient to consider the endpoint b. Assume that for some $\lambda_0 \in \mathbb{C} \setminus \mathbb{R}$ the equation $Jf' - Hf = \lambda_0 \Delta f$ has, up to scalar multiples, only one nontrivial solution that is square-integrable with respect to Δ at b, and suppose that for some $\lambda \in \mathbb{C} \setminus \mathbb{R}$ with $\lambda \neq \lambda_0$ the equation $Jf' - Hf = \lambda \Delta f$ does not have, up to scalar multiples, only one nontrivial solution which is square-integrable with respect to Δ at b. Since

$$1 \leq \dim(\operatorname{dom} D(b, \lambda)) \leq 2$$

by Theorem 7.4.2 there exist two linearly independent solutions of $Jf' - Hf = \lambda \Delta f$ which are square-integrable with respect to Δ at b. But then Theorem 7.4.1 implies that there also exist two linearly independent solutions of $Jf' - Hf = \lambda_0 \Delta f$ that are square-integrable with respect to Δ at b; a contradiction. $\square$

Theorem 7.4.1 and Corollary 7.4.3 yield the limit-point and limit-circle classification for real canonical systems in the next definition and corollary. The terminology is inspired by the terminology for Sturm–Liouville equations in Section 6.1.

Definition 7.4.4. For a real canonical system the endpoint a or b of the interval $\imath$ is said to be in the *limit-circle case* if for some, and hence for all $\lambda \in \mathbb{C}$ there exist two linearly independent solutions of $Jf' - Hf = \lambda\Delta f$ that are square-integrable with respect to Δ at a or b, respectively. The endpoint a or b of the interval $\imath$ is said to be in the *limit-point case* if for some, and hence for all $\lambda \in \mathbb{C} \setminus \mathbb{R}$ there exists, up to scalar multiples, only one nontrivial solution of $Jf' - Hf = \lambda\Delta f$ that is square-integrable with respect to Δ at a or b, respectively.

Note that, by Theorem 7.4.2, at any endpoint of the interval $\imath$ there is at least one nontrivial solution that is square-integrable with respect to Δ and there are at most two linearly independent solutions that are square-integrable with respect to Δ. This leads to *Weyl's alternative* for canonical systems.

Corollary 7.4.5. *For a real canonical system each of the endpoints of the interval is either in the limit-circle case or in the limit-point case.*

For completeness also the special case of regular and quasiregular endpoints is briefly discussed. The next corollary is an immediate consequence of Corollary 7.3.5.

Corollary 7.4.6. *A regular or quasiregular endpoint of a real canonical system is in the limit-circle case.*

A simple but useful characterization of the limit-point case is given in the following corollary. It is stated for the endpoint b, but clearly there is a similar statement for the endpoint a.

Corollary 7.4.7. *Let the canonical system be real and assume that the endpoint a is regular or quasiregular. Then the following statements hold:*

(i) *If the endpoint b is in the limit-point case, then for all $\lambda \in \mathbb{R}$ the equation $Jf' - Hf = \lambda\Delta f$ has, up to scalar multiples, at most one nontrivial solution that is square-integrable with respect to Δ at b.*

(ii) *If there exists $\lambda_0 \in \mathbb{R}$ such that the equation $Jf' - Hf = \lambda_0\Delta f$ has, up to scalar multiples, at most one nontrivial solution that is square-integrable with respect to Δ at b, then the endpoint b is in the limit-point case.*

Proof. (i) If there exists $\lambda \in \mathbb{R}$ for which the homogeneous equation has two linearly independent solutions that are square-integrable with respect to Δ at b, then by Theorem 7.4.1, for each $\lambda \in \mathbb{C}$ all nontrivial solutions are square-integrable with respect to Δ at b. Hence, b is in the limit-circle case; a contradiction.

(ii) If b is in the limit-circle case, then for all $\lambda \in \mathbb{C}$, and hence for $\lambda \in \mathbb{R}$, the homogeneous equation has two linearly independent solutions which are square-integrable with respect to Δ at b. This implies (ii). $\square$

If, for instance, the endpoint b is regular or quasiregular, then any solution of (7.2.3) with g square-integrable with respect to Δ at b has a limit at b by

Proposition 7.3.2 and b is in the limit-circle case by Corollary 7.4.6. However, if b is in the limit-circle case, then the solutions of (7.2.3) are square-integrable with respect to Δ at b, but they do not necessarily have a limit at b. It will be shown in this case that there exists a natural transformation which turns the system into one where b is quasiregular; cf. Lemma 7.2.5.

Corollary 7.4.8. *Assume that a is regular and that b is in the limit-circle case. Let $g \in \mathcal{L}^2_\Delta(a, b)$ and let $f(\cdot, \lambda)$ be a solution of*

$$Jf' - Hf = \lambda \Delta f + \Delta g.$$

Let $U(\cdot, \lambda_0)$, $\lambda_0 \in \mathbb{R}$, be a matrix function as in (7.2.17). Then the limit

$$\widetilde{f}(b) = \lim_{t \to b} U(t, \lambda_0)^{-1} f(t) \tag{7.4.6}$$

exists in $\mathbb{C}^2$. Moreover, for each $\gamma \in \mathbb{C}^2$ there exists a unique solution $f(\cdot, \lambda)$ of (7.2.3) such that $\widetilde{f}(b) = \gamma$ and the corresponding function

$$\lambda \mapsto \lim_{t \to b} U(t, \lambda_0)^{-1} f(t, \lambda)$$

is entire.

Proof. Let $g \in \mathcal{L}^2_\Delta(a, b)$ and let f be a solution of (7.2.3). Since b is in the limit-circle case, there exists for $\lambda_0 \in \mathbb{R}$ and $c_0 \in [a, b)$ a matrix function $U(\cdot, \lambda_0)$ satisfying (7.2.17) that is square-integrable with respect to Δ at b. Thus, the function $\widetilde{\Delta}$ defined in (7.2.18) is integrable at b, which means that the endpoint b for the system in (7.2.20) is quasiregular. Since g is square-integrable with respect to Δ at b, the function $\widetilde{g}$ in Lemma 7.2.5 is square-integrable with respect to $\widetilde{\Delta}$ at b. Therefore, the assertion is clear from Proposition 7.3.2 as $\widetilde{f}$ is a solution of (7.2.20). $\square$

Let the endpoint a be regular or quasiregular. Let $g, k \in \mathcal{L}^2_\Delta(\imath)$ and let f, h be solutions of the inhomogeneous equations (7.2.6) such that $f, h \in \mathcal{L}^2_\Delta(\imath)$. Then for $a \le t < b$ one has

$$h(t)^* Jf(t) - h(a)^* Jf(a) = \int_a^t \left(h(s)^* \Delta(s) g(s) - k(s)^* \Delta(s) f(s) \right) ds \tag{7.4.7}$$

by the Lagrange identity in Corollary 7.2.3. It follows from (7.4.7) that the limit

$$\lim_{t \to b} h(t)^* Jf(t)$$

exists. Of course, when b is regular or quasiregular, then the individual limits $\lim_{t \to b} f(t)$ and $\lim_{t \to b} h(t)$ exist by Proposition 7.3.2, see also Corollary 7.3.4. In general the existence of the individual limits $\lim_{t \to b} f(t)$ and $\lim_{t \to b} h(t)$ is not guaranteed. However, in the case where b is in the limit-circle case but not quasiregular the next corollary suggests to employ the limits in (7.4.6).

Corollary 7.4.9. *Assume that the endpoint* a *is regular and that* b *is in the limit-circle case. Let* $g, k \in L^2_\Delta(\imath)$ *and let* f, h *be solutions of the inhomogeneous equations (7.2.6) such that* $\widetilde{f}, \widetilde{h} \in L^2_\Delta(\imath)$. *Then*

$$\lim_{t \to b} h(t)^* J f(t) = \widetilde{h}(b)^* J \widetilde{f}(b), \tag{7.4.8}$$

where $\widetilde{f}(b)$ *and* $\widetilde{h}(b)$ *are as in (7.4.6). Moreover,*

$$\widetilde{h}(b)^* J \widetilde{f}(b) - h(a)^* J f(a) = \int_a^b \left(h(s)^* \Delta(s) g(s) - k(s)^* \Delta(s) f(s) \right) ds. \tag{7.4.9}$$

Proof. It follows by taking limits in (7.4.7) that

$$\lim_{t \to b} h(t)^* J f(t) - h(a)^* J f(a) = \int_a^b \left(h(s)^* \Delta(s) g(s) - k(s)^* \Delta(s) f(s) \right) ds.$$

Now apply Corollary 7.2.6 and Corollary 7.4.8. Take the limit $t \to b$ and (7.4.8) and (7.4.9) follow. □

7.5   Definite canonical systems

The general class of canonical differential equations as in (7.2.3) will now be narrowed down by imposing a definiteness condition; see Definition 7.5.5. This condition will be assumed in the rest of this chapter. In this section various equivalent formulations of the definiteness condition will be presented. Moreover, it will be shown that the solution of a definite canonical system (7.2.3) can be cut off near an endpoint of the interval $\imath$, in the sense that the solution is modified in such a way that it becomes trivial in a neighborhood of that endpoint.

It will be convenient to begin the discussion of definiteness of the canonical system (7.2.3) with the notion of definiteness when the system is restricted to an arbitrary subinterval $\jmath \subset \imath$.

Definition 7.5.1. Let $\jmath \subset \imath$ be a nonempty interval. The canonical system (7.2.3) is said to be *definite* on $\jmath$ if for each solution f of $J f' - H f = 0$ on $\jmath$ one has

$$\Delta(t) f(t) = 0, \ t \in \jmath \quad \Rightarrow \quad f(t) = 0, \ t \in \jmath.$$

Observe that if a solution f of the canonical system (7.2.3) vanishes on a nonempty subinterval $\jmath \subset \imath$, then $f(t) = 0$ for $t \in \imath$; cf. Theorem 7.2.1. Hence, it is clear that if the canonical system (7.2.3) is definite on $\jmath$, then it is also definite on every interval $\widetilde{\jmath}$ with the property that $\jmath \subset \widetilde{\jmath} \subset \imath$. Also observe that with the subinterval $\jmath \subset \imath$ and a continuous function f one has

$$\Delta(t) f(t) = 0, \ t \in \jmath \quad \Leftrightarrow \quad \int_\jmath f(s)^* \Delta(s) f(s) \, ds = 0. \tag{7.5.1}$$

Clearly, if $\Delta(t)$ has full rank for almost all $t \in \jmath$, then the canonical system is automatically definite on $\jmath$.

Lemma 7.5.2. *Let* $\jmath \subset \imath$ *be a nonempty interval. The canonical system* (7.2.3) *is definite on the interval* $\jmath \subset \imath$ *if and only if for all* $\lambda \in \mathbb{C}$ *and for each solution* f *of* $Jf' - Hf = \lambda \Delta f$ *on* $\jmath$ *one has*

$$\Delta(t)f(t) = 0, \ t \in \jmath \quad \Rightarrow \quad f(t) = 0, \ t \in \jmath.$$

Proof. Assume that the canonical system is definite on $\jmath$. Choose $\lambda \in \mathbb{C}$ and let f be a solution of $Jf' - Hf = \lambda \Delta f$ on $\jmath$ with $\Delta(t)f(t) = 0$ for almost all $t \in \jmath$. Thus, f is a solution of $Jf' - Hf = 0$ with $\Delta(t)f(t) = 0$ for almost all $t \in \jmath$. By assumption this implies that $f(t) = 0$ for $t \in \jmath$. The converse statement is trivial. $\square$

The following result is an alternative useful version of Lemma 7.5.2 in terms of a fundamental matrix $Y(\cdot, \lambda)$.

Corollary 7.5.3. *Let* $Y(\cdot, \lambda)$, $\lambda \in \mathbb{C}$, *be a fundamental matrix for* (7.2.3) *and let* $I \subset \imath$ *be a compact interval. Then the system* (7.2.3) *is definite on* I *if and only if the* 2×2 *matrix*

$$\int_I Y(s, \lambda)^* \Delta(s) Y(s, \lambda) \, ds \tag{7.5.2}$$

is invertible for some, and hence for all $\lambda \in \mathbb{C}$.

Proof. Assume that (7.2.3) is definite on I. If the (nonnegative) matrix in (7.5.2) is not invertible, then there exists a nontrivial $\gamma \in \mathbb{C}^2$ for which

$$\gamma^* \left(\int_I Y(s, \lambda)^* \Delta(s) Y(s, \lambda) \, ds \right) \gamma = 0, \tag{7.5.3}$$

or alternatively $\Delta(t)Y(t, \lambda)\gamma = 0$ for $t \in I$; cf. (7.5.1). Since $Y(\cdot, \lambda)\gamma$ is a solution of $Jf' - Hf = \lambda \Delta f$, it follows from the definiteness that $Y(t, \lambda)\gamma = 0$ for $t \in I$, which implies $\gamma = 0$. This contradiction shows that the matrix in (7.5.2) is invertible.

Conversely, assume that the (nonnegative) matrix in (7.5.2) is invertible. In order to show that (7.2.3) is definite, let

$$Jf'(t) - H(t)f(t) = \lambda \Delta(t)f(t), \quad \Delta(t)f(t) = 0, \quad t \in I.$$

Since $Y(\cdot, \lambda)$ is a fundamental matrix of $Jf' - Hf = \lambda \Delta f$, every solution of this equation can be written in the form $f = Y(\cdot, \lambda)\gamma$ with a unique $\gamma \in \mathbb{C}^2$. The condition $\Delta(t)f(t) = 0$, $t \in I$, implies that (7.5.3) holds. Therefore, $\gamma = 0$ and thus the system (7.2.3) is definite. $\square$

The next proposition shows that there is no difference between global definiteness and local definiteness.

Proposition 7.5.4. *The canonical system* (7.2.3) *is definite on $\imath$ if and only if there exists a compact interval $I \subset \imath$ such that the canonical system* (7.2.3) *is definite on the interval I.*

Proof. If the canonical system (7.2.3) is definite on the interval I, then it is clearly definite on the larger interval $\imath$.

To see the converse statement, let the canonical system (7.2.3) be definite on the interval $\imath$; in other words, assume that for each solution f of $Jf' - Hf = 0$ on $\imath$ one has

$$\Delta(t)f(t) = 0, \quad t \in \imath \quad \Rightarrow \quad f(t) = 0, \quad t \in \imath.$$

Introduce for each compact subinterval K of $\imath$ the subset $d(K)$ of $\mathbb{C}^2$ by

$$d(K) = \left\{ \phi \in \mathbb{C}^2 : |\phi| = 1, \int_K \phi^* Y(s,0)^* \Delta(s) Y(s,0) \phi \, ds = 0 \right\}.$$

Clearly, $d(K)$ is compact and $K \subset \widetilde{K}$ implies $d(\widetilde{K}) \subset d(K)$. Now choose an increasing sequence of compact intervals (K_n) such that their union equals the interval $\imath$. Then

$$\bigcap_{n \in \mathbb{N}} d(K_n) = \emptyset. \tag{7.5.4}$$

Indeed, assume that there exists an element $\phi \in \mathbb{C}^2$ with $|\phi| = 1$, such that

$$\int_{K_n} \phi^* Y(s,0)^* \Delta(s) Y(s,0) \phi \, ds = 0$$

for every $n \in \mathbb{N}$. Then, by monotone convergence,

$$\int_\imath \phi^* Y(s,0)^* \Delta(s) Y(s,0) \phi \, ds = 0.$$

As the canonical system (7.2.3) is definite, this implies by (7.5.1) that $Y(\cdot, 0)\phi = 0$, which leads to $\phi = 0$; a contradiction. Therefore, the identity (7.5.4) is valid. Since each of the sets $d(K_n)$ in (7.5.4) is compact, it follows that there exists a compact interval K_m such that $d(K_m) = \emptyset$. Hence, $I = K_m$ satisfies the requirements. To see this, let $Jf' - Hf = 0$ on K_m and assume that $\Delta(t)f(t) = 0$, $t \in K_m$, or, equivalently, $\int_{K_m} f(s)^* \Delta(s) f(s) = 0$; cf. (7.5.1). Since $d(K_m) = \emptyset$ one concludes that $f = 0$. □

In the rest of the text one often speaks of definite systems in the following sense.

Definition 7.5.5. *The canonical system* (7.2.3) *is said to be* definite *if it is* definite *on $\imath$.*

The next result is about smoothly cutting off the solution of a definite canonical system (7.2.3) near an endpoint of the interval $\imath$, i.e., modifying the solution so that it becomes trivial in a neighborhood of that endpoint. The following proposition and corollary will be used in Section 7.6.

Proposition 7.5.6. *Let the canonical system (7.2.3) be definite and choose a compact interval $[\alpha, \beta] \subset \imath$ such that the system is definite on $[\alpha, \beta]$. Let $g \in \mathcal{L}^2_{\Delta,\mathrm{loc}}(\imath)$ and let $f \in AC(\imath)$ be a solution of the inhomogeneous equation (7.2.3) for some $\lambda \in \mathbb{C}$. Then there exist functions $f_a \in AC(\imath)$ and $g_a \in \mathcal{L}^2_{\Delta,\mathrm{loc}}(\imath)$ satisfying*

$$J f_a'(t) - H(t) f_a(t) = \lambda \Delta(t) f_a(t) + \Delta(t) g_a(t)$$

such that

$$f_a(t) = \begin{cases} f(t), & t \in (a, \alpha], \\ 0, & t \in [\beta, b), \end{cases} \quad \text{and} \quad g_a(t) = \begin{cases} g(t), & t \in (a, \alpha], \\ 0, & t \in [\beta, b). \end{cases}$$

Similarly, there exist functions $f_b \in AC(\imath)$ and $g_b \in \mathcal{L}^2_{\Delta,\mathrm{loc}}(\imath)$ satisfying

$$J f_b'(t) - H(t) f_b(t) = \lambda \Delta(t) f_b(t) + \Delta(t) g_b(t)$$

such that

$$f_b(t) = \begin{cases} 0, & t \in (a, \alpha], \\ f(t), & t \in [\beta, b), \end{cases} \quad \text{and} \quad g_b(t) = \begin{cases} 0, & t \in (a, \alpha], \\ g(t), & t \in [\beta, b). \end{cases}$$

Proof. Let the functions f and g be as indicated. The result will be proved for the functions f_b and g_b; the proof for the functions f_a and g_a is similar.

Let $[\alpha, \beta] \subseteq \imath$ be a compact interval on which the canonical system (7.2.3) is definite; cf. Proposition 7.5.4. Let $k \in \mathcal{L}^2_\Delta(\alpha, \beta)$ and fix a fundamental system $Y(\cdot, \lambda)$ by the initial condition $Y(\alpha, \lambda) = I$. According to (7.2.15), the function defined by

$$h(t) = -Y(t, \lambda) \int_\alpha^t J Y(s, \bar{\lambda})^* \Delta(s) k(s) \, ds \qquad (7.5.5)$$

satisfies the inhomogeneous equation

$$J h'(t) - H(t) h(t) = \lambda \Delta(t) h(t) + \Delta(t) k(t), \quad \alpha < t < \beta,$$

and in the endpoints it has the values

$$h(\alpha) = 0 \quad \text{and} \quad h(\beta) = -Y(\beta, \lambda) \int_\alpha^\beta J Y(s, \bar{\lambda})^* \Delta(s) k(s) \, ds.$$

It will be shown that there exists a function $k \in \mathcal{L}^2_\Delta(\alpha, \beta)$ such that $h(\beta) = f(\beta)$.

In order to verify this, observe that $Y(\beta, \lambda)$ is invertible and that the integral operator

$$\ell \mapsto \int_\alpha^\beta J Y(s, \bar{\lambda})^* \Delta(s) \ell(s) \, ds$$

taking $\mathcal{L}^2_\Delta(\alpha, \beta)$ into $\mathbb{C}^2$ is surjective. To see this, assume that $\gamma \in \mathbb{C}^2$ is orthogonal to the range of this integral operator, that is,

$$0 = \gamma^* \int_\alpha^\beta JY(s, \bar{\lambda})^* \Delta(s)\ell(s)\, ds = \int_\alpha^\beta (Y(s, \bar{\lambda})J^*\gamma)^* \Delta(s)\ell(s)\, ds$$

for all $\ell \in \mathcal{L}^2_\Delta(\alpha, \beta)$. With $\ell(s) = Y(s, \bar{\lambda})J^*\gamma$ it then follows from (7.5.1) and Lemma 7.5.2 that $\ell(s) = 0$ for $s \in (\alpha, \beta)$, which implies that $\gamma = 0$. Thus, the integral operator is surjective.

Now choose $k \in \mathcal{L}^2_\Delta(\alpha, \beta)$ as above, so that h defined by (7.5.5) satisfies $h(\alpha) = 0$ and $h(\beta) = f(\beta)$. Hence, the functions f_b and g_b defined by

$$f_b(t) = \begin{cases} 0, & t \in (a, \alpha], \\ h(t), & t \in (\alpha, \beta), \\ f(t), & t \in [\beta, b), \end{cases} \quad \text{and} \quad g_b(t) = \begin{cases} 0, & t \in (a, \alpha], \\ k(t), & t \in (\alpha, \beta), \\ g(t), & t \in [\beta, b), \end{cases}$$

satisfy the appropriate inhomogeneous canonical equations on (a, α), (α, β), and (β, b). Since $f_b(\alpha) = h(\alpha)$ and $f_b(\beta) = h(\beta)$ it follows that $f_b \in AC(\imath)$. $\square$

In particular, if f is a solution of the homogeneous system (7.2.4), then f can be localized as indicated above. The following restatement of this fact in terms of matrix functions (groupings of column vector functions) is useful. Note that the modification of the solutions of the homogeneous equation involves a solution of the inhomogeneous equation.

Corollary 7.5.7. *Let the canonical system (7.2.3) be definite and choose a compact interval $[\alpha, \beta] \subset \imath$ such that the system is definite on $[\alpha, \beta]$. Let $Y(\cdot, \lambda)$ be a fundamental matrix of (7.2.4). Then there exist a 2×2 matrix function $Y_a(\cdot, \lambda) \in AC(\imath)$ and a 2×2 matrix function $Z_a(\cdot, \lambda)$ whose columns belong to $\mathcal{L}^2_\Delta(\imath)$, satisfying*

$$JY'_a(t, \lambda) - H(t)Y_a(t, \lambda) = \lambda\Delta(t)Y_a(t, \lambda) + \Delta(t)Z_a(t, \lambda)$$

such that

$$Y_a(t, \lambda) = \begin{cases} Y(t, \lambda), & t \in (a, \alpha], \\ 0, & t \in [\beta, b), \end{cases} \quad \text{and} \quad Z_a(t, \lambda) = \begin{cases} 0, & t \in (a, \alpha], \\ 0, & t \in [\beta, b). \end{cases}$$

Similarly, there exist a 2×2 matrix function $Y_b(\cdot, \lambda) \in AC(\imath)$ and a 2×2 matrix function $Z_b(\cdot, \lambda)$ whose columns belong to $\mathcal{L}^2_\Delta(\imath)$, satisfying

$$JY'_b(t, \lambda) - H(t)Y_b(t, \lambda) = \lambda\Delta(t)Y_b(t, \lambda) + \Delta(t)Z_b(t, \lambda)$$

such that

$$Y_b(t, \lambda) = \begin{cases} 0, & t \in (a, \alpha], \\ Y(t, \lambda), & t \in [\beta, b), \end{cases} \quad \text{and} \quad Z_b(t, \lambda) = \begin{cases} 0, & t \in (a, \alpha], \\ 0, & t \in [\beta, b). \end{cases}$$

With $\phi \in \mathbb{C}^2$ observe that the function $Y_a(\cdot, \lambda)\phi$ belongs to $\mathcal{L}_\Delta^2(\imath)$ if and only if $Y(\cdot, \lambda)\phi$ is square-integrable with respect to Δ at a, and, likewise, that the function $Y_b(\cdot, \lambda)\phi$ belongs to $\mathcal{L}_\Delta^2(\imath)$ if and only if $Y(\cdot, \lambda)\phi$ is square-integrable with respect to Δ at b.

It is useful to have a special notation for the elements that modify the pairs $\{Y(\cdot, \lambda), \lambda Y(\cdot, \lambda)\}$ in Corollary 7.5.7. Define the matrix functions $\mathcal{Y}_a(\cdot, \lambda)$ and $\mathcal{Y}_b(\cdot, \lambda)$ by

$$\begin{aligned}
\mathcal{Y}_a(\cdot, \lambda) &:= \{Y_a(\cdot, \lambda), \lambda Y_a(\cdot, \lambda) + Z_a(\cdot, \lambda)\}, \\
\mathcal{Y}_b(\cdot, \lambda) &:= \{Y_b(\cdot, \lambda), \lambda Y_b(\cdot, \lambda) + Z_b(\cdot, \lambda)\},
\end{aligned} \tag{7.5.6}$$

that is, for $\phi \in \mathbb{C}^2$ one has

$$\begin{aligned}
\mathcal{Y}_a(\cdot, \lambda)\phi &= \{Y_a(\cdot, \lambda)\phi, \lambda Y_a(\cdot, \lambda)\phi + Z_a(\cdot, \lambda)\phi\}, \\
\mathcal{Y}_b(\cdot, \lambda)\phi &= \{Y_b(\cdot, \lambda)\phi, \lambda Y_b(\cdot, \lambda)\phi + Z_b(\cdot, \lambda)\phi\}.
\end{aligned}$$

Note that $\mathcal{Y}_a(\cdot, \lambda)$ and $\mathcal{Y}_b(\cdot, \lambda)$ satisfy

$$\begin{aligned}
\mathcal{Y}_a(t, \lambda) &= \begin{cases} \{Y(t, \lambda), \lambda Y(t, \lambda)\}, & a < t \leq \alpha, \\ \{0, 0\}, & \beta \leq t < b, \end{cases} \\
\mathcal{Y}_b(t, \lambda) &= \begin{cases} \{0, 0\}, & a < t \leq \alpha, \\ \{Y(t, \lambda), \lambda Y(t, \lambda)\}, & \beta \leq t < b. \end{cases}
\end{aligned} \tag{7.5.7}$$

It is clear from the construction that the columns of $\mathcal{Y}_a(\cdot, \lambda)$ or $\mathcal{Y}_b(\cdot, \lambda)$ are square-integrable on (a, b) with respect to Δ if and only if the corresponding columns of $Y(\cdot, \lambda)$ have this property at a or b, respectively.

7.6 Maximal and minimal relations for canonical systems

In this and later sections it will be assumed that the canonical system (7.2.3) is real as in Definition 7.2.7 and definite as in Definition 7.5.5: such systems will be called *real definite canonical systems*. In this context the central Hilbert space will be $L_\Delta^2(\imath)$, in which the maximal and minimal relations associated with the real definite canonical system (7.2.3) will be defined. In principle, both these relations may be multivalued. The results from Section 7.4 and Section 7.5 make it possible to consider the limit-circle case and the limit-point case from the point of view of the maximal and minimal relations.

The real definite canonical system (7.2.3) induces the *maximal relation* $T_{\max}$ in $L_\Delta^2(\imath)$ defined by

$$T_{\max} = \{\{f, g\} \in L_\Delta^2(\imath) \times L_\Delta^2(\imath) : Jf' - Hf = \Delta g\}.$$

Since the elements of $L^2_\Delta(\imath)$ are equivalence classes, the definition of $T_{\max}$ needs the following explanation: an element $\{f, g\} \in L^2_\Delta(\imath) \times L^2_\Delta(\imath)$ belongs to $T_{\max}$ if and only if the equivalence class f contains an absolutely continuous representative $\widetilde{f}$ such that the inhomogeneous equation $J\widetilde{f}'(t) - H(t)\widetilde{f}(t) = \Delta(t)\widetilde{g}(t)$ is satisfied for almost every $t \in \imath$. Here $\widetilde{g}$ is any representative of $g \in L^2_\Delta(\imath)$; observe that the function $\Delta(t)\widetilde{g}(t)$ is independent of the representative. The above argument also shows that the relation $T_{\max}$ is linear.

Since the canonical system (7.2.3) is assumed to be definite, the absolutely continuous representative is unique.

Lemma 7.6.1. *If $\{f, g\} \in T_{\max}$, then the equivalence class f has a unique absolutely continuous representative.*

Proof. Let $\{f, g\} \in T_{\max}$ and let $\widetilde{f}_1$ and $\widetilde{f}_2$ be absolutely continuous representatives of f. Then $J(\widetilde{f}_1 - \widetilde{f}_2)' - H(\widetilde{f}_1 - \widetilde{f}_2) = 0$ holds and

$$\Delta(t)(\widetilde{f}_1 - \widetilde{f}_2)(t) = 0, \quad t \in \imath.$$

Therefore, by Definition 7.5.5, it follows that $\widetilde{f}_1(t) = \widetilde{f}_2(t)$ for all $t \in \imath$. $\square$

It will be shown that $T_{\max}$ is the adjoint of a symmetric relation whose defect numbers are equal and at most $(2, 2)$. Let T_0 be the *preminimal relation*, i.e., the restriction of the maximal relation $T_{\max}$ to the elements where the first component has compact support in $\imath$:

$$T_0 = \big\{\{f, g\} \in T_{\max} : f \text{ has compact support}\big\}.$$

More precisely, an element $\{f, g\} \in L^2_\Delta(\imath) \times L^2_\Delta(\imath)$ belongs to T_0 if and only if the equivalence class f contains an absolutely continuous representative $\widetilde{f}$ with compact support such that the inhomogeneous equation $J\widetilde{f}'(t) - H(t)\widetilde{f}(t) = \Delta(t)\widetilde{g}(t)$ is satisfied for almost every $t \in \imath$. Here $\widetilde{g}$ is any representative of $g \in L^2_\Delta(\imath)$. The *minimal relation* $T_{\min}$ is defined as $T_{\min} = \overline{T}_0$.

Theorem 7.6.2. *The closure $T_{\min} = \overline{T}_0$ of T_0 is a closed symmetric relation in $L^2_\Delta(\imath)$ and it satisfies*

$$T_{\min} \subset (T_{\min})^* = T_{\max},$$

and, consequently, $T_{\min} = (T_{\max})^*$.

Proof. Step 1. It will be shown that

$$T_{\max} \subset (T_0)^*. \tag{7.6.1}$$

For this purpose, let $\{f, g\} \in T_{\max}$, $\{h, k\} \in T_0$, and choose an interval $[\alpha, \beta] \subset \imath$ containing the support h (and hence the support of Δk). Then

$$
\begin{aligned}
(g, h)_\Delta - (f, k)_\Delta &= \int_\alpha^\beta h(s)^* \Delta(s) g(s)\, ds - \int_\alpha^\beta k(s)^* \Delta(s) f(s)\, ds \\
&= h(\beta)^* Jf(\beta) - h(\alpha)^* Jf(\alpha) \\
&= 0
\end{aligned}
$$

by Corollary 7.2.3. Here $(\cdot, \cdot)_\Delta$ denotes the scalar product in $L^2_\Delta(\imath)$, f and h are the uniquely defined absolutely continuous representatives, while g and k are arbitrary representatives. Observe that the integral does not depend on the particular choice of g and k. This shows $\{f, g\} \in (T_0)^*$ and hence (7.6.1) follows.

Step 2. It will be shown that

$$(T_0)^* \subset T_{\max}. \tag{7.6.2}$$

For this, let $\{f, g\} \in (T_0)^*$. By Theorem 7.2.1, there exists a nontrivial absolutely continuous function u on $\imath$ such that $Ju' - Hu = \Delta g$. The aim is to show that for any representative f the difference $f - u$ is absolutely continuous modulo an element whose $\mathcal{L}^2_\Delta(a, b)$-norm is zero. Recall that the system is assumed to be definite, and hence there exists a compact interval $[\alpha_0, \beta_0]$ on which it is definite; cf. Proposition 7.5.4. Choose an interval $[\alpha_1, \beta_1] \subset \imath$ which contains $[\alpha_0, \beta_0]$; then the system is also definite on $[\alpha_1, \beta_1]$.

It is convenient to introduce the subspace

$$\mathfrak{M}_1 := \left\{ k \in \mathcal{L}^2_\Delta(\alpha_1, \beta_1) : \begin{array}{c} Jh' - Hh = \Delta k \text{ for some } h \in AC[\alpha_1, \beta_1] \\ \text{such that } h(\alpha_1) = h(\beta_1) = 0 \end{array} \right\}.$$

Let $k \in \mathfrak{M}_1$ and let $h \in AC[\alpha_1, \beta_1]$ be a solution of $Jh' - Hh = \Delta k$ for which $h(\alpha_1) = h(\beta_1) = 0$. It follows from (7.2.15) that

$$h(t) = Y(t, 0)J^{-1} \int_{\alpha_1}^t Y(s, 0)^* \Delta(s)k(s)\, ds, \quad t \in [\alpha_1, \beta_1], \tag{7.6.3}$$

where the fundamental matrix $Y(\cdot, \lambda)$ is fixed by $Y(\alpha_1, \lambda) = I$. Note that the condition $h(\beta_1) = 0$ implies

$$\int_{\alpha_1}^{\beta_1} Y(s, 0)^* \Delta(s)k(s)\, ds = 0. \tag{7.6.4}$$

Conversely, if $k \in \mathcal{L}^2_\Delta(\alpha_1, \beta_1)$ satisfies (7.6.4), then $k \in \mathfrak{M}_1$ since $h \in AC[\alpha_1, \beta_1]$ in (7.6.3) satisfies $Jh' - Hh = \Delta k$ and $h(\alpha_1) = h(\beta_1) = 0$. In other words, one has

$$\mathfrak{M}_1 = \left\{ k \in \mathcal{L}^2_\Delta(\alpha_1, \beta_1) : \int_{\alpha_1}^{\beta_1} Y(s, 0)^* \Delta(s)k(s)\, ds = 0 \right\}.$$

Now let $k \in \mathfrak{M}_1$ and let h be defined by (7.6.3). Then the pair of functions $\{h, k\}$ can be trivially extended to all of $\imath$ and the extended pair, which will also be denoted by $\{h, k\}$, belongs to T_0. As $\{f, g\} \in (T_0)^*$, one has $(h, g)_\Delta = (k, f)_\Delta$, and since the supports of h and k are inside $[\alpha_1, \beta_1]$ it follows that

$$\int_{\alpha_1}^{\beta_1} g(s)^* \Delta(s)h(s)\, ds = \int_{\alpha_1}^{\beta_1} f(s)^* \Delta(s)k(s)\, ds. \tag{7.6.5}$$

Consider the pair $\{u, g\}$ on $[\alpha_1, \beta_1]$. Note that on this interval u is absolutely continuous and g is square-integrable with respect to Δ. It follows from the Lagrange identity in Corollary 7.2.3 applied to the pairs $\{h, k\}$ and $\{u, g\}$, and $h(\alpha_1) = h(\beta_1) = 0$ that

$$\int_{\alpha_1}^{\beta_1} g(s)^* \Delta(s) h(s) \, ds = \int_{\alpha_1}^{\beta_1} u(s)^* \Delta(s) k(s) \, ds. \tag{7.6.6}$$

Combining (7.6.5) and (7.6.6), one obtains that

$$\int_{\alpha_1}^{\beta_1} (f(s) - u(s))^* \Delta(s) k(s) \, ds = 0$$

for all $k \in \mathfrak{M}_1$. In other words, the restriction of $f - u$ to $[\alpha_1, \beta_1]$ is orthogonal to $\mathfrak{M}_1$ in the semidefinite Hilbert space $\mathcal{L}_\Delta^2(\alpha_1, \beta_1)$. Furthermore, by (7.6.4) one has that $Y(\cdot, 0)\gamma$ is orthogonal to $\mathfrak{M}_1$ in $\mathcal{L}_\Delta^2(\alpha_1, \beta_1)$ for all $\gamma \in \mathbb{C}^2$, and since the same is true for $f - u$ it follows that $f - u - Y(\cdot, 0)\gamma$ is orthogonal to $\mathfrak{M}_1$ in $\mathcal{L}_\Delta^2(\alpha_1, \beta_1)$ for all $\gamma \in \mathbb{C}^2$.

Next it will be shown that for some $\gamma_1 \in \mathbb{C}^2$ the function $f - u - Y(\cdot, 0)\gamma_1$ belongs to $\mathfrak{M}_1$. In fact, first of all it is clear from (7.2.15) that for any $\gamma \in \mathbb{C}^2$

$$h(t) = Y(t, 0) J^{-1} \int_{\alpha_1}^t Y(s, 0)^* \Delta(s) \big(f(s) - u(s) - Y(s, 0)\gamma \big) \, ds$$

satisfies $Jh' - Hh = \Delta(f - u - Y(\cdot, 0)\gamma)$ and $h(\alpha_1) = 0$. To satisfy the boundary condition $h(\beta_1) = 0$, choose $\gamma = \gamma_1 \in \mathbb{C}^2$ such that

$$\int_{\alpha_1}^{\beta_1} Y(s, 0)^* \Delta(s)(f(s) - u(s)) \, ds = \int_{\alpha_1}^{\beta_1} Y(s, 0)^* \Delta(s) Y(s, 0)\gamma_1 \, ds;$$

this is possible since the system is definite on $[\alpha_1, \beta_1]$ and hence the matrix

$$\int_{\alpha_1}^{\beta_1} Y(s, 0)^* \Delta(s) Y(s, 0) \, ds$$

is invertible; cf. Corollary 7.5.3. Therefore, $f - u - Y(\cdot, 0)\gamma_1 \in \mathfrak{M}_1$. Since the element $f - u - Y(\cdot, 0)\gamma_1$ is orthogonal to $\mathfrak{M}_1$ in $\mathcal{L}_\Delta^2(\alpha_1, \beta_1)$, this yields

$$\int_{\alpha_1}^{\beta_1} \big(f(s) - u(s) - Y(s, 0)\gamma_1 \big)^* \Delta(s) \big(f(s) - u(s) - Y(s, 0)\gamma_1 \big) \, ds = 0,$$

and hence there exists a function ω_1 on $[\alpha_1, \beta_1]$ such that

$$f(s) = u(s) + Y(s, 0)\gamma_1 + \omega_1(s), \quad \Delta(s)\omega_1(s) = 0, \quad s \in [\alpha_1, \beta_1].$$

Likewise, on any interval $[\alpha_2, \beta_2]$ extending $[\alpha_1, \beta_1]$ the same argument shows that there exist $\gamma_2 \in \mathbb{C}^2$ and a function ω_2 such that

$$f(s) = u(s) + \widetilde{Y}(s, 0)\gamma_2 + \omega_2(s), \quad \Delta(s)\omega_2(s) = 0, \quad s \in [\alpha_2, \beta_2];$$

here the fundamental matrix $\widetilde{Y}(\cdot, \lambda)$ is fixed by $\widetilde{Y}(\alpha_2, \lambda) = I$. Hence, on the smaller interval one obtains for $s \in [\alpha_1, \beta_1]$

$$\omega_1(s) - \omega_2(s) = \widetilde{Y}(s, 0)\gamma_2 - Y(s, 0)\gamma_1, \quad \Delta(s)(\omega_1(s) - \omega_2(s)) = 0.$$

Since the system is definite on the interval $[\alpha_1, \beta_1]$, this shows that $\omega_1(s) = \omega_2(s)$ and $Y(s, 0)\gamma_1 = \widetilde{Y}(s, 0)\gamma_2$ for $s \in [\alpha_1, \beta_1]$, and hence $Y(s, 0)\gamma_1 = \widetilde{Y}(s, 0)\gamma_2$ for $s \in \imath$. One concludes that there exists a function ω such that

$$f(s) = u(s) + Y(s, 0)\gamma_1 + \omega(s), \quad \Delta(s)\omega(s) = 0, \quad s \in \imath.$$

Thus, the functions f and $u + Y(\cdot, 0)\gamma_1$ belong to the same equivalence class in $L_\Delta^2(\imath)$. Since $J(u + Y(\cdot, 0)\gamma_1)' - H(u + Y(\cdot, 0)\gamma_1) = \Delta g$ it follows that $\{f, g\} \in T_{\max}$ and $u + Y(\cdot, 0)\gamma_1$ is the unique absolutely continuous representative of f. This implies (7.6.2).

Step 3. It follows from (7.6.1) and (7.6.2) that $T_{\max} = (T_0)^*$ and, in particular, this implies that $T_{\max}$ is closed. Hence, the fact that $T_0 \subset T_{\max}$ and the definition $T_{\min} = \overline{T}_0$ imply that

$$T_{\min} = \overline{T}_0 \subset T_{\max} = (T_0)^* = (T_{\min})^*.$$

Thus, $T_{\min}$ is a (closed) symmetric relation and $T_{\min} = (T_{\max})^*$.   □

At this stage note that $T_{\min}$ is a closed symmetric relation which need not be densely defined in $L_\Delta^2(\imath)$. Consider the orthogonal decomposition

$$L_\Delta^2(\imath) = (\operatorname{mul} T_{\min})^\perp \oplus \operatorname{mul} T_{\min} = \overline{\operatorname{dom} T_{\max}} \oplus \operatorname{mul} T_{\min} \qquad (7.6.7)$$

and recall from Theorem 1.4.11 that $T_{\min}$ admits the corresponding orthogonal sum decomposition

$$T_{\min} = (T_{\min})_{\mathrm{op}} \,\widehat{\oplus}\, (\{0\} \times \operatorname{mul} T_{\min}). \qquad (7.6.8)$$

The operator part $(T_{\min})_{\mathrm{op}}$ is not necessarily densely defined in $\overline{\operatorname{dom} T_{\max}}$ and $\{0\} \times \operatorname{mul} T_{\min}$ is the purely multivalued self-adjoint relation in $\operatorname{mul} T_{\min}$.

Since by Theorem 7.6.2 the relation $T_{\min}$ is closed and symmetric, while $(T_{\min})^* = T_{\max}$, it follows from the von Neumann decomposition, as given in Theorem 1.7.11, that the relation $T_{\max}$ has the componentwise sum decomposition

$$T_{\max} = T_{\min} \,\widehat{+}\, \widehat{\mathfrak{N}}_\lambda(T_{\max}) \,\widehat{+}\, \widehat{\mathfrak{N}}_\mu(T_{\max}), \quad \lambda \in \mathbb{C}^+, \, \mu \in \mathbb{C}^-,$$

where the sums are direct. Now assume that $f \in \mathfrak{N}_\zeta(T_{\max})$ with $\zeta \in \mathbb{C} \setminus \mathbb{R}$. Then $\{f, \zeta f\} \in T_{\max}$, which is equivalent to $f \in L^2_\Delta(\imath)$ having an absolutely continuous representative $\widetilde{f}$ such that $J\widetilde{f}' - H\widetilde{f} = \zeta \Delta \widetilde{f}$. Since the system consists of 2×2 matrix functions, there are precisely two linearly independent solutions of this homogeneous equation and at most two linearly independent solutions that are square-integrable with respect to Δ. Furthermore, since the canonical system is assumed to be real, the number of solutions at $\zeta \in \mathbb{C}^+$ that are square-integrable with respect to Δ coincides with the number of solutions at $\bar{\zeta} \in \mathbb{C}^-$ that are square-integrable with respect to Δ by Corollary 7.2.9. Taking into account Corollary 7.4.5 and Corollary 7.4.6 one then obtains the following statement. The case where both endpoints of the interval $\imath$ are in the limit-point case will be dealt with in Corollary 7.6.9.

Corollary 7.6.3. *Let $T_{\min}$ be the minimal symmetric relation associated with the real definite canonical system (7.2.3) in $L^2_\Delta(\imath)$. Then the following statements hold:*

(i) *If both endpoints of $\imath$ are regular or quasiregular, then the defect numbers of $T_{\min}$ are $(2, 2)$.*

(ii) *If one endpoint of $\imath$ is regular or quasiregular and one endpoint is in the limit-point case, then the defect numbers of $T_{\min}$ are $(1, 1)$.*

Recall that elements $\{f, g\} \in T_{\max}$ satisfy the equation $Jf' - Hf = \Delta g$ and that the entries also satisfy the integrability condition $f, g \in L^2_\Delta(\imath)$. These two ingredients make it possible to extend the usual Lagrange identity in Corollary 7.2.3 on a compact subinterval to all of $\imath$. This new Lagrange identity for the elements in $T_{\max}$ will play an important role in the rest of this chapter.

Lemma 7.6.4. *Let $T_{\max}$ be the maximal relation associated with the real definite canonical system (7.2.3) in $L^2_\Delta(\imath)$. Then for all $\{f, g\}, \{h, k\} \in T_{\max}$ one has*

$$(g, h)_\Delta - (f, k)_\Delta = \lim_{t \to b} h(t)^* Jf(t) - \lim_{t \to a} h(t)^* Jf(t), \qquad (7.6.9)$$

where $f(t)$ and $h(t)$ denote the values of the unique absolutely continuous representatives of f and h, respectively.

Proof. First observe that for all elements $\{f, g\}, \{h, k\} \in T_{\max}$ and every compact subinterval $[\alpha, \beta] \subset \imath$ one has

$$\int_\alpha^\beta \left(h(s)^* \Delta(s) g(s) - k(s)^* \Delta(s) f(s) \right) ds = h(\beta)^* Jf(\beta) - h(\alpha)^* Jf(\alpha)$$

by the Lagrange identity in Corollary 7.2.3; here $f(t)$ and $h(t)$ denote the values of the unique absolutely continuous representatives of f and h, and $g(t)$ and $k(t)$ are the values of some representatives of g and k. Observe that the integral on the left-hand side does not depend on the choice of the representatives of g and

k. The limit of the left-hand side exists as $\beta \to b$ and $\alpha \to a$, respectively, since $f, g, h, k \in L^2_\Delta(\imath)$. As a consequence, one sees that each of the limits

$$\lim_{t \to a} h(t)^* J f(t) \quad \text{and} \quad \lim_{t \to b} h(t)^* J f(t)$$

exists and hence the Lagrange identity takes the limit form (7.6.9).   □

Remark 7.6.5. Observe that in (7.6.9) one uses the values $f(t)$ and $h(t)$ of the unique absolutely continuous representatives of f and h in $\operatorname{dom} T_{\max}$, respectively. For instance, if $\{0, g\} \in T_{\max}$ and $\{h, k\} \in T_{\max}$, then there exists an absolutely continuous function $\widetilde{f}$ such that $\Delta(t)\widetilde{f}(t) = 0$ and $J\widetilde{f}'(t) - H(t)\widetilde{f}(t) = \Delta(t)\widetilde{g}(t)$ is satisfied for almost every $t \in \imath$, where $\widetilde{g}$ is any representative of $g \in L^2_\Delta(\imath)$. In this situation the identity (7.6.9) has the form

$$(g, h)_\Delta - (0, k)_\Delta = \lim_{t \to b} h(t)^* J\widetilde{f}(t) - \lim_{t \to a} h(t)^* J\widetilde{f}(t).$$

The elements in the minimal symmetric relation $T_{\min} = \overline{T}_0$ can be easily characterized in terms of these limits.

Corollary 7.6.6. *Let $T_{\min}$ and $T_{\max}$ be the minimal and maximal relations associated with the real definite canonical system (7.2.3) in $L^2_\Delta(\imath)$ and let $\{f, g\} \in T_{\max}$. Then $\{f, g\} \in T_{\min}$ if and only if*

$$\lim_{t \to a} h(t)^* J f(t) = 0 \quad \text{and} \quad \lim_{t \to b} h(t)^* J f(t) = 0 \tag{7.6.10}$$

for all $\{h, k\} \in T_{\max}$, where $f(t)$ and $h(t)$ denote the values of the unique absolutely continuous representatives of f and h, respectively.

Proof. Observe first that since $T_{\min} = (T_{\max})^*$ one has $\{f, g\} \in T_{\min}$ if and only if $(g, h)_\Delta = (f, k)_\Delta$ for all $\{h, k\} \in T_{\max}$. Hence, it follows from the Lagrange identity (7.6.9) that $\{f, g\} \in T_{\min}$ if and only if

$$\lim_{t \to b} h(t)^* J f(t) = \lim_{t \to a} h(t)^* J f(t) \tag{7.6.11}$$

for all $\{h, k\} \in T_{\max}$. To see that for $\{f, g\} \in T_{\min}$ each of the limits in (7.6.11) is zero, consider $\{h, k\} \in T_{\max}$ and use Proposition 7.5.6 (with $\lambda = 0$) to obtain an element $\{h_a, k_a\} \in T_{\max}$ that coincides with $\{h, k\}$ in a neighborhood of a and with $\{0, 0\}$ in a neighborhood of b. Then (7.6.11) implies

$$\lim_{t \to a} h(t)^* J f(t) = \lim_{t \to a} h_a(t)^* J f(t) = \lim_{t \to b} h_a(t)^* J f(t) = 0,$$

and hence (7.6.10) follows together with (7.6.11). Conversely, if (7.6.10) holds for some $\{f, g\} \in T_{\max}$ and all $\{h, k\} \in T_{\max}$, then the identity (7.6.11) holds for all $\{h, k\} \in T_{\max}$ and hence $\{f, g\} \in T_{\min}$.   □

The main difficulty when dealing with the boundary value problems associated with the system (7.2.3) is to break the limits in (7.6.9) and in (7.6.10) into limits of the separate factors. The case where the endpoints are regular, quasiregular, or in the limit-circle case will be pursued in Section 7.7. If one of the endpoints is in the limit-point case, the situation is somewhat simpler, since one of the limits in (7.6.9) automatically vanishes, as will be shown now. A further discussion of the remaining limit will be pursued in Section 7.8.

Lemma 7.6.7. *Let $T_{\max}$ be the maximal relation associated with the real definite canonical system* (7.2.3) *in* $L^2_\Delta(\imath)$, *let $\lambda \in \mathbb{C}$, and let $\mathcal{Y}_a(\cdot, \lambda)$ and $\mathcal{Y}_b(\cdot, \lambda)$ be as in* (7.5.6). *Then the following statements hold:*

(i) *Let a be a regular or quasiregular endpoint and let b be in the limit-point case. Then for $\lambda \in \mathbb{C}$*

$$T_{\max} = T_{\min} \,\widehat{+}\, \{\mathcal{Y}_a(\cdot, \lambda)\phi : \phi \in \mathbb{C}^2\}, \qquad (7.6.12)$$

where the sum is direct.

(ii) *Let b be a regular or quasiregular endpoint and let a be in the limit-point case. Then for $\lambda \in \mathbb{C}$*

$$T_{\max} = T_{\min} \,\widehat{+}\, \{\mathcal{Y}_b(\cdot, \lambda)\phi : \phi \in \mathbb{C}^2\},$$

where the sum is direct.

Proof. It suffices to consider the case (i) since the case (ii) can be proved in a similar way. Let $Y(\cdot, \lambda)$ be a fundamental matrix of (7.2.4). Note that if a is regular or quasiregular, then it follows from Corollary 7.3.3 and (7.5.7) that $\mathcal{Y}_a(\cdot, \lambda)\phi \in L^2_\Delta(\imath) \times L^2_\Delta(\imath)$, and Corollary 7.5.7 implies that $\mathcal{Y}_a(\cdot, \lambda)\phi \in T_{\max}$ for all $\phi \in \mathbb{C}^2$.

As $T_{\min} \subset T_{\max}$, it is clear that the right-hand side of (7.6.12) is contained in $T_{\max}$. By assumption and Corollary 7.6.3 (ii) $T_{\max}$ is a two-dimensional extension of $T_{\min}$ and hence it suffices to show that the elements $\mathcal{Y}_a(\cdot, \lambda)\phi$, $\phi \in \mathbb{C}^2$, span a two-dimensional subspace of $T_{\max}$ which has a trivial intersection with $T_{\min}$. In other words, it remains to check that $\mathcal{Y}_a(\cdot, \lambda)\phi \in T_{\min}$ if and only if $\phi = 0$. Suppose that $\mathcal{Y}_a(\cdot, \lambda)\phi \in T_{\min}$ for some $\phi \in \mathbb{C}^2$. For all $\psi \in \mathbb{C}^2$ one has $\mathcal{Y}_a(\cdot, \bar{\lambda})\psi \in T_{\max}$ and therefore, by Corollary 7.6.6,

$$0 = \lim_{t \to a} \psi^* Y_a(t, \bar{\lambda})^* J Y_a(t, \lambda)\phi.$$

Since $Y_a(\cdot, \lambda) = Y(\cdot, \lambda)$ and $Y_a(\cdot, \bar{\lambda}) = Y(\cdot, \bar{\lambda})$ in a neighborhood of a, it follows that

$$0 = \lim_{t \to a} \psi^* Y(t, \bar{\lambda})^* J Y(t, \lambda)\phi$$

for all $\psi \in \mathbb{C}^2$. Fix the fundamental matrix $Y(\cdot, \lambda)$ by $Y(a, \lambda) = I$. This leads to $\psi^* J \phi = 0$ for all $\psi \in \mathbb{C}^2$, which implies $\phi = 0$. Hence, the right-hand side of (7.6.12) is a two-dimensional extension of $T_{\min}$ which is contained in $T_{\max}$, and therefore coincides with $T_{\max}$. $\square$

In the next lemma the case of a singular endpoint in the limit-point case is discussed.

Lemma 7.6.8. *Let $T_{\max}$ be the maximal relation associated with the real definite canonical system (7.2.3) in $L^2_\Delta(\imath)$. Then the endpoint a or b of the interval $\imath$ is in the limit-point case if and only if for all $\{f, g\}, \{h, k\} \in T_{\max}$ one has*

$$\lim_{t \to a} h(t)^* J f(t) = 0 \quad or \quad \lim_{t \to b} h(t)^* J f(t) = 0,$$

respectively. Here $f(t)$ and $h(t)$ denote the values of the unique absolutely continuous representatives of f and h, respectively.

Proof. It suffices to consider the case that the endpoint a is regular. The proof of the case where b is regular is similar. As usual the fundamental matrix is fixed by $Y(a, \lambda) = I$.

Assume that b is in the limit-point case. In this case one has the decomposition (7.6.12) in Lemma 7.6.7. Let $\{f, g\}, \{h, k\} \in T_{\max}$ be decomposed in the form

$$\{f, g\} = \{f_0, g_0\} + \mathcal{Y}_a(\cdot, \lambda)\phi \quad \text{and} \quad \{h, k\} = \{h_0, k_0\} + \mathcal{Y}_a(\cdot, \lambda)\psi,$$

where $\{f_0, g_0\}, \{h_0, k_0\} \in T_{\min}$ and $\phi, \psi \in \mathbb{C}^2$. Then it follows from (7.5.7) that

$$\lim_{t \to b} h(t)^* J f(t) = \lim_{t \to b} h_0(t)^* J f_0(t) = 0,$$

where Corollary 7.6.6 was used in the last step.

Conversely, assume that for all $\{f, g\}, \{h, k\} \in T_{\max}$

$$\lim_{t \to b} h^*(t) J f(t) = 0. \tag{7.6.13}$$

Then b is in the limit-point case. To see this, assume that b is not in the limit-point case, so that b is in the limit-circle case by Corollary 7.4.5. It then follows that for $\lambda_0 \in \mathbb{R}$ the columns of the matrix function $\mathcal{Y}_b(\cdot, \lambda_0)$ are square-integrable with respect to Δ at b. Consider $\{f, g\} = \{h, k\} = \mathcal{Y}_b(\cdot, \lambda_0)\phi \in T_{\max}$ for some $\phi \in \mathbb{C}^2$ such that $\phi^* J \phi \neq 0$. Using (7.5.7) and $Y(t, \lambda_0)^* J Y(t, \lambda_0)$ (see (7.2.8)), one computes

$$\lim_{t \to b} h^*(t) J f(t) = \lim_{t \to b} \phi^* Y(t, \lambda_0)^* J Y(t, \lambda_0)\phi = \phi^* J \phi \neq 0,$$

which contradicts (7.6.13). $\square$

Corollary 7.6.9. *Let $T_{\min}$ be the minimal symmetric relation associated with the real definite canonical system (7.2.3) in $L^2_\Delta(\imath)$ and assume that both endpoints of $\imath$ are in the limit-point case. Then the defect numbers of $T_{\min}$ are $(0, 0)$ and $T_{\min} = T_{\max}$ is self-adjoint in $L^2_\Delta(\imath)$.*

Proof. Assume that the system is definite on the interval $[\alpha, \beta] \subset \imath$. Then the system is also definite on the intervals (a, β') and (α', b), with $\beta' \in (\beta, b)$ and $\alpha' \in (a, \alpha)$, respectively. Denote the maximal relation in $L^2_\Delta(\alpha', b)$ associated with the canonical system by $T_{\max}(\alpha', b)$. It follows that the endpoint α' is regular and that the endpoint b is in the limit-point case for the canonical system on (α', b). To see the last assertion, assume that there are two linearly independent solutions on (α', b) which are square-integrable with respect to Δ at b. Since these solutions admit unique extensions to solutions on (a, b), one obtains a contradiction. In particular, for all $\{f_b, g_b\}, \{h_b, k_b\} \in T_{\max}(\alpha', b)$ one concludes from Lemma 7.6.8 that

$$\lim_{t \to b} h_b(t)^* J f_b(t) = 0.$$

Now consider $\{f, g\}, \{h, k\} \in T_{\max}$ and let f_b, g_b, h_b, k_b be the restrictions of f, g, h, k to the interval (α', b). Then one has $\{f_b, g_b\}, \{h_b, k_b\} \in T_{\max}(\alpha', b)$, and hence

$$\lim_{t \to b} h(t)^* J f(t) = \lim_{t \to b} h_b(t)^* J f_b(t) = 0.$$

A similar argument applies to the canonical system on (a, β') and shows that

$$\lim_{t \to a} h(t)^* J f(t) = 0$$

for all $\{f, g\}, \{h, k\} \in T_{\max}$. Therefore, Lemma 7.6.4 implies

$$(g, h)_\Delta - (f, k)_\Delta = \lim_{t \to b} h(t)^* J f(t) - \lim_{t \to a} h(t)^* J f(t) = 0$$

for all $\{f, g\}, \{h, k\} \in T_{\max}$, and hence $T_{\max} \subset T^*_{\max}$. From Theorem 7.6.2 one now concludes $T_{\min} = T_{\max}$ and thus it follows that the defect numbers of $T_{\min}$ are $(0, 0)$. □

7.7   Boundary triplets for the limit-circle case

Assume that the system (7.2.3) is real and definite, and assume that the endpoints of the system are both in the limit-circle case. A boundary triplet will be presented for $T_{\max} = (T_{\min})^*$ and the self-adjoint extensions of $T_{\min}$ will be described in terms of the boundary triplet. For a straightforward presentation the case where the endpoints are regular or quasiregular is discussed first. At the end of the section it will be explained what modifications are necessary for endpoints which are in the limit-circle case and which are not regular or quasiregular.

The symmetric relation $T_{\min} = \overline{T}_0$ will now be described when a and b are regular or quasiregular.

Lemma 7.7.1. *Assume that a and b are regular or quasiregular endpoints for the canonical system (7.2.3). Then the minimal relation $T_{\min}$ is given by*

$$T_{\min} = \left\{ \{f, g\} \in T_{\max} : f(a) = f(b) = 0 \right\},$$

where $f(a)$ and $f(b)$ denote the boundary values of the unique absolutely continuous representatives of f.

Proof. According to Corollary 7.6.6, the element $\{f, g\} \in T_{\max}$ belongs to $T_{\min}$ if and only if

$$\lim_{t \to a} h(t)^* J f(t) = 0 \quad \text{and} \quad \lim_{t \to b} h(t)^* J f(t) = 0$$

for all $\{h, k\} \in T_{\max}$. Since the endpoints are regular or quasiregular, these conditions are the same as

$$h(a)^* J f(a) = 0 \quad \text{and} \quad h(b)^* J f(b) = 0$$

for all $\{h, k\} \in T_{\max}$. Now observe that for any $\gamma \in \mathbb{C}^2$ and $k \in L^2_\Delta(\imath)$ there exists an element $h \in L^2_\Delta(\imath)$ such that $\{h, k\} \in T_{\max}$ and $h(a) = \gamma$ or $h(b) = \gamma$. Hence, it follows that $f(a) = 0$ and $f(b) = 0$. $\qquad\square$

When the endpoints of the interval $\imath = (a, b)$ are regular or quasiregular for the canonical system, then the solutions of $Jf' - Hf = \lambda \Delta f$, $\lambda \in \mathbb{C}$, automatically belong to $L^2_\Delta(\imath)$ and thus $\dim \ker (T_{\max} - \lambda) = 2$, so that the defect numbers of $T_{\min}$ are $(2, 2)$; cf. Corollary 7.6.3. In the next theorem a boundary triplet for $(T_{\min})^* = T_{\max}$ is provided and the corresponding γ-field and Weyl function are obtained in terms of an arbitrary fundamental matrix $Y(\cdot, \lambda)$ fixed by $Y(c, \lambda) = I$ for some $c \in [a, b]$.

Theorem 7.7.2. *Assume that a and b are regular or quasiregular endpoints for the canonical system (7.2.3) and let the fundamental matrix $Y(\cdot, \lambda)$ of (7.2.4) be fixed by $Y(c, \lambda) = I$ for some $c \in [a, b]$. Then $\{\mathbb{C}^2, \Gamma_0, \Gamma_1\}$, with*

$$\Gamma_0\{f, g\} = \frac{1}{\sqrt{2}}(f(a) + f(b)) \quad \text{and} \quad \Gamma_1\{f, g\} = -\frac{J}{\sqrt{2}}(f(a) - f(b)),$$

where $\{f, g\} \in T_{\max}$, is a boundary triplet for $(T_{\min})^* = T_{\max}$; here $f(a)$ and $f(b)$ denote the boundary values of the unique absolutely continuous representative of f. The corresponding γ-field and Weyl function are given by

$$\gamma(\lambda) = \sqrt{2} Y(\cdot, \lambda)\big(Y(a, \lambda) + Y(b, \lambda)\big)^{-1}, \quad \lambda \in \rho(A_0),$$

and

$$M(\lambda) = -J\big(Y(a, \lambda) - Y(b, \lambda)\big)\big(Y(a, \lambda) + Y(b, \lambda)\big)^{-1}, \quad \lambda \in \rho(A_0).$$

Proof. Let $\{f, g\}, \{h, k\} \in T_{\max}$. Since the endpoints a and b are regular or quasiregular, one has the Lagrange identity

$$(g, h)_\Delta - (f, k)_\Delta = \int_a^b \big(h(s)^* \Delta(s) g(s) - k(s)^* \Delta(s) f(s)\big) \, ds$$
$$= h(b)^* J f(b) - h(a)^* J f(a);$$

cf. Corollary 7.3.4. On the other hand, a straightforward calculation shows that

$$
\begin{aligned}
&\big(\Gamma_1\{f,g\},\Gamma_0\{h,k\}\big) - \big(\Gamma_0\{f,g\},\Gamma_1\{h,k\}\big) \\
&= -\frac{1}{2}\big(h(a)+h(b)\big)^* J\big(f(a)-f(b)\big) + \frac{1}{2}\big(h(a)-h(b)\big)^* J^*\big(f(a)+f(b)\big) \\
&= h(b)^* J f(b) - h(a)^* J f(a),
\end{aligned}
$$

and hence the boundary mappings Γ_0 and Γ_1 satisfy the abstract Green identity (2.1.1). Furthermore, the mapping $(\Gamma_0,\Gamma_1)^\top : T_{\max} \to \mathbb{C}^4$ is surjective. To see this, observe first that

$$
\begin{pmatrix} \Gamma_0\{f,g\} \\ \Gamma_1\{f,g\} \end{pmatrix} = \frac{1}{\sqrt{2}} \begin{pmatrix} I & I \\ -J & J \end{pmatrix} \begin{pmatrix} f(a) \\ f(b) \end{pmatrix}, \quad \{f,g\} \in T_{\max},
$$

and that the 4×4-matrix on the right-hand side is invertible. Hence, it suffices to check that for any $\gamma_a, \gamma_b \in \mathbb{C}^2$ there exists $\{f,g\} \in T_{\max}$ such that

$$
\begin{pmatrix} f(a) \\ f(b) \end{pmatrix} = \begin{pmatrix} \gamma_a \\ \gamma_b \end{pmatrix}. \tag{7.7.1}
$$

Choose a solution of the equation $Jh' - Hh = 0$ such that $h(a) = \gamma_a$ and modify h as in Proposition 7.5.6, so that it becomes a solution h_a of an inhomogeneous equation $Jh'_a - Hh_a = \Delta k_a$ which coincides with h in a neighborhood of a and vanishes in a neighborhood of the endpoint b. Then one has $\{h_a, k_a\} \in T_{\max}$ and $h_a(a) = \gamma_a$ and $h_a(b) = 0$. The same argument shows that there exists an element $\{h_b, k_b\} \in T_{\max}$ such that $h_b(b) = \gamma_b$ and $h_b(a) = 0$. Thus, for $f = h_a + h_b$ and $g = k_a + k_b$ one has $\{f,g\} \in T_{\max}$ and (7.7.1) holds. It follows that the mapping $(\Gamma_0,\Gamma_1)^\top : T_{\max} \to \mathbb{C}^4$ is surjective, as claimed. Therefore, $\{\mathbb{C}^2, \Gamma_0, \Gamma_1\}$ is a boundary triplet for $(T_{\min})^* = T_{\max}$.

To obtain the expressions for the associated γ-field and Weyl function, let $\lambda \in \rho(A_0)$, where $A_0 = \ker \Gamma_0$, and note that

$$
\mathfrak{N}_\lambda(T_{\max}) = \{ Y(\cdot,\lambda)\phi : \phi \in \mathbb{C}^2 \}, \quad \lambda \in \mathbb{C}.
$$

Hence, for $\widehat{f}_\lambda = \{ Y(\cdot,\lambda)\phi, \lambda Y(\cdot,\lambda)\phi \}$, $\phi \in \mathbb{C}^2$, and $\lambda \in \rho(A_0)$ one has

$$
\Gamma_0 \widehat{f}_\lambda = \frac{1}{\sqrt{2}}\big(Y(a,\lambda) + Y(b,\lambda) \big)\phi \quad \text{and} \quad \Gamma_1 \widehat{f}_\lambda = -\frac{J}{\sqrt{2}}\big(Y(a,\lambda) - Y(b,\lambda) \big)\phi,
$$

which leads to

$$
\gamma(\lambda) = \left\{ \left\{ \frac{1}{\sqrt{2}}\big(Y(a,\lambda) + Y(b,\lambda) \big)\phi, Y(\cdot,\lambda)\phi \right\} : \phi \in \mathbb{C}^2 \right\}
$$

and

$$
M(\lambda) = \left\{ \left\{ \frac{1}{\sqrt{2}}\big(Y(a,\lambda) + Y(b,\lambda) \big)\phi, -\frac{J}{\sqrt{2}}\big(Y(a,\lambda) - Y(b,\lambda) \big)\phi \right\} : \phi \in \mathbb{C}^2 \right\};
$$

cf. Definition 2.3.1 and Definition 2.3.4. Now observe that for $\lambda \in \rho(A_0)$ the matrix $Y(a, \lambda) + Y(b, \lambda)$ is invertible, as otherwise $(Y(a, \lambda) + Y(b, \lambda))\psi = 0$ for some nontrivial $\psi \in \mathbb{C}^2$ would imply that λ is an eigenvalue of the self-adjoint relation $A_0 = \ker \Gamma_0$ with corresponding eigenfunction $Y(\cdot, \lambda)\psi$; a contradiction. Therefore, the formulas for the γ-field and the Weyl function follow from the above identities for $\gamma(\lambda)$ and $M(\lambda)$. $\qquad\qquad\qquad\qquad\qquad\qquad\qquad\square$

Before formulating the next proposition some terminology is recalled. Let T be an integral operator of the form

$$Tf(t) = \int_a^b K(t, s)f(s)\, ds,$$

where f is a $\mathbb{C}^2$-valued function and K is a $\mathbb{C}^{2\times 2}$-valued measurable matrix kernel. If K is square-integrable with respect to the Lebesgue measure on $\imath \times \imath$, that is,

$$\int_a^b \int_a^b \|K(t, s)\|_2^2 \, ds\, dt < \infty,$$

where $\| \cdot \|_2$ is the Hilbert–Schmidt matrix norm in (7.1.4), then T is a bounded linear operator from $L^2(\imath)$ into itself, which belongs to the Hilbert–Schmidt class. Recall that a bounded linear operator from $L^2(\imath)$ into itself belongs to the *Hilbert–Schmidt class* if for some, and hence for all orthonormal bases (φ_i) in $L^2(\imath)$ one has

$$\sum_{i,j} |(T\varphi_i, \varphi_j)|^2 < \infty.$$

Proposition 7.7.3. *Assume that a and b are regular or quasiregular endpoints for the canonical system (7.2.3) and let $\{\mathbb{C}^2, \Gamma_0, \Gamma_1\}$ be the boundary triplet for $T_{\max}$ in Theorem 7.7.2 with corresponding Weyl function M. Let the fundamental matrix $Y(\cdot, \lambda)$ be fixed by $Y(a, \lambda) = I$. Then the self-adjoint relation $A_0 = \ker \Gamma_0$ is given by*

$$A_0 = \ker \Gamma_0 = \{\{f, g\} \in T_{\max} : f(a) + f(b) = 0\},$$

where $f(a)$ and $f(b)$ denote the boundary values of the unique absolutely continuous representative of f. The resolvent of A_0 is an integral operator

$$((A_0 - \lambda)^{-1}g)(t) = \int_a^b G_0(t, s, \lambda)\Delta(s)g(s)\, ds, \quad \lambda \in \rho(A_0), \qquad (7.7.2)$$

which belongs to the Hilbert–Schmidt class. The Green function $G_0(t, s, \lambda)$ is given by

$$G_0(t, s, \lambda) = G_{0,e}(t, s, \lambda) + G_{0,i}(t, s, \lambda), \qquad (7.7.3)$$

where the entire part $G_{0,e}$ is given by

$$G_{0,e}(t, s, \lambda) = Y(t, \lambda)\left[\frac{1}{2}J\,\mathrm{sgn}\,(s - t)\right]Y(s, \bar{\lambda})^*$$

$$= \frac{1}{2}\begin{cases} -Y(t, \lambda)JY(s, \bar{\lambda})^*, & s < t, \\ Y(t, \lambda)JY(s, \bar{\lambda})^*, & s > t, \end{cases} \qquad (7.7.4)$$

and

$$G_{0,\mathrm{i}}(t, s, \lambda) = Y(t, \lambda) \left[-\frac{1}{2} J M(\lambda) J \right] Y(s, \bar{\lambda})^*. \tag{7.7.5}$$

Proof. Step 1. The resolvent of A_0 has the form (7.7.2) with G_0 as in (7.7.3). To see this, let $\lambda \in \rho(A_0)$ and $g \in L^2_\Delta(\imath)$, and define the function

$$f(t) = \int_a^b G_0(t, s, \lambda) \Delta(s) g(s) \, ds.$$

From the structure of the Green function in (7.7.3), (7.7.4), and (7.7.5), it follows that

$$f(t) = \frac{1}{2} Y(t, \lambda) J \left(-\int_a^t Y(s, \bar{\lambda})^* \Delta(s) g(s) \, ds + \int_t^b Y(s, \bar{\lambda})^* \Delta(s) g(s) \, ds \right)$$
$$+ Y(t, \lambda) E_0(\lambda) \int_a^b Y(s, \bar{\lambda})^* \Delta(s) g(s) \, ds,$$

with $E_0(\lambda) = -\frac{1}{2} J M(\lambda) J$. Hence, f is well defined and absolutely continuous. A straightforward computation together with (7.2.10) shows that

$$J f'(t) = \frac{1}{2} J Y'(t, \lambda) J \left(-\int_a^t Y(s, \bar{\lambda})^* \Delta(s) g(s) \, ds + \int_t^b Y(s, \bar{\lambda})^* \Delta(s) g(s) \, ds \right)$$
$$+ \Delta(t) g(t) + J Y'(t, \lambda) E_0(\lambda) \int_a^b Y(s, \bar{\lambda})^* \Delta(s) g(s) \, ds.$$

This implies that

$$J f' - H f = \lambda \Delta f + \Delta g = \Delta(g + \lambda f),$$

and thus one has $\{f, g + \lambda f\} \in T_{\max}$. Furthermore, it is clear from the definition of f and $Y(a, \lambda) = I$ that

$$f(a) = \left[\frac{1}{2} J + E_0(\lambda) \right] \int_a^b Y(s, \bar{\lambda})^* \Delta(s) g(s) \, ds$$

and

$$f(b) = Y(b, \lambda) \left[-\frac{1}{2} J + E_0(\lambda) \right] \int_a^b Y(s, \bar{\lambda})^* \Delta(s) g(s) \, ds.$$

Since $E_0(\lambda) = -\frac{1}{2} J M(\lambda) J = -\frac{1}{2}(I - Y(b, \lambda))(I + Y(b, \lambda))^{-1} J$, observe that

$$\frac{1}{2} J + E_0(\lambda) = Y(b, \lambda)(I + Y(b, \lambda))^{-1} J,$$

$$-\frac{1}{2} J + E_0(\lambda) = -(I + Y(b, \lambda))^{-1} J.$$

Thus,

$$\left[\frac{1}{2} J + E_0(\lambda) \right] + Y(b, \lambda) \left[-\frac{1}{2} J + E_0(\lambda) \right] = 0,$$

and hence $\Gamma_0\{f, g + \lambda f\} = \frac{1}{\sqrt{2}}(f(a) + f(b)) = 0$, which implies $\{f, g + \lambda f\} \in A_0$. Therefore, $f = (A_0 - \lambda)^{-1}g$ and the resolvent of A_0 is given by (7.7.2).

Step 2. The kernel $G_\Delta(\cdot, \cdot, \lambda)$ defined by

$$G_\Delta(t, s, \lambda) = \Delta(t)^{\frac{1}{2}} G_0(t, s, \lambda) \Delta(s)^{\frac{1}{2}}, \qquad (7.7.6)$$

where the kernel $G_0(\cdot, \cdot, \lambda)$ is given in (7.7.3), satisfies

$$\int_a^b \int_a^b \left\| G_\Delta(t, s, \lambda) \right\|_2^2 ds\, dt < \infty; \qquad (7.7.7)$$

here $\|\cdot\|_2$ is the Hilbert–Schmidt matrix norm. In fact, from (7.7.3), (7.7.4), (7.7.5), and (7.1.5) it follows that

$$\int_a^b \int_a^b \left\| G_\Delta(t, s, \lambda) \right\|_2^2 ds\, dt = \int_a^b \int_a^b \left\| \Delta(t)^{\frac{1}{2}} G_0(t, s, \lambda) \Delta(s)^{\frac{1}{2}} \right\|_2^2 ds\, dt$$

$$\leq C \int_a^b \int_a^b \left\| \Delta(t)^{\frac{1}{2}} Y(t, \lambda) \right\|_2^2 \left\| Y(s, \bar{\lambda})^* \Delta(s)^{\frac{1}{2}} \right\|_2^2 ds\, dt.$$

To show that the right-hand side is finite, note that with $Y(\cdot, \lambda) = (Y_1(\cdot, \lambda)\, Y_2(\cdot, \lambda))$ one has

$$\int_a^b \left\| \Delta(t)^{\frac{1}{2}} Y(t, \lambda) \right\|_2^2 dt = \int_a^b \left| \Delta(t)^{\frac{1}{2}} Y_1(t, \lambda) \right|^2 dt + \int_a^b \left| \Delta(t)^{\frac{1}{2}} Y_2(t, \lambda) \right|^2 dt < \infty,$$

as the columns $Y_1(\cdot, \lambda)$ and $Y_2(\cdot, \lambda)$ of $Y(\cdot, \lambda)$ are square-integrable with respect to Δ. Due to the identity $\|A\|_2 = \|A^*\|_2$, it follows that

$$\int_a^b \left\| Y(s, \bar{\lambda})^* \Delta(s)^{\frac{1}{2}} \right\|_2^2 ds = \int_a^b \left\| \Delta(s)^{\frac{1}{2}} Y(s, \lambda) \right\|_2^2 dt < \infty.$$

Therefore, the kernel G_Δ is square-integrable with respect to the Lebesgue measure on $[a, b] \times [a, b]$ and hence (7.7.7) holds. Consequently, the integral operator T_Δ, defined by

$$T_\Delta f(t) = \int_a^b G_\Delta(t, s, \lambda) f(s)\, ds, \quad f \in L^2(\imath), \qquad (7.7.8)$$

belongs to the Hilbert–Schmidt class in $L^2(\imath)$.

Step 3. The operator $(A_0 - \lambda)^{-1}$ belongs to the Hilbert–Schmidt class or, equivalently,

$$\sum_{i,j} \left| ((A_0 - \lambda)^{-1} u_i, u_j) \right|^2 < \infty, \quad \lambda \in \rho(A_0), \qquad (7.7.9)$$

for some, and hence for any orthonormal basis (u_i) in $L^2_\Delta(\imath)$. To see (7.7.9), observe that $(\Delta^{\frac{1}{2}} u_i)$ is an orthonormal system in $L^2(\imath)$ and that (7.7.2) and (7.7.6) give

$$
\begin{aligned}
\left((A_0 - \lambda)^{-1} u_i, u_j\right)_\Delta &= \int_a^b u_j(t)^* \Delta(t) \left(\int_a^b G_0(t, s, \lambda) \Delta(s) u_i(s) \, ds \right) dt \\
&= \int_a^b \int_a^b \left(\Delta(t)^{\frac{1}{2}} u_j(t)\right)^* G_\Delta(t, s, \lambda) \left(\Delta(s)^{\frac{1}{2}} u_i(s)\right) \, ds \, dt \\
&= (T_\Delta \, \Delta^{\frac{1}{2}} u_i, \Delta^{\frac{1}{2}} u_j)_{L^2(\imath)},
\end{aligned}
$$

where T_Δ is the Hilbert–Schmidt operator (7.7.8) in $L^2(\imath)$ whose kernel is given by (7.7.6). Hence, by Step 2 it follows that (7.7.9) holds, which implies that $(A_0 - \lambda)^{-1}$ is a Hilbert–Schmidt operator. $\qquad\square$

Since the resolvent of the self-adjoint relation A_0 is a Hilbert–Schmidt operator, the spectrum of A_0 is discrete. As the minimal relation $T_{\min}$ has no eigenvalues, the next statement follows immediately from Proposition 3.4.8.

Theorem 7.7.4. Let $\{\mathbb{C}^2, \Gamma_0, \Gamma_1\}$ be the boundary triplet for $T_{\max}$ as in Theorem 7.7.2. Then the operator part $(T_{\min})_{\mathrm{op}}$ is simple in $L^2_\Delta(\imath) \ominus \operatorname{mul} T_{\min}$.

Theorem 7.7.4 together with the considerations in Section 3.5 and Section 3.6 ensure that the Weyl function M in Theorem 7.7.2 contains the complete spectral data of A_0. In the present situation the eigenvalues of A_0 coincide with the poles of the Weyl function and the multiplicities of the eigenvalues of A_0 coincide with the multiplicities of the poles of M.

Let $\{\mathbb{C}^2, \Gamma_0, \Gamma_1\}$ be the boundary triplet in Theorem 7.7.2 with corresponding γ-field γ and Weyl function M. The self-adjoint (maximal dissipative, maximal accumulative) extensions $A_\Theta \subset T_{\max}$ of $T_{\min}$ are in a one-to-one correspondence to the self-adjoint (maximal dissipative, maximal accumulative) relations Θ in $\mathbb{C}^2$ via

$$
\begin{aligned}
A_\Theta &= \left\{ \{f, g\} \in T_{\max} : \{\Gamma_0\{f, g\}, \Gamma_1\{f, g\}\} \in \Theta \right\} \\
&= \left\{ \{f, g\} \in T_{\max} : \{f(a) + f(b), -Jf(a) + Jf(b)\} \in \Theta \right\},
\end{aligned}
\tag{7.7.10}
$$

where $f(a)$ and $f(b)$ denote the boundary values of the unique absolutely continuous representative of f. Recall from Theorem 2.6.1 that for $\lambda \in \rho(A_\Theta) \cap \rho(A_0)$ the Kreĭn formula for the corresponding resolvents reads

$$
(A_\Theta - \lambda)^{-1} = (A_0 - \lambda)^{-1} + \gamma(\lambda)\left(\Theta - M(\lambda)\right)^{-1} \gamma(\bar\lambda)^*.
\tag{7.7.11}
$$

Assume in the following that Θ is a self-adjoint relation in $\mathbb{C}^2$. Since the spectrum of A_0 is discrete and the difference of the resolvents of A_0 and A_Θ is an operator of rank ≤ 2, it is clear that the spectrum of the self-adjoint relation

A_Θ is also discrete. Note that $\lambda \in \rho(A_0)$ is an eigenvalue of A_Θ if and only if $\ker(\Theta - M(\lambda))$ is nontrivial, and that

$$\ker(A_\Theta - \lambda) = \gamma(\lambda)\ker(\Theta - M(\lambda)).$$

For the self-adjoint relation Θ one may use a parametric representation with the help of 2×2 matrices $\mathcal{A}$ and $\mathcal{B}$ as in Section 1.10 and give a complete description of the (discrete) spectrum of A_Θ via poles of a transform of the Weyl function M; cf. Section 3.8 and Section 6.3.

In the following paragraph and corollary it is assumed for simplicity that the relation Θ in (7.7.10) is a self-adjoint 2×2 matrix. In this case the self-adjoint relation A_Θ in (7.7.10) is given by

$$A_\Theta = \{\{f, g\} \in T_{\max} : \Theta(f(a) + f(b)) = -Jf(a) + Jf(b)\} \qquad (7.7.12)$$

and according to Section 3.8 the spectral properties of A_Θ can also be described with the help of the function

$$\lambda \mapsto (\Theta - M(\lambda))^{-1}; \qquad (7.7.13)$$

that is, the poles of the matrix function (7.7.13) coincide with the (discrete) spectrum of A_Θ and the dimension of the eigenspace $\ker(A_\Theta - \lambda)$ coincides with the dimension of the range of the residue of the function in (7.7.13) at λ. Now fix a fundamental matrix $Y(\cdot, \lambda)$ by $Y(a, \lambda) = I$ as in Proposition 7.7.3. By Proposition 7.7.3, the resolvent $(A_0 - \lambda)^{-1}$ in the Kreĭn formula (7.7.11) is an integral operator. Since $I + JM(\lambda) = 2(I + Y(b, \lambda))^{-1}$, the γ-field and Weyl function in Theorem 7.7.2 are connected in the present situation via

$$\gamma(\cdot, \lambda) = \frac{1}{\sqrt{2}}Y(\cdot, \lambda)(I + JM(\lambda)), \quad \lambda \in \mathbb{C} \setminus \mathbb{R}.$$

One verifies that

$$\gamma(\bar\lambda)^* g = \frac{1}{\sqrt{2}}(I - M(\lambda)J)\int_a^b Y(s, \bar\lambda)^* \Delta(s)g(s)\,ds, \quad g \in L^2_\Delta(\imath),$$

and this implies that the second term on the right-hand side of (7.7.11) applied to $g \in L^2_\Delta(\imath)$ can be written as

$$\frac{1}{2}Y(\cdot, \lambda)(I + JM(\lambda))(\Theta - M(\lambda))^{-1}(I - M(\lambda)J)\int_a^b Y(s, \bar\lambda)^* \Delta(s)g(s)\,ds.$$

Combining this expression with the entire part $G_{0,e}$ in (7.7.4) and the part $G_{0,i}$ in (7.7.5) for $(A_0 - \lambda)^{-1}$, one sees that the resolvent of the self-adjoint extension A_Θ is an integral operator in $L^2_\Delta(\imath)$ of the form

$$((A_\Theta - \lambda)^{-1}g)(t) = \int_a^b G_\Theta(t, s, \lambda)\Delta(s)g(s)\,ds, \quad \lambda \in \rho(A_\Theta) \cap \rho(A_0), \qquad (7.7.14)$$

where $g \in L_\Delta^2(\imath)$. The Green function $G_\Theta(t, s, \lambda)$ in (7.7.14) is given by

$$G_\Theta(t, s, \lambda) = Y(t, \lambda) \left[\frac{1}{2} J \operatorname{sgn}(s - t) + E_\Theta(\lambda) \right] Y(s, \bar{\lambda})^*, \qquad (7.7.15)$$

where

$$E_\Theta(\lambda) = -\frac{1}{2} J \left[M(\lambda) + (M(\lambda) - J)(\Theta - M(\lambda))^{-1}(M(\lambda) + J) \right] J. \qquad (7.7.16)$$

In the next corollary the Green function in (7.7.14) is further decomposed in the case that the self-adjoint relation Θ in $\mathbb{C}^2$ is a self-adjoint matrix.

Corollary 7.7.5. *Let a and b be regular or quasiregular endpoints for the canonical system (7.2.3) and let $\{\mathbb{C}^2, \Gamma_0, \Gamma_1\}$ be the boundary triplet for $T_{\max}$ in Theorem 7.7.2 with corresponding Weyl function M. Assume that the fundamental matrix $Y(\cdot, \lambda)$ is fixed by $Y(a, \lambda) = I$, let Θ be a self-adjoint matrix in $\mathbb{C}^2$, and let A_Θ be the self-adjoint extension in (7.7.12). Then the Green function $G_\Theta(t, s, \lambda)$ in (7.7.14) has the decomposition*

$$G_\Theta(t, s, \lambda) = G_{\Theta,\mathrm{e}}(t, s, \lambda) + G_{\Theta,\mathrm{i}}(t, s, \lambda),$$

where the entire part $G_{\Theta,\mathrm{e}}$ is given by

$$G_{\Theta,\mathrm{e}}(t, s, \lambda) = Y(t, \lambda) \left[\frac{1}{2} J \operatorname{sgn}(s - t) + \frac{1}{2} J\Theta J \right] Y(s, \bar{\lambda})^*,$$

and

$$G_{\Theta,\mathrm{i}}(t, s, \lambda) = Y_\Theta(t, \lambda) \left[\frac{1}{2}(\Theta - M(\lambda))^{-1} \right] Y_\Theta(s, \bar{\lambda})^*,$$

where $Y_\Theta(t, \lambda) = Y(t, \lambda)(I + J\Theta)$.

Proof. Since Θ is a self-adjoint 2×2-matrix, one sees that

$$(M(\lambda) - J)(\Theta - M(\lambda))^{-1}(M(\lambda) + J)$$
$$= -M(\lambda) - \Theta + (\Theta - J)(\Theta - M(\lambda))^{-1}(\Theta + J).$$

Therefore, $E_\Theta(\lambda)$ in (7.7.16) has the form

$$E_\Theta(\lambda) = -\frac{1}{2} J \left[-\Theta + (\Theta - J)(\Theta - M(\lambda))^{-1}(\Theta + J) \right] J$$
$$= \frac{1}{2} J\Theta J + (I + J\Theta) \left[\frac{1}{2}(\Theta - M(\lambda))^{-1} \right] (I - \Theta J).$$

The assertion now follows from this identity combined with (7.7.15). $\qquad \square$

At the end of this section the assumption is that the endpoints a and b are in the general limit-circle case, so that the assumption that a and b are regular or quasiregular is abandoned. The transformation in Lemma 7.2.5 will be useful as for any $\lambda_0 \in \mathbb{R}$ the solution matrix $U(\cdot, \lambda_0)$ is now square-integrable with respect to Δ. This implies that the transformed equation (7.2.20) is in the quasiregular case at a and b. Then most of the above results remain true once the (limit) values $f(a)$ and $f(b)$ are replaced by the limits in (7.4.6). The next proposition is the counterpart of Theorem 7.7.2.

Proposition 7.7.6. *Assume that a and b are in the limit-circle case and let $Y(\cdot, \lambda)$ be a fundamental matrix. Let λ_0 in $\mathbb{R}$, let $U(\cdot, \lambda_0)$ be a solution matrix as in Lemma 7.2.5, and consider the limits*

$$\widetilde{f}(a) = \lim_{t \to a} U(t, \lambda_0)^{-1} f(t) \quad and \quad \widetilde{f}(b) = \lim_{t \to b} U(t, \lambda_0)^{-1} f(t)$$

for $\{f, g\} \in T_{\max}$; cf. Corollary 7.4.8. Then $\{\mathbb{C}^2, \Gamma_0, \Gamma_1\}$, with

$$\Gamma_0\{f, g\} = \frac{1}{\sqrt{2}} (\widetilde{f}(a) + \widetilde{f}(b)) \quad and \quad \Gamma_1\{f, g\} = -\frac{J}{\sqrt{2}} (\widetilde{f}(a) - \widetilde{f}(b)),$$

where $\{f, g\} \in T_{\max}$, is a boundary triplet for $(T_{\min})^* = T_{\max}$. The corresponding γ-field and Weyl function are given by

$$\gamma(\lambda) = \sqrt{2} Y(\cdot, \lambda) (\widetilde{Y}(a, \lambda) + \widetilde{Y}(b, \lambda))^{-1}, \quad \lambda \in \rho(A_0),$$

and

$$M(\lambda) = -J (\widetilde{Y}(a, \lambda) - \widetilde{Y}(b, \lambda))(\widetilde{Y}(a, \lambda) + \widetilde{Y}(b, \lambda))^{-1}, \quad \lambda \in \rho(A_0),$$

where $\widetilde{Y}(\cdot, \lambda)\phi = U(\cdot, \lambda_0)^{-1} Y(\cdot, \lambda)\phi$ for $\phi \in \mathbb{C}^2$ and $\lambda \in \rho(A_0)$.

Proof. Recall that due to Corollary 7.4.9 the Lagrange formula takes the form

$$\widetilde{h}(b)^* J \widetilde{f}(b) - \widetilde{h}(a)^* J \widetilde{f}(a) = \int_a^b (h(s)^* \Delta(s) g(s) - k(s)^* \Delta(s) f(s)) \, ds$$

for $\{f, g\}, \{h, k\} \in T_{\max}$. Now the same computation as in the proof of Theorem 7.7.2 shows that the abstract Green identity (2.1.1) is satisfied. The surjectivity of the map $(\Gamma_0, \Gamma_1)^\top : T_{\max} \to \mathbb{C}^4$, and the form of the γ-field and Weyl function follow in the same way as in the proof of Theorem 7.7.2. $\square$

7.8 Boundary triplets for the limit-point case

Assume that the system (7.2.3) is real and definite, and assume that the endpoint a is in the limit-circle case and the endpoint b is in the limit-point case. A boundary triplet will be presented for $T_{\max} = (T_{\min})^*$ and will be used to describe the self-adjoint extensions of $T_{\min}$. To make the presentation straightforward, the case where the endpoint a is regular or quasiregular is dealt with first. At the end of the section it will be explained what modifications are necessary if the endpoint a is in the limit-circle case.

The symmetric relation $T_{\min} = \overline{T}_0$ will now be described when a is a regular or quasiregular endpoint and b is in the limit-point case.

Lemma 7.8.1. *Assume that the endpoint a is regular or quasiregular and the endpoint b is in the limit-point case. Then the minimal relation $T_{\min}$ is given by*

$$T_{\min} = \{\{f, g\} \in T_{\max} : f(a) = 0\},$$

where $f(a)$ denotes the boundary value of the unique absolutely continuous representative of f.

Proof. According to Corollary 7.6.6 and Lemma 7.6.8, an element $\{f, g\} \in T_{\max}$ belongs to $T_{\min}$ if and only if

$$\lim_{t \to a} h(t)^* J f(t) = 0$$

for all $\{h, k\} \in T_{\max}$. Since the endpoint a is regular or quasiregular this condition is the same as

$$h(a)^* J f(a) = 0$$

for all $\{h, k\} \in T_{\max}$. Now observe that for any $\gamma \in \mathbb{C}^2$ there exists $\{h, k\} \in T_{\max}$ such that $h(a) = \gamma$. In fact, choose a solution $J u' - H u = 0$ such that $u(a) = \gamma$ and use Proposition 7.5.6 to modify u to a function $h \in L^2_\Delta(\imath)$ which coincides with u in a neighborhood of a, vanishes in a neighborhood of b, and satisfies $J h' - H h = \Delta k$ with some $k \in L^2_\Delta(\imath)$, that is, $\{h, k\} \in T_{\max}$. Since $\gamma^* J f(a) = 0$ for all $\gamma \in \mathbb{C}^2$, it follows that $f(a) = 0$. $\qquad\square$

Let the endpoint a be regular or quasiregular and let b be in the limit-point case. Then there exists for some, and hence for all $\lambda \in \mathbb{C} \setminus \mathbb{R}$, up to scalar multiples, one nontrivial solution of $J f' - H f = \lambda \Delta f$, which is square-integrable with respect to Δ at b and thus $\dim \ker (T_{\max} - \lambda) = 1$ for $\lambda \in \mathbb{C} \setminus \mathbb{R}$. This implies that the defect numbers are $(1, 1)$; cf. Corollary 7.6.3. In the next theorem a boundary triplet is provided in this case. To avoid confusion, recall that $Y_1(\cdot, \lambda)$ and $Y_2(\cdot, \lambda)$ are the columns of a fundamental matrix $Y(\cdot, \lambda)$, whereas f_1 and f_2 stand for the components of the 2×1 vector function f.

Theorem 7.8.2. *Assume that the endpoint a is regular or quasiregular and that the endpoint b is in the limit-point case. Let $Y(\cdot, \lambda)$ be a fundamental matrix fixed by $Y(a, \lambda) = I$. Then $\{\mathbb{C}, \Gamma_0, \Gamma_1\}$, where*

$$\Gamma_0\{f, g\} = f_1(a) \quad and \quad \Gamma_1\{f, g\} = f_2(a), \quad \{f, g\} \in T_{\max},$$

is a boundary triplet for $(T_{\min})^* = T_{\max}$; here $f_1(a)$ and $f_2(a)$ denote the boundary values of the components of the unique absolutely continuous representative of f. Moreover, if $\lambda \in \mathbb{C} \setminus \mathbb{R}$ and $\chi(\cdot, \lambda)$ is a nontrivial element in $\mathfrak{N}_\lambda(T_{\max})$, then one has $\chi_1(a, \lambda) \neq 0$. For all $\lambda \in \mathbb{C} \setminus \mathbb{R}$ the corresponding γ-field and Weyl function are given by

$$\gamma(\cdot, \lambda) = Y_1(\cdot, \lambda) + M(\lambda) Y_2(\cdot, \lambda) \quad and \quad M(\lambda) = \frac{\chi_2(a, \lambda)}{\chi_1(a, \lambda)}.$$

Proof. Since the endpoint a is assumed to be regular or quasiregular, the elements $\{f, g\}, \{h, k\} \in T_{\max}$ have boundary values $f(a), h(a) \in \mathbb{C}^2$. Due to Lemma 7.6.8, the Lagrange identity in Corollary 7.3.4 takes the form

$$(g, h)_\Delta - (f, k)_\Delta = \int_a^b \left(h(s)^* \Delta(s) g(s) - k(s)^* \Delta(s) f(s) \right) ds$$
$$= -h(a)^* J f(a)$$
$$= f_2(a)\overline{h_1(a)} - f_1(a)\overline{h_2(a)}$$
$$= \left(\Gamma_1\{f, g\}, \Gamma_0\{h, k\} \right) - \left(\Gamma_0\{f, g\}, \Gamma_1\{h, k\} \right).$$

Hence, the boundary mappings Γ_0 and Γ_1 satisfy the abstract Green identity (2.1.1). In the proof of Lemma 7.8.1 it was shown that for $\gamma \in \mathbb{C}^2$ there exists $\{h, k\} \in T_{\max}$ such that $h(a) = \gamma$, and so the mapping $(\Gamma_0, \Gamma_1)^\top : T_{\max} \to \mathbb{C}^2$ is surjective. It follows that $\{\mathbb{C}, \Gamma_0, \Gamma_1\}$ is a boundary triplet for $(T_{\min})^* = T_{\max}$.

Due to the assumption that the endpoint b is in the limit-point case, each eigenspace $\mathfrak{N}_\lambda(T_{\max})$, $\lambda \in \mathbb{C} \setminus \mathbb{R}$, has dimension 1. Hence, $\widehat{f}_\lambda \in \widehat{\mathfrak{N}}_\lambda(T_{\max})$ has the form $\widehat{f}_\lambda = \{\chi(\cdot, \lambda)c, \lambda\chi(\cdot, \lambda)c\}$ for some $c \in \mathbb{C}$, where $\chi(\cdot, \lambda)$ is a nontrivial element in $\mathfrak{N}_\lambda(T_{\max})$, $\lambda \in \mathbb{C} \setminus \mathbb{R}$. It follows from Definition 2.3.1 and Definition 2.3.4 that

$$\gamma(\lambda) = \left\{ \{\chi_1(a, \lambda)c, \chi(\cdot, \lambda)c\} : c \in \mathbb{C} \right\}, \quad \lambda \in \mathbb{C} \setminus \mathbb{R},$$

and

$$M(\lambda) = \left\{ \{\chi_1(a, \lambda)c, \chi_2(a, \lambda)c\} : c \in \mathbb{C} \right\}, \quad \lambda \in \mathbb{C} \setminus \mathbb{R}.$$

Observe that $\chi_1(a, \lambda) \neq 0$ for $\lambda \in \mathbb{C} \setminus \mathbb{R}$, as otherwise $\lambda \in \mathbb{C} \setminus \mathbb{R}$ would be an eigenvalue of the self-adjoint relation $A_0 = \ker \Gamma_0$ and $\chi(\cdot, \lambda)$ would be a corresponding eigenfunction. Thus, one concludes that

$$\gamma(\lambda) = \frac{\chi(\cdot, \lambda)}{\chi_1(a, \lambda)} \quad \text{and} \quad M(\lambda) = \frac{\chi_2(a, \lambda)}{\chi_1(a, \lambda)}, \quad \lambda \in \mathbb{C} \setminus \mathbb{R}.$$

Note that $\chi(\cdot, \lambda) = \alpha_1 Y_1(\cdot, \lambda) + \alpha_2 Y_2(\cdot, \lambda)$ for some $\alpha_1, \alpha_2 \in \mathbb{C}$ and that the assumption $Y(a, \lambda) = I$ yields

$$\begin{pmatrix} \chi_1(a, \lambda) \\ \chi_2(a, \lambda) \end{pmatrix} = \chi(a, \lambda) = \begin{pmatrix} \alpha_1 \\ \alpha_2 \end{pmatrix}.$$

This implies

$$\gamma(\lambda) = \frac{\alpha_1 Y_1(\cdot, \lambda) + \alpha_2 Y_2(\cdot, \lambda)}{\chi_1(a, \lambda)} = Y_1(\cdot, \lambda) + M(\lambda) Y_2(\cdot, \lambda), \quad \lambda \in \mathbb{C} \setminus \mathbb{R},$$

establishing the formulas for the γ-field and Weyl function. $\square$

Note that the γ-field and Weyl function corresponding to the boundary triplet $\{\mathbb{C}, \Gamma_0, \Gamma_1\}$ in Theorem 7.8.2 are defined and analytic on the resolvent set of the self-adjoint relation $A_0 = \ker \Gamma_0$. It follows in the same way as in Section 6.4 (see the discussion after the proof of Proposition 6.4.1) that the expressions for γ and M in Theorem 7.8.2 extend to points in $\rho(A_0) \cap \mathbb{R}$.

In the next proposition the resolvent of the self-adjoint relation A_0 is expressed as an integral operator.

Proposition 7.8.3. *Assume that the endpoint a is regular or quasiregular and that the endpoint b is in the limit-point case. Let $Y(\cdot, \lambda)$ be a fundamental matrix fixed by $Y(a, \lambda) = I$. Let $\{\mathbb{C}, \Gamma_0, \Gamma_1\}$ be the boundary triplet for $T_{\max}$ in Theorem 7.8.2 with corresponding Weyl function M. Then the self-adjoint relation $A_0 = \ker \Gamma_0$ is given by*

$$A_0 = \ker \Gamma_0 = \{\{f, g\} \in T_{\max} : f_1(a) = 0\},$$

where $f_1(a)$ denotes the boundary value of the first component of the unique absolutely continuous representative of f. The resolvent of A_0 is an integral operator

$$\left((A_0 - \lambda)^{-1}g\right)(t) = \int_a^b G_0(t, s, \lambda)\Delta(s)g(s)\,ds, \quad \lambda \in \mathbb{C} \setminus \mathbb{R}, \tag{7.8.1}$$

where $g \in L_\Delta^2(\imath)$. The Green function $G_0(t, s, \lambda)$ is given by

$$G_0(t, s, \lambda) = G_{0,e}(t, s, \lambda) + G_{0,i}(t, s, \lambda), \tag{7.8.2}$$

where the entire part $G_{0,e}$ is given by

$$G_{0,e}(t, s, \lambda) = \begin{cases} Y_1(t, \lambda)Y_2(s, \bar{\lambda})^*, & s < t, \\ Y_2(t, \lambda)Y_1(s, \bar{\lambda})^*, & s > t, \end{cases} \tag{7.8.3}$$

and

$$G_{0,i}(t, s, \lambda) = Y_2(t, \lambda)M(\lambda)Y_2(s, \bar{\lambda})^*. \tag{7.8.4}$$

Proof. To prove the identity (7.8.1), consider $g \in L_\Delta^2(\imath)$ and define the function f by the right-hand side of (7.8.1) with G_0 as in (7.8.2). In view of (7.8.3) and (7.8.4), this means that

$$f(t) = \left(Y_1(t, \lambda) + Y_2(t, \lambda)M(\lambda)\right) \int_a^t Y_2(s, \bar{\lambda})^*\Delta(s)g(s)\,ds$$
$$+ Y_2(t, \lambda) \int_t^b \left(Y_1(s, \bar{\lambda})^* + M(\bar{\lambda})^*Y_2(s, \bar{\lambda})^*\right)\Delta(s)g(s)\,ds. \tag{7.8.5}$$

Observe that, indeed, the integral near b exists, since one has

$$\gamma(\cdot, \bar{\lambda}) = Y_1(\cdot, \bar{\lambda}) + M(\bar{\lambda})Y_2(\cdot, \bar{\lambda}) \in L_\Delta^2(\imath)$$

and $g \in L^2_\Delta(i)$. It follows that the function f in (7.8.5) is well defined and absolutely continuous. Rewrite (7.8.5) in the form

$$f(t) = Y(t, \lambda) \begin{pmatrix} 1 \\ M(\lambda) \end{pmatrix} \int_a^t (0 \ \ 1) \, Y(s, \bar{\lambda})^* \Delta(s) g(s) \, ds$$
$$+ Y(t, \lambda) \begin{pmatrix} 0 \\ 1 \end{pmatrix} \int_t^b (1 \ \ M(\bar{\lambda})^*) \, Y(s, \bar{\lambda})^* \Delta(s) g(s) \, ds. \tag{7.8.6}$$

Then a straightforward calculation using the identity

$$\begin{pmatrix} 1 \\ M(\lambda) \end{pmatrix} (0 \ \ 1) - \begin{pmatrix} 0 \\ 1 \end{pmatrix} (1 \ \ M(\bar{\lambda})^*) = \begin{pmatrix} 0 & 1 \\ -1 & 0 \end{pmatrix} = -J \tag{7.8.7}$$

and (7.2.10) shows that f satisfies the inhomogenenous equation

$$Jf' - Hf = \lambda \Delta f + \Delta g. \tag{7.8.8}$$

Moreover, one sees from (7.8.5) that f satisfies

$$f(a) = \begin{pmatrix} 0 \\ 1 \end{pmatrix} \int_a^b (Y_1(s, \bar{\lambda}) + Y_2(s, \bar{\lambda}) M(\bar{\lambda}))^* \Delta(s) g(s) \, ds = \begin{pmatrix} 0 \\ (g, \gamma(\bar{\lambda}))_\Delta \end{pmatrix}.$$

Now denote the function on the left-hand side of (7.8.1) by $h = (A_0 - \lambda)^{-1} g$. Then it is clear that

$$\{h, \lambda h + g\} = \{(A_0 - \lambda)^{-1} g, g + \lambda (A_0 - \lambda)^{-1} g\} \in A_0 \subset T_{\max}, \tag{7.8.9}$$

so that h also satisfies (7.8.8). Moreover, by (7.8.9) and Proposition 2.3.2 one obtains

$$h_1(a) = \Gamma_0\{h, \lambda h + g\} = 0,$$
$$h_2(a) = \Gamma_1\{h, \lambda h + g\} = \gamma(\bar{\lambda})^* g = (g, \gamma(\bar{\lambda}))_\Delta.$$

Thus, for a fixed $\lambda \in \mathbb{C} \setminus \mathbb{R}$ the functions h and f satisfy the same inhomogeneous equation (7.8.8) and they have the same initial value $f(a) = h(a)$. Since the solution is unique, $h = f$. One concludes that $(A_0 - \lambda)^{-1} g$ is given by the right-hand side of (7.8.1). $\qquad \square$

Note that there exist canonical systems whose Weyl functions are of the form $M(\lambda) = \alpha + \beta \lambda$ where $\alpha \in \mathbb{R}$ and $\beta \geq 0$; cf. Example 7.10.4 for a special case. Hence, the functions M and $G_{0,i}$ in (7.8.4) may be entire.

Theorem 7.8.4. *Assume that the endpoint a is regular or quasiregular and that the endpoint b is in the limit-point case. Then the operator part $(T_{\min})_{op}$ is simple in the Hilbert space $L^2_\Delta(i) \ominus \mathrm{mul}\, T_{\min}$.*

Proof. *Step* 1. Let $g \in L^2_\Delta(\imath)$ and define $f = (A_0 - \lambda)^{-1}g$. Then it is clear that $\{f, g + \lambda f\} \in A_0$ and one has

$$f(t) = -Y(t, \lambda)J \int_a^t Y(s, \bar{\lambda})^* \Delta(s)g(s)\, ds + Y(t, \lambda) \begin{pmatrix} 0 \\ (g, \gamma(\bar{\lambda}))_\Delta \end{pmatrix}, \qquad (7.8.10)$$

where $Y(\cdot, \lambda)$ is the fundamental matrix fixed by $Y(a, \lambda) = I$. In fact, it follows from Proposition 7.8.3 and its proof that f is given by (7.8.1) or, equivalently, by (7.8.6). Now on the right-hand side of (7.8.6) substract and add the term

$$Y(t, \lambda) \begin{pmatrix} 0 \\ 1 \end{pmatrix} \int_a^t \begin{pmatrix} 1 & M(\bar{\lambda})^* \end{pmatrix} Y(s, \bar{\lambda})^* \Delta(s)g(s)\, ds,$$

and use (7.8.7) and $\gamma(\cdot, \bar{\lambda}) = Y_1(\cdot, \bar{\lambda}) + M(\bar{\lambda})Y_2(\cdot, \bar{\lambda})$. This yields (7.8.10).

Step 2. The multivalued part $\operatorname{mul} T_{\min}$ is given by

$$\left(\operatorname{span} \{ \gamma(\lambda) : \lambda \in \mathbb{C} \setminus \mathbb{R} \} \right)^\perp = \operatorname{mul} T_{\min}, \qquad (7.8.11)$$

which, in view of Corollary 3.4.6, is equivalent to $(T_{\min})_{\mathrm{op}}$ being simple in the Hilbert space $L^2_\Delta(\imath) \ominus \operatorname{mul} T_{\min}$.

The identity (7.8.11) will be verified by exhibiting the corresponding inclusions. For the inclusion ($\supset$) in (7.8.11), let $g \in \operatorname{mul} T_{\min}$. Since $\{0, g\} \in T_{\min}$ and $\{\gamma(\lambda), \lambda\gamma(\lambda)\} \in T_{\max}$ for all $\lambda \in \mathbb{C} \setminus \mathbb{R}$ one sees that

$$(g, \gamma(\lambda))_\Delta = (g, \gamma(\lambda))_\Delta - (0, \lambda\gamma(\lambda))_\Delta = 0.$$

Hence, $g \in \operatorname{mul} T_{\min}$ is orthogonal to all $\gamma(\lambda)$, $\lambda \in \mathbb{C} \setminus \mathbb{R}$.

For the inclusion ($\subset$) in (7.8.11), assume that $g \in L^2_\Delta(\imath)$ is orthogonal to all $\gamma(\lambda)$, $\lambda \in \mathbb{C} \setminus \mathbb{R}$. Then it follows from (7.8.10) that

$$f(t) = \left((A_0 - \lambda)^{-1}g \right)(t) = -Y(t, \lambda)J \int_a^t Y(s, \bar{\lambda})^* \Delta(s)g(s)\, ds. \qquad (7.8.12)$$

Clearly, $f(a) = 0$, so that, in fact,

$$\{f, g + \lambda f\} \in T_{\min}. \qquad (7.8.13)$$

Let $h \in L^2_\Delta(\imath)$ have compact support, say in $[a', b'] \subset \imath$. Then it follows from (7.8.12) that

$$\left((A_0 - \lambda)^{-1}g, h \right)_\Delta = -\int_a^b \left(\int_a^t h(t)^* \Delta(t) Y(t, \lambda) J Y(s, \bar{\lambda})^* \Delta(s)g(s)\, ds \right) dt$$

and due to the structure of the double integral the integration takes place only on the square $[a', b'] \times [a', b']$.

Now consider a bounded interval $\delta \subset \mathbb{R}$ such that the endpoints of δ are not eigenvalues of A_0. Then the spectral projection of A_0 corresponding to the interval δ is given by Stone's formula (1.5.7) (see also Example A.1.4),

$$(E(\delta)g, h)_\Delta = \lim_{\varepsilon \downarrow 0} \frac{1}{2\pi i} \int_\delta \left(((A_0 - (\mu + i\varepsilon))^{-1} - (A_0 - (\mu - i\varepsilon))^{-1})g, h\right)_\Delta d\mu.$$

Making use of the above integral for $((A_0 - \lambda)^{-1}g, h)_\Delta$ one has that

$$(E(\delta)g, h)_\Delta = \lim_{\varepsilon \downarrow 0} \frac{1}{2\pi i} \int_\delta \int_a^b \int_a^t h(t)^* \Delta(t) F_\varepsilon(t, s, \mu) \Delta(s) g(s) \, ds \, dt \, d\mu$$

$$= \lim_{\varepsilon \downarrow 0} \frac{1}{2\pi i} \int_a^b \int_a^t h(t)^* \Delta(t) \left(\int_\delta F_\varepsilon(t, s, \mu) \, d\mu\right) \Delta(s) g(s) \, ds \, dt,$$

where

$$F_\varepsilon(t, s, \mu) = Y(t, \mu - i\varepsilon) JY(s, \overline{\mu - i\varepsilon})^* - Y(t, \mu + i\varepsilon) JY(s, \overline{\mu + i\varepsilon})^*.$$

To justify the application of Fubini's theorem above note that each of the functions

$$s \mapsto \Delta(s) g(s) \quad \text{and} \quad t \mapsto \Delta(t) h(t)$$

is integrable on $[a', b']$, due to $g, h \in L^2_\Delta(a, b)$ and Lemma 7.1.4, and that the function

$$(s, t, \lambda) \mapsto Y(t, \lambda) JY(s, \bar{\lambda})^* - Y(t, \bar{\lambda}) JY(s, \lambda)^*, \quad s, t \in [a', b'], \ \lambda \in K,$$

where $K \subset \mathbb{C}$ is some compact set, is continuous and hence bounded on the set $[a', b'] \times [a', b'] \times K$. Since the mapping $\lambda \mapsto Y(t, \lambda)$ is entire, it follows that

$$\lim_{\varepsilon \downarrow 0} \int_\delta F_\varepsilon(t, s, \mu) \, d\mu = 0$$

and dominated convergence implies that $(E(\delta)g, h)_\Delta = 0$ for any $h \in L^2_\Delta(\imath)$ with compact support. Therefore, $E(\delta)g = 0$ for any bounded interval δ with endpoints not in $\sigma_p(A_0)$. With $\delta \to \mathbb{R}$ one concludes $E(\mathbb{R})g = 0$ and this implies $g \in \text{mul}\, A_0$. Since $\{f, g + \lambda f\} \in A_0$ and $\{0, g\} \in A_0$, it follows that $\{f, \lambda f\} \in A_0$ and hence $f = 0$, as $\lambda \in \mathbb{C} \setminus \mathbb{R}$ is not an eigenvalue of A_0. Since $\{f, g + \lambda f\} \in T_{\min}$ by (7.8.13), one concludes $\{0, g\} \in T_{\min}$, that is, $g \in \text{mul}\, T_{\min}$. This shows the inclusion ($\subset$) in (7.8.11). $\qquad \square$

Let $\{\mathbb{C}, \Gamma_0, \Gamma_1\}$ be the boundary triplet in Theorem 7.8.2. Then the self-adjoint extensions of $T_{\min}$ are in a one-to-one correspondence to the numbers $\tau \in \mathbb{R} \cup \{\infty\}$ via

$$A_\tau = \{\{f, g\} \in T_{\max} : \Gamma_1\{f, g\} = \tau \Gamma_0\{f, g\}\}. \tag{7.8.14}$$

Note that A_0 corresponds to $\tau = \infty$. For a given $\tau \in \mathbb{R} \cup \{\infty\}$ one can transform the boundary triplet $\{\mathbb{C}, \Gamma_0, \Gamma_1\}$ as follows:

$$\begin{pmatrix} \Gamma_0^\tau \\ \Gamma_1^\tau \end{pmatrix} = \frac{1}{\sqrt{\tau^2 + 1}} \begin{pmatrix} \tau & -1 \\ 1 & \tau \end{pmatrix} \begin{pmatrix} \Gamma_0 \\ \Gamma_1 \end{pmatrix} \tag{7.8.15}$$

(see (2.5.19)), so that $A_\tau = \ker \Gamma_0^\tau$. Then, by (2.5.20), the γ-field and Weyl function corresponding to the new boundary triplet are given by

$$M_\tau(\lambda) = \frac{1 + \tau M(\lambda)}{\tau - M(\lambda)} \quad \text{and} \quad \gamma_\tau(\lambda) = \frac{\sqrt{\tau^2 + 1}}{\tau - M(\lambda)} \gamma(\lambda), \quad \lambda \in \mathbb{C} \setminus \mathbb{R}. \tag{7.8.16}$$

The Weyl function M_τ and the γ-field γ_τ are connected by

$$\frac{M_\tau(\lambda) - M_\tau(\mu)^*}{\lambda - \bar{\mu}} = \gamma_\tau(\mu)^* \gamma_\tau(\lambda), \quad \lambda, \mu \in \mathbb{C} \setminus \mathbb{R}.$$

Let $Y(\cdot, \lambda)$ be a fundamental matrix fixed by $Y(a, \lambda) = I$. In a similar fashion one can transform $Y(\cdot, \lambda)$ to a fundamental matrix $V(\cdot, \lambda)$ given by

$$\begin{pmatrix} V_1(\cdot, \lambda) & V_2(\cdot, \lambda) \end{pmatrix} = \begin{pmatrix} Y_1(\cdot, \lambda) & Y_2(\cdot, \lambda) \end{pmatrix} \frac{1}{\sqrt{\tau^2 + 1}} \begin{pmatrix} \tau & 1 \\ -1 & \tau \end{pmatrix}. \tag{7.8.17}$$

Note that $V(a, \bar{\lambda})^* JV(a, \lambda) = J$ holds; cf. (7.2.8) and (7.2.11). Due to this transformation the γ-field γ_τ can be written in terms of the new fundamental system as

$$\gamma_\tau(\lambda) = V_1(\cdot, \lambda) + M_\tau(\lambda) V_2(\cdot, \lambda),$$

which belongs to $L^2_\Delta(\imath)$, while the second column of $V(\cdot, \lambda)$ satisfies the (formal) boundary condition which determines $A_\tau = \ker \Gamma_0^\tau$:

$$V_2(a, \lambda) = \frac{1}{\sqrt{\tau^2 + 1}} \begin{pmatrix} 1 \\ \tau \end{pmatrix}. \tag{7.8.18}$$

The next proposition is the counterpart of Proposition 7.8.3 for the self-adjoint extensions A_τ.

Proposition 7.8.5. *Assume that the endpoint a is regular or quasiregular and that the endpoint b is in the limit-point case. Let $Y(\cdot, \lambda)$ be a fundamental matrix fixed by $Y(a, \lambda) = I$. Let $\{\mathbb{C}, \Gamma_0, \Gamma_1\}$ be the boundary triplet for $T_{\max}$ in Theorem 7.8.2. For $\tau \in \mathbb{R}$ let A_τ be a self-adjoint extension of $T_{\min}$ given by (7.8.14) and let M_τ be as in (7.8.16). Then the resolvent of A_τ is an integral operator*

$$((A_\tau - \lambda)^{-1} g)(t) = \int_a^b G_\tau(t, s, \lambda) \Delta(s) g(s) \, ds, \quad \lambda \in \mathbb{C} \setminus \mathbb{R}, \tag{7.8.19}$$

where $g \in L^2_\Delta(\imath)$. The Green's function $G_\tau(t, s, \lambda)$ is given by

$$G_\tau(t, s, \lambda) = G_{\tau, \mathrm{e}}(t, s, \lambda) + G_{\tau, \mathrm{i}}(t, s, \lambda), \tag{7.8.20}$$

where the entire part $G_{\tau,e}$ is given by

$$G_{\tau,e}(t,s,\lambda) = \begin{cases} V_1(t,\lambda)V_2(s,\bar{\lambda})^*, & s < t, \\ V_2(t,\lambda)V_1(s,\bar{\lambda})^*, & s > t, \end{cases} \tag{7.8.21}$$

and

$$G_{\tau,i}(t,s,\lambda) = V_2(t,\lambda)M_\tau(\lambda)V_2(s,\bar{\lambda})^*. \tag{7.8.22}$$

Proof. The proof of Proposition 7.8.5 is similar to the proof of Proposition 7.8.3. In order to show the identity (7.8.19), consider $g \in L^2_\Delta(\imath)$ and define the function f by the right-hand side of (7.8.19). Then one has

$$f(t) = \left(V_1(t,\lambda) + V_2(t,\lambda)M_\tau(\lambda)\right) \int_a^t V_2(s,\bar{\lambda})^* \Delta(s)g(s)\,ds$$
$$+ V_2(t,\lambda) \int_t^b \left(V_1(s,\bar{\lambda})^* + M_\tau(\bar{\lambda})^* V_2(s,\bar{\lambda})^*\right)\Delta(s)g(s)\,ds \tag{7.8.23}$$

and the same arguments as in the proof of Proposition 7.8.3 show that f is well defined, absolutely continuous, and satisfies the inhomogenenous equation

$$Jf' - Hf = \lambda\Delta f + \Delta g. \tag{7.8.24}$$

Moreover, one sees from (7.8.23) that f satisfies

$$f(a) = \frac{1}{\sqrt{\tau^2+1}}\begin{pmatrix}1\\\tau\end{pmatrix}\int_a^b \left(V_1(s,\bar{\lambda}) + V_2(s,\bar{\lambda})M_\tau(\bar{\lambda})\right)^*\Delta(s)g(s)\,ds$$
$$= \frac{1}{\sqrt{\tau^2+1}}\begin{pmatrix}1\\\tau\end{pmatrix}(g,\gamma_\tau(\bar{\lambda}))_\Delta,$$

and hence

$$\frac{\tau f_1(a) - f_2(a)}{\sqrt{\tau^2+1}} = 0 \quad \text{and} \quad \frac{f_1(a) + \tau f_2(a)}{\sqrt{\tau^2+1}} = (g,\gamma_\tau(\bar{\lambda}))_\Delta.$$

Now denote the function on the left-hand side of (7.8.19) by $h = (A_\tau - \lambda)^{-1}g$. Then h also satisfies (7.8.24) and from (7.8.15) and Proposition 2.3.2 one obtains

$$\frac{\tau h_1(a) - h_2(a)}{\sqrt{\tau^2+1}} = \Gamma_0^\tau\{h, \lambda h + g\} = 0,$$
$$\frac{h_1(a) + \tau h_2(a)}{\sqrt{\tau^2+1}} = \Gamma_1^\tau\{h, \lambda h + g\} = \gamma_\tau(\bar{\lambda})^* g = (g, \gamma_\tau(\bar{\lambda}))_\Delta.$$

Thus, for a fixed $\lambda \in \mathbb{C} \setminus \mathbb{R}$ the functions h and f satisfy the same initial value problem and hence it follows that $h = f$. Therefore, $(A_\tau - \lambda)^{-1}g$ is given by the right-hand side of (7.8.19). $\qquad\square$

Let $\{\mathbb{C}, \Gamma_0, \Gamma_1\}$ be the boundary triplet for $T_{\max}$ in Theorem 7.8.2 and consider the self-adjoint extension $A_\tau = \ker \Gamma_\tau$; cf. (7.8.14). Assume that the Weyl function M_τ corresponding to $\{\mathbb{C}, \Gamma_0^\tau, \Gamma_1^\tau\}$ has the integral representation

$$M_\tau(\lambda) = \alpha_\tau + \beta_\tau \lambda + \int_{\mathbb{R}} \left(\frac{1}{t - \lambda} - \frac{t}{t^2 + 1} \right) d\sigma_\tau(t), \qquad (7.8.25)$$

with $\alpha_\tau \in \mathbb{R}$, $\beta_\tau \geq 0$, and σ_τ a nondecreasing function with

$$\int_{\mathbb{R}} \frac{1}{t^2 + 1} d\sigma_\tau(t) < \infty.$$

Recall from Theorem 3.5.10 and Lemma A.2.6 that $\operatorname{mul} A_\tau \ominus \operatorname{mul} T_{\min}$ is nontrivial if and only if $\beta_\tau > 0$.

Lemma 7.8.6. *Let $\tau \in \mathbb{R} \cup \{\infty\}$ and let $E_\tau(\cdot)$ be the spectral measure of the self-adjoint relation A_τ. For $h \in L^2_\Delta(\imath)$ with compact support define the Fourier transform $\widehat{f}$ by*

$$\widehat{f}(\mu) = \int_a^b V_2(s, \mu)^* \Delta(s) f(s) \, ds, \quad \mu \in \mathbb{R},$$

where $V_2(\cdot, \mu)$ is the formal solution in (7.8.17). Let σ_τ be the function in the integral representation (7.8.25) of the Weyl function M_τ. Then for every bounded open interval $\delta \subset \mathbb{R}$ such that its endpoints are not eigenvalues of A_τ one has

$$(E_\tau(\delta) f, f)_\Delta = \int_\delta \widehat{f}(\mu) \, \overline{\widehat{f}(\mu)} \, d\sigma_\tau(\mu). \qquad (7.8.26)$$

Proof. Recall that $(A_\tau - \lambda)^{-1}$ is given by (7.8.19), where the Green function $G_\tau(t, s, \lambda)$ in (7.8.20) is given by (7.8.21) and (7.8.22). Assume that the function $h \in L^2_\Delta(\imath)$ has compact support in $[a', b'] \subset \imath$. Then

$$((A_\tau - \lambda)^{-1} f, f)_\Delta = \int_a^b f(t)^* \Delta(t) \left(\int_a^b G_\tau(t, s, \lambda) \Delta(s) f(s) \, ds \right) dt$$

for each $\lambda \in \mathbb{C} \setminus \mathbb{R}$, where, in fact, the integration takes place only on the square $[a', b'] \times [a', b']$.

Let $\delta \subset \mathbb{R}$ be a bounded interval such that the endpoints of δ are not eigenvalues of A_τ. Then the spectral projection of A_τ corresponding to the interval δ is given by Stone's formula (1.5.7) (see also Example A.1.4)

$$(E(\delta) f, f)_\Delta = \lim_{\varepsilon \downarrow 0} \frac{1}{2\pi i} \int_\delta \left(((A_\tau - (\mu + i\varepsilon))^{-1} - (A_\tau - (\mu - i\varepsilon))^{-1}) f, f \right)_\Delta d\mu$$

$$= \lim_{\varepsilon \downarrow 0} \frac{1}{2\pi i} \int_\delta \left(\int_a^b \int_a^b f(t)^* \Delta(t) \big(G_\tau(t, s, \mu + i\varepsilon) \right.$$

$$\left. - G_\tau(t, s, \mu - i\varepsilon) \big) \Delta(s) f(s) \, ds \, dt \right) d\mu.$$

Decompose the Green function in (7.8.20) as in (7.8.21) and (7.8.22). Since the function $\lambda \mapsto V(t, \lambda)$ is entire, one verifies in the same way as in the proof of Theorem 7.8.4 that

$$\lim_{\varepsilon \downarrow 0} \frac{1}{2\pi i} \int_\delta \left(\int_a^b \int_a^b f(t)^* \Delta(t) \big(G_{\tau,\mathrm{e}}(t, s, \mu + i\varepsilon) \right.$$
$$\left. - G_{\tau,\mathrm{e}}(t, s, \mu - i\varepsilon) \big) \Delta(s) f(s) \, ds \, dt \right) d\mu = 0.$$

Therefore, it remains to consider the corresponding integral with $G_{\tau,\mathrm{i}}$, which takes the form

$$\lim_{\varepsilon \downarrow 0} \frac{1}{2\pi i} \int_\delta \left(\int_a^b \int_a^b f(t)^* \Delta(t) \big(V_2(t, \mu + i\varepsilon) M_\tau(\mu + i\varepsilon) V_2(s, \mu - i\varepsilon)^* \right.$$
$$\left. - V_2(t, \mu - i\varepsilon) M_\tau(\mu - i\varepsilon) V_2(s, \mu + i\varepsilon)^* \big) \Delta(s) f(s) \, ds \, dt \right) d\mu$$

$$= \lim_{\varepsilon \downarrow 0} \frac{1}{2\pi i} \int_\delta \left(\int_a^b \int_a^b f(t)^* \Delta(t) \big[(g_{t,s} M_\tau)(\mu + i\varepsilon) \right.$$
$$\left. - (g_{t,s} M_\tau)(\mu - i\varepsilon) \big] \Delta(s) f(s) \, ds \, dt \right) d\mu, \tag{7.8.27}$$

where $g_{t,s}$ stands for the 2×2 matrix function

$$g_{t,s}(\eta) = V_2(t, \eta) \, V_2(s, \bar{\eta})^*.$$

For $t, s \in [a', b']$ this function is entire in η. For $\varepsilon_0 > 0$ and $A < B$ such that $\bar{\delta} \subset (A, B)$ consider the rectangle $R = [A, B] \times [-i\varepsilon_0, i\varepsilon_0]$. Then the function $\{t, s, \eta\} \mapsto g_{t,s}(\eta)$ is bounded on $[a', b'] \times [a', b'] \times R$, and since $\Delta h \in L^1(a', b')$, it follows that for each fixed ε such that $0 < \varepsilon \le \varepsilon_0$

$$\frac{1}{2\pi i} \int_\delta \left(\int_a^b \int_a^b f(t)^* \Delta(t) \big[(g_{t,s} M_\tau)(\mu + i\varepsilon) - (g_{t,s} M_\tau)(\mu - i\varepsilon) \big] \Delta(s) f(s) \, ds \, dt \right) d\mu$$

$$= \frac{1}{2\pi i} \int_a^b \int_a^b f(t)^* \Delta(t) \left(\int_\delta \big[(g_{t,s} M_\tau)(\mu + i\varepsilon) - (g_{t,s} M_\tau)(\mu - i\varepsilon) \big] d\mu \right) \Delta(s) f(s) \, ds \, dt.$$

By the Stieltjes inversion formula in Lemma A.2.7 and Remark A.2.10, one sees

$$\lim_{\varepsilon \downarrow 0} \frac{1}{2\pi i} \int_\delta \big[(g_{t,s} M_\tau)(\mu + i\varepsilon) - (g_{t,s} M_\tau)(\mu - i\varepsilon) \big] d\mu = \int_\delta g_{t,s}(\mu) \, d\sigma_\tau(\mu)$$

for all $t, s \in [a', b']$. To justify taking the limit $\varepsilon \downarrow 0$ inside the integral (7.8.27) one needs dominated convergence. For this purpose recall from Lemma A.2.7 and

Remark A.2.10 that there exists $m \geq 0$ such that for $0 < \varepsilon \leq \varepsilon_0$ one has

$$\left| \int_{\delta} \left[(g_{t,s} M_\tau)(\mu + i\varepsilon) - (g_{t,s} M_\tau)(\mu - i\varepsilon) \right] d\mu \right| \tag{7.8.28}$$
$$\leq m \, \sup\{ |g_{t,s}(\eta)|, |g'_{t,s}(\eta)| : t, s \in [a', b'], \, \lambda \in R \},$$

where $R = [A, B] \times [-i\varepsilon_0, i\varepsilon_0]$. Since the functions

$$\{t, s, \eta\} \mapsto g_{t,s}(\eta) \quad \text{and} \quad \{t, s, \eta\} \mapsto g'_{t,s}(\eta)$$

are bounded on $[a', b'] \times [a', b'] \times R$, it follows that the integral in (7.8.28), regarded as a function in $\{t, s\}$ on $[a', b'] \times [a', b']$, is bounded by some constant for all $0 < \varepsilon \leq \varepsilon_0$. Furthermore, $\Delta f \in L^1(a', b')$ implies that there is an integrable majorant for (7.8.27), and so dominated convergence and Fubini's theorem show that

$$(E(\delta)f, f)_\Delta = \int_a^b \int_a^b f(t)^* \Delta(t) \int_\delta g_{t,s}(\mu) \, d\sigma_\tau(\mu) \, \Delta(s) f(s) \, ds \, dt$$
$$= \int_\delta \left(\int_a^b (V_2(t, \mu)^* \Delta(t) f(t))^* \, dt \right) \left(\int_a^b V_2(s, \mu)^* \Delta(s) f(s) \, ds \right) d\sigma_\tau(\mu)$$

for every open interval δ such that the endpoints are not eigenvalues of A_τ. This gives the formula in (7.8.26). □

The next theorem is a consequence of Lemma 7.8.6 and Theorem B.2.3.

Theorem 7.8.7. *Let $\tau \in \mathbb{R} \cup \{\infty\}$, let $V_2(\cdot, \mu)$ be the formal solution in (7.8.17), and let σ_τ be the function in the integral representation of the Weyl function M_τ. Then the Fourier transform*

$$f \mapsto \widehat{f}, \qquad \widehat{f}(\mu) = \int_a^b V_2(s, \mu)^* \Delta(s) f(s) \, ds, \quad \mu \in \mathbb{R},$$

extends by continuity from compactly supported functions $f \in L^2_\Delta(\imath)$ to a surjective partial isometry $\mathcal{F}$ from $L^2_\Delta(\imath)$ to $L^2_{d\sigma_\tau}(\mathbb{R})$ with $\ker \mathcal{F} = \operatorname{mul} A_\tau$. The restriction $\mathcal{F}_{\mathrm{op}} : L^2_\Delta(\imath) \ominus \operatorname{mul} A_\tau \to L^2_{d\sigma_\tau}(\mathbb{R})$ is a unitary mapping, such that the self-adjoint operator $(A_\tau)_{\mathrm{op}}$ in $L^2_\Delta(\imath) \ominus \operatorname{mul} A_\tau$ is unitarily equivalent to multiplication by the independent variable in $L^2_{d\sigma_\tau}(\mathbb{R})$.

Proof. It follows from Lemma 7.8.6 that the condition (B.2.2) is satisfied. Furthermore, for every $\mu \in \mathbb{R}$ there exists a compactly supported function $f \in L^2_\Delta(\imath)$ such that

$$\widehat{f}(\mu) = \int_a^b V_2(s, \mu)^* \Delta(s) f(s) \, ds \neq 0.$$

To see this, assume that for some $\mu \in \mathbb{R}$

$$\int_a^b V_2(s, \mu)^* \Delta(s) f(s) \, ds = 0$$

for all compactly supported $f \in L^2_\Delta(\imath)$. This implies that $V_2(s,\mu)^* \Delta(s) = 0$ for a.e. $s \in (a,b)$. By definiteness one has $V_2(s,\mu) = 0$ for a.e. $s \in (a,b)$, which is a contradiction. Therefore, condition (B.2.9) is satisfied and the result follows from Theorem B.2.3. □

In the next lemma the Fourier transform $\mathcal{F}\gamma_\tau$ of the γ-field in (7.8.16) corresponding to the boundary triplet $\{\mathbb{C}, \Gamma^\tau_0, \Gamma^\tau_1\}$ is computed; this allows to identify the model in Theorem 7.8.7 with the model for scalar Nevanlinna functions in Section 4.3.

Lemma 7.8.8. *Let $\tau \in \mathbb{R} \cup \{\infty\}$ and let γ_τ be the γ-field in (7.8.16). Then for all $\lambda \in \mathbb{C} \setminus \mathbb{R}$ one has almost everywhere in the sense of $d\sigma_\tau$:*

$$[\mathcal{F}\gamma_\tau(\lambda)](\mu) = \frac{1}{\mu - \lambda}, \quad \mu \in \mathbb{R},$$

where $\mathcal{F}$ is the Fourier transform from $L^2_\Delta(\imath)$ onto $L^2_{d\sigma_\tau}(\mathbb{R})$ in Theorem 7.8.7.

Proof. Recall first that for $g \in L^2_\Delta(\imath)$ and $\lambda \in \mathbb{C} \setminus \mathbb{R}$ the function $(A_\tau - \lambda)^{-1} g$ is given by the identity (7.8.19), which holds for all $t \in \imath$. In fact, the Green function G_τ in (7.8.19) is a 2×2 matrix function and the following notation will be useful:

$$G_\tau(t, s, \lambda) = \begin{pmatrix} G_{\tau,1}(t, s, \lambda) \\ G_{\tau,2}(t, s, \lambda) \end{pmatrix},$$

where each of these components is a 1×2 matrix function. The fundamental matrix $V(\cdot, \lambda)$ in (7.8.17) is written in the form

$$\begin{pmatrix} V_1(\cdot, \lambda) & V_2(\cdot, \lambda) \end{pmatrix} = \begin{pmatrix} V_{11}(\cdot, \lambda) & V_{12}(\cdot, \lambda) \\ V_{21}(\cdot, \lambda) & V_{22}(\cdot, \lambda) \end{pmatrix}.$$

Now observe that

$$G_{\tau,1}(t, s, \lambda) = \begin{cases} (V_{11}(t, \lambda) + M_\tau(\lambda) V_{12}(t, \lambda)) V_2(s, \bar{\lambda})^*, & s < t, \\ V_{12}(t, \lambda) \gamma_\tau(s, \bar{\lambda})^*, & s > t, \end{cases} \quad (7.8.29)$$

and

$$G_{\tau,2}(t, s, \lambda) = \begin{cases} (V_{21}(t, \lambda) + M_\tau(\lambda) V_{22}(t, \lambda)) V_2(s, \bar{\lambda})^*, & s < t, \\ V_{22}(t, \lambda) \gamma_\tau(s, \bar{\lambda})^*, & s > t, \end{cases} \quad (7.8.30)$$

which follows easily from (7.8.21) and (7.8.22); cf. (7.8.23). Note also that

$$V_{ij}(\cdot, \bar{\lambda}) = \overline{V_{ij}(\cdot, \lambda)}, \quad i, j = 1, 2,$$

since the system is real (see Lemma 7.2.8), and that (7.2.10) implies the useful identity

$$V_{11}(t, \lambda) V_{22}(t, \lambda) - V_{21}(t, \lambda) V_{12}(t, \lambda) = 1, \quad t \in \imath. \quad (7.8.31)$$

For $\lambda \in \mathbb{C} \setminus \mathbb{R}$ one has

$$
\big((A_\tau - \lambda)^{-1}g\big)(t) = \begin{pmatrix} \int_a^b G_{\tau,1}(t,s,\lambda)\Delta(s)g(s)\,ds \\ \int_a^b G_{\tau,2}(t,s,\lambda)\Delta(s)g(s)\,ds \end{pmatrix} = \begin{pmatrix} (g, G_{\tau,1}(t,\cdot,\lambda)^*)_\Delta \\ (g, G_{\tau,2}(t,\cdot,\lambda)^*)_\Delta \end{pmatrix}.
$$

Since the Fourier transform $\mathcal{F} : L^2_\Delta(\imath) \to L^2_{d\sigma_\tau}(\mathbb{R})$ in Theorem 7.8.7 is a partial isometry with $\ker \mathcal{F} = \operatorname{mul} A_\tau$, it follows from (1.1.10) that

$$
\begin{aligned}
\big((A_\tau - \lambda)^{-1}g\big)(t) &= \begin{pmatrix} (\mathcal{F}g, \mathcal{F}G_{\tau,1}(t,\cdot,\lambda)^*)_{L^2_{d\sigma_\tau}(\mathbb{R})} \\ (\mathcal{F}g, \mathcal{F}G_{\tau,2}(t,\cdot,\lambda)^*)_{L^2_{d\sigma_\tau}(\mathbb{R})} \end{pmatrix} \\[2mm]
&= \begin{pmatrix} \int_\mathbb{R} \mathcal{F}g(\mu)\,\overline{[\mathcal{F}G_{\tau,1}(t,\cdot,\lambda)^*](\mu)}\,d\sigma_\tau(\mu) \\ \int_\mathbb{R} \mathcal{F}g(\mu)\,\overline{[\mathcal{F}G_{\tau,2}(t,\cdot,\lambda)^*](\mu)}\,d\sigma_\tau(\mu) \end{pmatrix}
\end{aligned} \tag{7.8.32}
$$

is valid for all $g \in (\operatorname{mul} A_\tau)^\perp$. It is clear that (7.8.32) is also true for all $g \in \operatorname{mul} A_\tau$, since in this case $(A_\tau - \lambda)^{-1}g = 0$ and $\mathcal{F}g = 0$. Therefore, (7.8.32) is valid for all $g \in L^2_\Delta(\imath)$. Moreover, if $g \in L^2_\Delta(\imath)$ one has for almost all $t \in \imath$

$$
\begin{aligned}
\big((A_\tau - \lambda)^{-1}g\big)(t) &= \int_\mathbb{R} \frac{V_2(t,\mu)}{\mu - \lambda}\,\mathcal{F}g(\mu)\,d\sigma_\tau(\mu) \\[2mm]
&= \begin{pmatrix} \int_\mathbb{R} \frac{V_{12}(t,\mu)}{\mu - \lambda}\,\mathcal{F}g(\mu)\,d\sigma_\tau(\mu) \\ \int_\mathbb{R} \frac{V_{22}(t,\mu)}{\mu - \lambda}\,\mathcal{F}g(\mu)\,d\sigma_\tau(\mu) \end{pmatrix};
\end{aligned} \tag{7.8.33}
$$

see (B.2.5). Furthermore, if $\mathcal{F}g$ has compact support, then the right-hand side of (7.8.33) is absolutely continuous, and hence in this case the equality holds for all $t \in \imath$. Therefore, when $\mathcal{F}g$ has compact support the right-hand sides of (7.8.32) and (7.8.33) are equal for all $t \in \imath$, and hence one has

$$
\frac{V_{12}(t,\mu)}{\mu - \lambda} = \overline{[\mathcal{F}G_{\tau,1}(t,\cdot,\lambda)^*](\mu)} \quad \text{and} \quad \frac{V_{22}(t,\mu)}{\mu - \lambda} = \overline{[\mathcal{F}G_{\tau,2}(t,\cdot,\lambda)^*](\mu)}
$$

for all $t \in \imath$. These identities hold for all $\mu \in \mathbb{R} \setminus \Omega(t)$, where the set $\Omega(t) \subset \mathbb{R}$ has $d\sigma_\tau$-measure 0. Hence, (by replacing λ with $\bar{\lambda}$ and taking conjugates) one has for all $t \in \imath$ and all $\mu \in \mathbb{R} \setminus \Omega(t)$

$$
\frac{V_{12}(t,\mu)}{\mu - \lambda} = [\mathcal{F}G_{\tau,1}(t,\cdot,\bar{\lambda})^*](\mu) \quad \text{and} \quad \frac{V_{22}(t,\mu)}{\mu - \lambda} = [\mathcal{F}G_{\tau,2}(t,\cdot,\bar{\lambda})^*](\mu). \tag{7.8.34}
$$

By means of (7.8.29), (7.8.30), (7.8.31), and (7.8.34) it is straightforward to verify that for all $t \in \imath$ and all $\mu \in \mathbb{R} \setminus \Omega(t)$,

$$
\begin{aligned}
\frac{1}{\mu - \lambda}&\big(V_{11}(t,\lambda)V_{22}(t,\mu) - V_{21}(t,\lambda)V_{12}(t,\mu)\big) \\
&= V_{11}(t,\lambda)\big[\mathcal{F}G_{\tau,2}(t,\cdot,\bar{\lambda})^*\big](\mu) - V_{21}(t,\lambda)\big[\mathcal{F}G_{\tau,1}(t,\cdot,\bar{\lambda})^*\big](\mu) \\
&= \mathcal{F}\big[V_{11}(t,\lambda)\,G_{\tau,2}(t,\cdot,\bar{\lambda})^* - V_{21}(t,\lambda)\,G_{\tau,1}(t,\cdot,\bar{\lambda})^*\big](\mu) \\
&= \mathcal{F}\big[W(t,\cdot,\lambda)\big](\mu),
\end{aligned} \tag{7.8.35}
$$

where the 2×1 matrix function $W(\cdot, \cdot, \lambda)$ is given by

$$W(t, s, \lambda) = \begin{cases} M_\tau(\lambda) V_2(s, \lambda), & s < t, \\ \gamma_\tau(s, \lambda), & s > t. \end{cases} \qquad (7.8.36)$$

The above identity and a limit process will give the desired result. In fact, first observe that according to the definition of $W(\cdot, \cdot, \lambda)$ in (7.8.36) one has

$$\|\gamma_\tau(\cdot, \lambda) - W(t, \cdot, \lambda)\|_\Delta^2 = \int_a^t |\Delta(s)^{\frac{1}{2}} V_1(s, \lambda)|^2 \, ds \to 0 \quad \text{as} \quad t \to a,$$

and hence the continuity of $\mathcal{F}: L_\Delta^2(\imath) \to L_{d\sigma_\tau}^2(\mathbb{R})$ implies that

$$\left\| [\mathcal{F}\gamma_\tau(\cdot, \lambda)] - [\mathcal{F}W(t, \cdot, \lambda)] \right\|_{L_{d\sigma_\tau}^2(\mathbb{R})} \to 0 \quad \text{as} \quad t \to a.$$

Now approximate a by a sequence $t_n \in \imath$. Then there exists a subsequence, again denoted by t_n, such that pointwise

$$[\mathcal{F}\gamma_\tau(\cdot, \lambda)](\mu) = \lim_{n \to \infty} [\mathcal{F}W(t_n, \cdot, \lambda)](\mu), \quad \mu \in \mathbb{R} \setminus \Omega,$$

where Ω is a set of measure 0 in the sense of $d\sigma_\tau$. Observe that (7.8.35) gives

$$[\mathcal{F}W(t_n, \cdot, \lambda)](\mu) = \frac{1}{\mu - \lambda} (V_{11}(t_n, \lambda) V_{22}(t_n, \mu) - V_{21}(t_n, \lambda) V_{12}(t_n, \mu))$$

for all $\mu \in \mathbb{R} \setminus \Omega(t_n)$. The limit on the right-hand side as $n \to \infty$ gives

$$\frac{1}{\mu - \lambda} (V_{11}(a, \lambda) V_{22}(a, \mu) - V_{21}(a, \lambda) V_{12}(a, \mu)) = \frac{1}{\mu - \lambda},$$

which follows from the form of the fundamental matrix $(V_1(\cdot, \lambda) \, V_2(\cdot, \lambda))$ in (7.8.17). Hence,

$$[\mathcal{F}\gamma_\tau(\cdot, \lambda)](\mu) = \frac{1}{\mu - \lambda}, \quad \mu \in \mathbb{R} \setminus \left(\Omega \cup \bigcup_{n=1}^\infty \Omega(t_n) \right),$$

which completes the proof. $\qquad \qquad \square$

Lemma 7.8.8 will be used to explain the model in Theorem 7.8.7 with the model for scalar Nevanlinna functions discussed in Section 4.3. Without loss of generality it is assumed that $T_{\min}$ is simple; cf. Section 3.4. The Weyl function M_τ of the boundary triplet $\{\mathbb{C}, \Gamma_0^\tau, \Gamma_1^\tau\}$ for $T_{\max}$ has the integral representation (7.8.25). If $\beta = 0$, then the discussion in Chapter 6 following Lemma 6.4.8 applies in this case as well. Hence, assume $\beta > 0$ in (7.8.25). Then by Theorem 4.3.4 there is a closed simple symmetric operator S in $L_{d\sigma_\tau}^2(\mathbb{R}) \oplus \mathbb{C}$ such that the Nevanlinna function M_τ in (7.8.25) is the Weyl function corresponding to the boundary triplet $\{\mathbb{C}, \Gamma_0', \Gamma_1'\}$ for S^* in Theorem 4.3.4. The γ-field corresponding to $\{\mathbb{C}, \Gamma_0', \Gamma_1'\}$ is denoted by γ' and it is given by (4.3.16). Furthermore, the restriction A'_0 corresponding to the boundary mapping Γ'_0 is a self-adjoint relation in $L^2_{d\sigma_\tau}(\mathbb{R}) \oplus \mathbb{C}$ whose operator part $(A'_0)_{\mathrm{op}}$ is the maximal multiplication operator by the independent variable in $L^2_{d\sigma_\tau}(\mathbb{R})$. By comparing with (4.3.16) one sees that, according to Lemma 7.8.8, the Fourier transform $\mathcal{F}_{\mathrm{op}}$ from $L^2_\Delta(\imath) \ominus \mathrm{mul}\, A_\tau$ onto $L^2_{d\sigma_\tau}(\mathbb{R})$ as a unitary mapping satisfies

$$\mathcal{F}_{\mathrm{op}}\, P\gamma_\tau(\lambda) = P'\gamma'(\lambda),$$

where P and P' stand for the orthogonal projections from $L^2_\Delta(\imath)$ onto $(\mathrm{mul}\, A_\tau)^\perp$ and from $L^2_{d\sigma_\tau}(\mathbb{R}) \oplus \mathbb{C}$ onto $L^2_{d\sigma_\tau}(\mathbb{R}) = (\mathrm{mul}\, A'_0)^\perp$. Recall that

$$(I - P)\gamma_\tau(\lambda) \quad \text{and} \quad (I - P')\gamma'(\lambda)$$

are independent of λ and belong to $\mathrm{mul}\, A_\tau$ and $\mathrm{mul}\, A'_0$, respectively; cf. Corollary 2.5.16. Hence, the mapping $\mathcal{F}_{\mathrm{m}}$ from $\mathrm{mul}\, A_\tau$ to $\mathrm{mul}\, A'_0$ defined by

$$\mathcal{F}_{\mathrm{m}}(I - P)\gamma_\tau(\lambda) = \beta^{\frac{1}{2}} = (I - P')\gamma'(\lambda)$$

is a one-to-one correspondence. In fact, $\mathcal{F}_{\mathrm{m}}$ is an isometry due to Proposition 3.5.7. Define the mapping U from the space $L^2_\Delta(\imath)$ to the model space $L^2_{d\sigma_\tau}(\mathbb{R}) \oplus \mathbb{C}$ by

$$U = \begin{pmatrix} \mathcal{F}_{\mathrm{op}} & 0 \\ 0 & \mathcal{F}_{\mathrm{m}} \end{pmatrix} : \begin{pmatrix} (\mathrm{mul}\, A_\tau)^\perp \\ \mathrm{mul}\, A_\tau \end{pmatrix} \rightarrow \begin{pmatrix} (\mathrm{mul}\, A'_0)^\perp \\ \mathrm{mul}\, A'_0 \end{pmatrix}.$$

Then it is clear that U is unitary and that

$$U\gamma_\tau(\lambda) = \gamma'(\lambda).$$

Hence, by Theorem 4.2.6, it follows that the boundary triplet $\{\mathbb{C}, \Gamma^\tau_0, \Gamma^\tau_1\}$ for $T_{\max}$ and the boundary triplet $\{\mathbb{C}, \Gamma'_0, \Gamma'_1\}$ for S^* are unitarily equivalent under the mapping U, in particular, one has

$$(A'_0)_{\mathrm{op}} = \mathcal{F}_{\mathrm{op}}(A_\tau)_{\mathrm{op}}\mathcal{F}^{-1}_{\mathrm{op}} \quad \text{and} \quad A'_0 = U A_\tau U^{-1}.$$

At the end of this section the case where the endpoint a is in the limit-circle case and the endpoint b is in the limit-point case is briefly discussed. In a similar way as in the end of Section 7.7 one makes use of the transformation in Lemma 7.2.5. The next proposition is the counterpart of Theorem 7.8.2; it is proved in the same way.

Proposition 7.8.9. *Assume that a is in the limit-circle case and that b is in the limit-point case. Let λ_0 in $\mathbb{R}$, let $U(\cdot, \lambda_0)$ be a solution matrix as in Lemma 7.2.5, and consider the limit*

$$\widetilde{f}(a) = \lim_{t \to a} U(t, \lambda_0)^{-1} f(t)$$

for $\{f,g\} \in T_{\max}$; *cf. Corollary 7.4.8. Then* $\{\mathbb{C},\Gamma_0,\Gamma_1\}$, *where*

$$\Gamma_0\{f,g\} = \widetilde{f}_1(a) \quad and \quad \Gamma_1\{f,g\} = \widetilde{f}_2(a), \quad \{f,g\} \in T_{\max},$$

is a boundary triplet for $(T_{\min})^* = T_{\max}$. *Let* $Y(\cdot,\lambda)$ *be a fundamental matrix fixed in such a way that* $\widetilde{Y}(\cdot,\lambda) = U(\cdot,\lambda_0)^{-1}Y(\cdot,\lambda)$ *satisfies* $\widetilde{Y}(a,\lambda) = I$. *Then for all* $\lambda \in \mathbb{C} \setminus \mathbb{R}$ *the* γ *-field* γ *and Weyl function* M *corresponding to* $\{\mathbb{C},\Gamma_0,\Gamma_1\}$ *are given by*

$$\gamma(\lambda) = Y_1(\cdot,\lambda) + M(\lambda)Y_2(\cdot,\lambda) \quad and \quad M(\lambda) = \frac{\widetilde{\chi}_2(a,\lambda)}{\widetilde{\chi}_1(a,\lambda)},$$

where $\widetilde{\chi}(\cdot,\lambda) = U(\cdot,\lambda_0)^{-1}\chi(\cdot,\lambda)$ *and* $\chi(\cdot,\lambda)$ *is a nontrivial element in* $\mathfrak{N}_\lambda(T_{\max})$.

7.9 Weyl functions and subordinate solutions

Consider the real definite canonical system (7.2.3) on the interval $\imath = (a,b)$ and assume that the endpoint a is regular and that the endpoint b is in the limit-point case. Let $\{\mathbb{C},\Gamma_0,\Gamma_1\}$ be the boundary triplet for $T_{\max}$ in Theorem 7.7.2 with γ-field γ and Weyl function M. The spectrum of the self-adjoint extension

$$A_0 = \ker \Gamma_0 = \{\{f,g\} \in T_{\max} : f_1(a) = 0\}$$

will be studied by means of *subordinate solutions* of the equation $Jy' - Hy = \lambda\Delta y$. The discussion in this section is parallel to the discussion in Section 6.7.

It is useful to take into account all self-adjoint extensions of $T_{\min}$. As in Section 7.8, there is a one-to-one correspondence to the numbers $\tau \in \mathbb{R} \cup \{\infty\}$ as restrictions of $T_{\max}$ via

$$A_\tau = \{\{f,g\} \in T_{\max} : \Gamma_1\{f,g\} = \tau\Gamma_0\{f,g\}\}, \tag{7.9.1}$$

with the understanding that A_0 corresponds to $\tau = \infty$. As before, let the boundary triplet $\{\mathbb{C},\Gamma_0^\tau,\Gamma_1^\tau\}$ be defined by the transformation (7.8.15) with the Weyl function and γ-field given by

$$M_\tau(\lambda) = \frac{1 + \tau M(\lambda)}{\tau - M(\lambda)} \quad and \quad \gamma_\tau(\lambda) = \frac{\sqrt{\tau^2 + 1}}{\tau - M(\lambda)}\gamma(\lambda), \quad \lambda \in \mathbb{C} \setminus \mathbb{R}. \tag{7.9.2}$$

Recall that the Weyl function and the γ-field are connected via

$$\frac{M_\tau(\lambda) - M_\tau(\mu)^*}{\lambda - \bar{\mu}} = \gamma_\tau(\mu)^*\gamma_\tau(\lambda), \quad \lambda,\mu \in \mathbb{C} \setminus \mathbb{R}. \tag{7.9.3}$$

The transformation (7.8.15) also induces a transformation of the fundamental matrix $Y(\cdot,\lambda)$ with $Y(a,\lambda) = I$ as in (7.8.17):

$$\begin{pmatrix} V_1(\cdot,\lambda) & V_2(\cdot,\lambda) \end{pmatrix} = \begin{pmatrix} Y_1(\cdot,\lambda) & Y_2(\cdot,\lambda) \end{pmatrix} \frac{1}{\sqrt{\tau^2 + 1}} \begin{pmatrix} \tau & 1 \\ -1 & \tau \end{pmatrix}. \tag{7.9.4}$$

Recall that
$$\gamma_\tau(\lambda) = V_1(\cdot, \lambda) + M_\tau(\lambda)V_2(\cdot, \lambda)$$
is square-integrable with respect to Δ, while the second column of $V(\cdot, \lambda)$ satisfies the (formal) boundary condition which determines A_τ; cf. (7.8.18).

In the next estimates it is more convenient to work in the semi-Hilbert space $\mathcal{L}^2_\Delta(\imath)$ rather than in the Hilbert space $L^2_\Delta(\imath)$. In fact, for each $x > a$ the notation $\mathcal{L}^2_\Delta(a, x)$ stands for the semi-Hilbert space with the semi-inner product
$$(f, g)_x = \int_a^x g(t)^* \Delta(t) f(t)\, dt, \quad f, g \in \mathcal{L}^2_\Delta(a, x),$$
and the seminorm corresponding to $(\cdot, \cdot)_x$ will be denoted by $\|\cdot\|_x$; here the index Δ is omitted. Hence, for fixed $f, g \in \mathcal{L}^2_\Delta(a, c)$, $a < c < b$, the function $x \mapsto (f, g)_x$ is absolutely continuous and
$$\frac{d}{dx}(f, g)_x = g(x)^* \Delta(x) f(x) \tag{7.9.5}$$
holds almost everywhere on (a, c).

Definition 7.9.1. Let $\lambda \in \mathbb{C}$. Then a solution $h(\cdot, \lambda)$ of $Jh' - Hh = \lambda\Delta h$ is said to be *subordinate* at b if
$$\lim_{x \to b} \frac{\|h(\cdot, \lambda)\|_x}{\|k(\cdot, \lambda)\|_x} = 0$$
for every nontrivial solution $k(\cdot, \lambda)$ of $Jk' - Hk = \lambda\Delta k$ which is not a scalar multiple of $h(\cdot, \lambda)$.

The spectrum of the self-adjoint extension A_0 will be studied in terms of solutions of the canonical system $Jy' - Hy = \xi\Delta y$, $\xi \in \mathbb{R}$, which do not necessarily belong to $\mathcal{L}^2_\Delta(a, b)$. Observe that if a solution $h(\cdot, \xi)$ of $Jy' - Hy = \xi\Delta y$ belongs to $\mathcal{L}^2_\Delta(a, b)$, then it is subordinate at b since b is in the limit-point case, and hence any other nontrivial solution which is not a multiple does not belong to $\mathcal{L}^2_\Delta(a, b)$.

By means of the fundamental system $(V_1(\cdot, \lambda)\ V_2(\cdot, \lambda))$ in (7.8.17) define for any $\lambda \in \mathbb{C}$ and $h \in \mathcal{L}^2_\Delta(a, x)$, $a < x < b$,
$$(\mathcal{H}(\lambda)h)(t) = V_1(t, \lambda) \int_a^t V_2(s, \bar{\lambda})^* \Delta(s) h(s)\, ds$$
$$- V_2(t, \lambda) \int_a^t V_1(s, \bar{\lambda})^* \Delta(s) h(s)\, ds, \quad t \in (a, x).$$

Thus, $\mathcal{H}(\lambda)$ is a well-defined integral operator and it is clear that the function $\mathcal{H}(\lambda)h$ is absolutely continuous. Using the identity (7.2.10) for $V(\cdot, \lambda)$ (which holds because $V(a, \bar{\lambda})^* JV(a, \lambda) = J$) one sees in the same way as in (7.2.14)–(7.2.16) that $f = \mathcal{H}(\lambda)h$ satisfies
$$Jf' - Hf = \lambda\Delta f + \Delta h, \quad f(a) = 0.$$

In particular, $\mathcal{H}(\lambda)$ maps $\mathcal{L}_\Delta^2(a, x)$ into itself. It follows directly that for $\lambda, \mu \in \mathbb{C}$ one has

$$V_i(\cdot, \lambda) - V_i(\cdot, \mu) = (\lambda - \mu)\mathcal{H}(\lambda)V_i(\cdot, \mu), \quad i = 1, 2, \tag{7.9.6}$$

since the functions on the left-hand side and the right-hand side both satisfy the same equation $Jy' - Hy = \lambda\Delta y + (\lambda - \mu)\Delta V_i(\cdot, \mu)$ and both functions vanish at the endpoint a.

Lemma 7.9.2. *Let $a < x < b$ and let $h \in \mathcal{L}_\Delta^2(a, x)$. Then the operator $\mathcal{H}(\lambda)$ satisfies*

$$\|\mathcal{H}(\lambda)h\|_x^2 \leq 2\|V_1(\cdot, \lambda)\|_x^2 \|V_2(\cdot, \lambda)\|_x^2 \|h\|_x^2$$

for each $x > a$.

Proof. The definition of $\mathcal{H}(\lambda)$ may be written as

$$(\mathcal{H}(\lambda)h)(t) = V_1(t, \lambda)g_2(t, \lambda) - V_2(t, \lambda)g_1(t, \lambda), \tag{7.9.7}$$

with the functions $g_i(\cdot, \lambda)$, $i = 1, 2$, defined by

$$g_i(t, \lambda) = \int_a^t V_i(s, \bar{\lambda})^* \Delta(s)h(s)\, ds.$$

The Cauchy–Schwarz inequality and Corollary 7.2.9 show that

$$|g_i(t, \lambda)|^2 \leq \|V_i(\cdot, \bar{\lambda})\|_t^2\|h\|_t^2 = \|V_i(\cdot, \lambda)\|_t^2\|h\|_t^2, \quad i = 1, 2. \tag{7.9.8}$$

Multiplying $(\mathcal{H}(\lambda)h)(t)$ in (7.9.7) on the left by the matrix $\Delta(t)^{\frac{1}{2}}$, using the inequality $|a + b|^2 \leq 2(|a|^2 + |b|^2)$, and using (7.9.8) one obtains

$$|\Delta(t)^{\frac{1}{2}}(\mathcal{H}(\lambda)h)(t)|^2$$
$$\leq 2\left(|\Delta(t)^{\frac{1}{2}}V_1(t, \lambda)|^2 |g_2(t, \lambda)|^2 + |\Delta(t)^{\frac{1}{2}}V_2(t, \lambda)|^2 |g_1(t, \lambda)|^2\right)$$
$$\leq 2\left(|\Delta(t)^{\frac{1}{2}}V_1(t, \lambda)|^2 \|V_2(\cdot, \lambda)\|_t^2\|h\|_t^2 + |\Delta(t)^{\frac{1}{2}}V_2(t, \lambda)|^2 \|V_1(\cdot, \lambda)\|_t^2\|h\|_t^2\right).$$

Integration of this inequality over (a, x) and (7.9.5) lead to

$$\|\mathcal{H}(\lambda)h\|_x^2 \leq 2\int_a^x \left(|\Delta(t)^{\frac{1}{2}}V_1(t, \lambda)|^2\|V_2(\cdot, \lambda)\|_t^2\|h\|_t^2\right.$$
$$\left. + |\Delta(t)^{\frac{1}{2}}V_2(t, \lambda)|^2\|V_1(\cdot, \lambda)\|_t^2\|h\|_t^2\right) dt$$
$$= 2\int_a^x \left(\frac{d}{dt}\|V_1(\cdot, \lambda)\|_t^2\|V_2(\cdot, \lambda)\|_t^2\right)\|h\|_t^2\, dt$$
$$\leq 2\|h\|_x^2\int_a^x \left(\frac{d}{dt}\|V_1(\cdot, \lambda)\|_t^2\|V_2(\cdot, \lambda)\|_t^2\right) dt,$$

which implies the desired result. $\qquad\qquad\qquad\qquad\qquad\qquad\qquad\qquad\square$

Since the system is assumed to be definite on $\imath = (a, b)$, there is a compact subinterval $[\alpha, \beta]$ such that the system is definite on $[\alpha, \beta]$ and hence on any interval (a, x) with $x > \beta$. This implies that for $x > \beta$ both functions

$$x \mapsto \|V_1(\cdot, \lambda)\|_x \quad \text{and} \quad x \mapsto \|V_2(\cdot, \lambda)\|_x$$

have positive values; cf. (7.5.1) and Lemma 7.5.2.

Lemma 7.9.3. *Let $\xi \in \mathbb{R}$ be a fixed number. The function $x \mapsto \varepsilon_\tau(x, \xi)$ given by*

$$\sqrt{2}\, \varepsilon_\tau(x, \xi) \|V_1(\cdot, \xi)\|_x \|V_2(\cdot, \xi)\|_x = 1, \quad x > \beta,$$

is well defined, continuous, nonincreasing, and satisfies

$$\lim_{x \to b} \varepsilon_\tau(x, \xi) = 0.$$

Proof. It is clear that $\varepsilon_\tau(x, \xi) > 0$ is well defined due to the assumption that $x > \beta$. Note that $x \mapsto \|V_1(\cdot, \xi)\|_x \|V_2(\cdot, \xi)\|_x$ is continuous and nondecreasing. The assumption that b is in the limit-point case implies that not both $V_1(\cdot, \xi)$ and $V_2(\cdot, \xi)$ belong to $L^2_\Delta(\imath)$. Thus, the limit result follows. □

The function $x \mapsto \varepsilon_\tau(x, \xi)$ appears in the estimate in the following theorem.

Theorem 7.9.4. *Let M_τ be the Weyl function in (7.9.2) corresponding to the boundary triplet $\{\mathbb{C}, \Gamma_0^\tau, \Gamma_1^\tau\}$. Assume that $\xi \in \mathbb{R}$ and let $\varepsilon_\tau(x, \xi)$ be as in Lemma 7.9.3. Then for $a < \beta < x < b$*

$$\frac{1}{d_0} \le \frac{\|V_2(\cdot, \xi)\|_x}{\|V_1(\cdot, \xi)\|_x} |M_\tau(\xi + i\varepsilon_\tau(x, \xi))| \le d_0,$$

where $d_0 = 1 + 2\big(\sqrt{2} + \sqrt{2 + \sqrt{2}}\,\big)$.

Proof. Assume that $\xi \in \mathbb{R}$ and let $\varepsilon > 0$. Define the function $\psi(\cdot, \xi, \varepsilon)$ by

$$\psi(\cdot, \xi, \varepsilon) = V_1(\cdot, \xi) + M_\tau(\xi + i\varepsilon)V_2(\cdot, \xi). \tag{7.9.9}$$

For any $a < x < b$ this leads to

$$\big| \|V_2(\cdot, \xi)\|_x |M_\tau(\xi + i\varepsilon)| - \|V_1(\cdot, \xi)\|_x \big| \le \|\psi(\cdot, \xi, \varepsilon)\|_x$$

or, equivalently, when $\beta < x < a$

$$\left| \frac{\|V_2(\cdot, \xi)\|_x}{\|V_1(\cdot, \xi)\|_x} |M_\tau(\xi + i\varepsilon)| - 1 \right| \le \frac{\|\psi(\cdot, \xi, \varepsilon)\|_x}{\|V_1(\cdot, \xi)\|_x}. \tag{7.9.10}$$

The term on the right-hand side of (7.9.10) will now be estimated in a suitable way. First note that it follows from (7.9.6) that for $\lambda \in \mathbb{C}$ and $\mu \in \mathbb{C} \setminus \mathbb{R}$ one obtains

$$V_1(\cdot, \lambda) + M_\tau(\mu)V_2(\cdot, \lambda) - \gamma_\tau(\mu) = (\lambda - \mu)\mathcal{H}(\lambda)\gamma_\tau(\cdot, \mu). \tag{7.9.11}$$

Applying the identity in (7.9.11) with $\lambda = \xi$ and $\mu = \xi + i\varepsilon$ one sees that

$$\psi(\cdot, \xi, \varepsilon) = \gamma_\tau(\cdot, \xi + i\varepsilon) - i\varepsilon\mathcal{H}(\xi)\gamma_\tau(\cdot, \xi + i\varepsilon),$$

which expresses the function $\psi(\cdot, \xi, \varepsilon)$ in (7.9.9) in terms of the γ-field γ_τ. Hence, it follows from Lemma 7.9.2 that

$$\|\psi(\cdot, \xi, \varepsilon)\|_x \leq \left(1 + \sqrt{2}\,\varepsilon\,\|V_1(\cdot, \xi)\|_x\|V_2(\cdot, \xi)\|_x\right)\|\gamma_\tau(\cdot, \xi + i\varepsilon)\|_x.$$

Therefore, the right-hand side of (7.9.10) is estimated by

$$\frac{\left(1 + \sqrt{2}\,\varepsilon\,\|V_1(\cdot, \xi)\|_x\|V_2(\cdot, \xi)\|_x\right)\|\gamma_\tau(\cdot, \xi + i\varepsilon)\|_x}{\|V_1(\cdot, \xi)\|_x}$$

$$= \frac{1 + \sqrt{2}\,\varepsilon\,\|V_1(\cdot, \xi)\|_x\|V_2(\cdot, \xi)\|_x}{(\|V_1(\cdot, \xi)\|_x\|V_2(\cdot, \xi)\|_x)^{\frac{1}{2}}}\,\frac{\|V_2(\cdot, \xi)\|_x^{\frac{1}{2}}}{\|V_1(\cdot, \xi)\|_x^{\frac{1}{2}}}\,\|\gamma_\tau(\cdot, \xi + i\varepsilon)\|_x.$$

Now observe that $\|\gamma_\tau(\cdot, \xi + i\varepsilon)\|_x \leq \|\gamma_\tau(\cdot, \xi + i\varepsilon)\|_b$ and it follows from (7.9.3) that

$$\|\gamma_\tau(\cdot, \xi + i\varepsilon)\|_b \leq \sqrt{\frac{\operatorname{Im} M_\tau(\xi + i\varepsilon)}{\varepsilon}} \leq \sqrt{\frac{|M_\tau(\xi + i\varepsilon)|}{\varepsilon}}.$$

Thus, for any $\varepsilon > 0$ and $\beta < x < b$ one obtains the inequality

$$\left|\frac{\|V_2(\cdot, \xi)\|_x}{\|V_1(\cdot, \xi)\|_x}\,|M_\tau(\xi + i\varepsilon)| - 1\right|$$

$$\leq \frac{1 + \sqrt{2}\,\varepsilon\,\|V_1(\cdot, \xi)\|_x\|V_2(\cdot, \xi)\|_x}{(\varepsilon\,\|V_1(\cdot, \xi)\|_x\|V_2(\cdot, \xi)\|_x)^{\frac{1}{2}}}\left(\frac{\|V_2(\cdot, \xi)\|_x}{\|V_1(\cdot, \xi)\|_x}\,|M_\tau(\xi + i\varepsilon)|\right)^{\frac{1}{2}}.$$

Now for $\xi \in \mathbb{R}$ and $\beta < x < b$ choose $\varepsilon = \varepsilon_\tau(x, \xi)$ in this estimate. This choice minimizes the first factor on the right-hand side to $2^{5/4}$. Hence, the nonnegative quantity

$$Q = \frac{\|V_2(\cdot, \xi)\|_x}{\|V_1(\cdot, \xi)\|_x}\,|M_\tau(\xi + i\varepsilon_\tau(x, \xi))|$$

satisfies the inequality

$$|Q - 1| \leq 2^{5/4}Q^{\frac{1}{2}}$$

or, equivalently, $Q^2 - 2Q + 1 \leq 4\sqrt{2}Q$. Therefore, $1/d_0 \leq Q \leq d_0$, which completes the proof. $\qquad\square$

The following result is now a direct consequence of Theorem 7.9.4.

Theorem 7.9.5. *Let M be the Weyl function corresponding to the boundary triplet $\{\mathbb{C}, \Gamma_0, \Gamma_1\}$ and let $\xi \in \mathbb{R}$. Then the following statements hold:*

(i) *If $\tau \in \mathbb{R}$, then the solution $Y_1(\cdot, \xi) + \tau Y_2(\cdot, \xi)$ of the boundary value problem*

$$Jf' - Hf = \xi \Delta f, \quad f_2(a) = \tau f_1(a),$$

which is unique up to scalar multiples, is subordinate if and only if

$$\lim_{\varepsilon \downarrow 0} M(\xi + i\varepsilon) = \tau.$$

(ii) *If $\tau = \infty$, then the solution $Y_2(\cdot, \xi)$ of the boundary value problem*

$$Jf' - Hf = \xi \Delta f, \quad f_1(a) = 0,$$

which is unique up to scalar multiples, is subordinate if and only if

$$\lim_{\varepsilon \downarrow 0} M(\xi + i\varepsilon) = \infty.$$

Proof. Since $x \mapsto \varepsilon_\tau(x, \xi)$ is continuous, nonincreasing, and has limit 0 as $x \to b$, one obtains the identity

$$\lim_{\varepsilon \downarrow 0} M_\tau(\xi + i\varepsilon) = \lim_{x \to b} M_\tau(\xi + i\varepsilon_\tau(x, \xi)).$$

(i) Assume that $\tau \in \mathbb{R}$ and note that

$$V_2(\cdot, \xi) = \frac{1}{\sqrt{\tau^2 + 1}} \left(Y_1(\cdot, \xi) + \tau Y_2(\cdot, \xi) \right)$$

by (7.9.4). It will be shown that $|M_\tau(\xi + i\varepsilon)| \to \infty$ for $\varepsilon \downarrow 0$ if and only if the solution $V_2(\cdot, \xi)$ is subordinate. To see this, assume first that $|M_\tau(\xi + i\varepsilon)| \to \infty$. Then, by Theorem 7.9.4, it follows that

$$\lim_{x \to b} \frac{\|V_2(\cdot, \xi)\|_x}{\|V_1(\cdot, \xi)\|_x} = 0. \tag{7.9.12}$$

Hence, for any $c_1, c_2 \in \mathbb{R}$, $c_1 \neq 0$, one obtains from (7.9.12) that

$$\lim_{x \to b} \frac{\|V_2(\cdot, \xi)\|_x}{\|c_1 V_1(\cdot, \xi) + c_2 V_2(\cdot, \xi)\|_x} = 0, \tag{7.9.13}$$

and therefore the solution $V_2(\cdot, \xi)$ is subordinate. Conversely, assume that $V_2(\cdot, \xi)$ is subordinate, so that (7.9.13) holds for all $c_1, c_2 \in \mathbb{R}$, $c_1 \neq 0$. Then clearly (7.9.12) holds, and therefore it follows from Theorem 7.9.4 that $|M_\tau(\xi + i\varepsilon)| \to \infty$.

It is a consequence of (7.9.2) that for $\varepsilon \downarrow 0$ one has

$$|M_\tau(\xi + i\varepsilon)| \to \infty \quad \Leftrightarrow \quad M(\xi + i\varepsilon) \to \tau.$$

This equivalence leads to the assertion for $\tau \in \mathbb{R}$.

(ii) The case $\tau = \infty$ can be treated in the same way as (i).   □

The self-adjoint extension $A_0 = \ker \Gamma_0$ of $T_{\min}$ is given by

$$A_0 = \{\{f, g\} \in T_{\max} : f_1(a) = 0\};\tag{7.9.14}$$

cf. (7.9.1). The boundary condition

$$f_1(a) = 0\tag{7.9.15}$$

plays a central role in the following definition, which is based on Theorem 7.9.5; cf. Definition 6.7.6.

Definition 7.9.6. With the canonical system $Jf' - Hf = \xi \Delta f$, $\xi \in \mathbb{R}$, the following subsets of $\mathbb{R}$ are associated:

(i) $\mathfrak{M}$ is the complement of the set of all $\xi \in \mathbb{R}$ for which a subordinate solution exists that does not satisfy (7.9.15);

(ii) $\mathfrak{M}_{ac}$ is the set of all $\xi \in \mathbb{R}$ for which no subordinate solution exists;

(iii) $\mathfrak{M}_s$ is the set of all $\xi \in \mathbb{R}$ for which a subordinate solution exists that satisfies (7.9.15);

(iv) $\mathfrak{M}_{sc}$ is the set of all $\xi \in \mathbb{R}$ for which a subordinate solution exists that satisfies (7.9.15) and does not belong to $L^2_\Delta(\imath)$;

(v) $\mathfrak{M}_p$ is the set of all $\xi \in \mathbb{R}$ for which a subordinate solution exists that satisfies (7.9.15) and belongs to $L^2_\Delta(\imath)$.

It is a direct consequence of Definition 7.9.6 that

$$\mathbb{R} = \mathfrak{M}^c \sqcup \mathfrak{M}_{ac} \sqcup \mathfrak{M}_s, \quad \mathfrak{M} = \mathfrak{M}_{ac} \sqcup \mathfrak{M}_s, \quad \text{and} \quad \mathfrak{M}_s = \mathfrak{M}_{sc} \sqcup \mathfrak{M}_p,$$

where $\sqcup$ stands for disjoint union.

Let the Weyl function M of the boundary triplet $\{\mathbb{C}, \Gamma_0, \Gamma_1\}$ have the integral representation

$$M(\lambda) = \alpha + \beta\lambda + \int_{\mathbb{R}} \left(\frac{1}{t - \lambda} - \frac{t}{t^2 + 1} \right) d\sigma(t),\tag{7.9.16}$$

where $\alpha \in \mathbb{R}$, $\beta \geq 0$, and the measure σ satisfies

$$\int_{\mathbb{R}} \frac{1}{t^2 + 1} \, d\sigma(t) < \infty;$$

cf. Theorem A.2.5. The following proposition is based on Corollary 3.1.8, where minimal supports for the various parts of the measure σ in the integral representation of M are described in terms of the boundary behavior of the Nevanlinna function M. The proof of Proposition 7.9.7 is the same as the proof of Proposition 6.7.7 and will not be repeated.

Proposition 7.9.7. *Let M be the Weyl function associated with the boundary triplet $\{\mathbb{C}, \Gamma_0, \Gamma_1\}$ and let σ be the corresponding measure in (7.9.16). Then the sets*

$$\mathcal{M}, \mathcal{M}_{\mathrm{ac}}, \mathcal{M}_{\mathrm{s}}, \mathcal{M}_{\mathrm{sc}}, \mathcal{M}_{\mathrm{p}},$$

are minimal supports for the measures

$$\sigma, \sigma_{\mathrm{ac}}, \sigma_{\mathrm{s}}, \sigma_{\mathrm{sc}}, \sigma_{\mathrm{p}},$$

respectively.

The minimal supports in Proposition 7.9.7 are intimately connected with the spectrum of A_0. For the absolutely continuous spectrum one obtains in the same way as in Theorem 6.7.8 the following result, where the notion of the absolutely continuous closure of a Borel set from Definition 3.2.4 is used. Similar statements (with an inclusion) can be formulated for the singular parts of the spectrum; cf. Section 3.6.

Theorem 7.9.8. *Let A_0 be the self-adjoint relation in (7.9.14) and let $\mathcal{M}_{\mathrm{ac}}$ be as in Definition 7.9.6. Then*

$$\sigma_{\mathrm{ac}}(A_0) = \mathrm{clos}_{\mathrm{ac}}(\mathcal{M}_{\mathrm{ac}}).$$

7.10   Special classes of canonical systems

In this section two particular types of canonical systems are studied. First it is shown how a class of Sturm–Liouville problems, which are slightly more general than the equations treated in Chapter 6, fit in the framework of canonical systems. In this context the results from the previous sections can be carried over to Sturm–Liouville equations. The second class of canonical systems which is discussed here consists of systems of the form (7.2.3) with $H = 0$. In this situation a simple limit-point/limit-circle criterion is provided.

Weighted Sturm–Liouville equations

Let $\imath \subset \mathbb{R}$ be an open interval. Let $1/p, q, r, s \in \mathcal{L}^1_{\mathrm{loc}}(\imath)$ be real functions, assume $r(t) \geq 0$ for almost all $t \in \imath$, and define the 2×2 matrix functions H and Δ by

$$H(t) = \begin{pmatrix} -q(t) & -s(t) \\ -s(t) & 1/p(t) \end{pmatrix} \quad \text{and} \quad \Delta(t) = \begin{pmatrix} r(t) & 0 \\ 0 & 0 \end{pmatrix}, \tag{7.10.1}$$

respectively. Let the 2×2 matrix J be as in (7.2.2). If the vector functions f and g satisfy the canonical system $Jf' - Hf = \Delta g$, then their first components $\mathfrak{f} = f_1$ and $\mathfrak{g} = g_1$ satisfy the weighted Sturm–Liouville equation

$$-(\mathfrak{f}^{[1]})' + s\mathfrak{f}^{[1]} + q\mathfrak{f} = r\mathfrak{g}, \quad \text{where} \quad \mathfrak{f}^{[1]} = p(\mathfrak{f}' + s\mathfrak{f}), \tag{7.10.2}$$

and, since $f \in AC(\imath)$, it follows that $\mathfrak{f}, \mathfrak{f}^{[1]} \in AC(\imath)$. Conversely, if $\mathfrak{f}, \mathfrak{f}^{[1]} \in AC(\imath)$ and $\mathfrak{f}, \mathfrak{g}$ satisfy (7.10.2), then the vector functions

$$f = \begin{pmatrix} \mathfrak{f} \\ \mathfrak{f}^{[1]} \end{pmatrix} \quad \text{and} \quad g = \begin{pmatrix} \mathfrak{g} \\ 0 \end{pmatrix},$$

satisfy the canonical system (7.2.3) with the coefficients given by (7.10.1). Since the functions $1/p, q, r, s$ are assumed to be real, the system $Jf' - Hf = \Delta g$ is real. Moreover, the canonical system corresponding to (7.10.1) is definite, when any solution f of $Jf' - Hf = 0$ which satisfies $\Delta f = 0$ vanishes or, equivalently,

$$-f_2' + qf_1 + sf_2 = 0, \quad f_1' + sf_1 - (1/p)f_2 = 0, \quad rf_1 = 0 \quad \Rightarrow \quad f = 0.$$

In accordance with the definition of definiteness for canonical equations, the Sturm–Liouville equation (7.10.2) is said to be *definite* if

$$-(\mathfrak{f}^{[1]})' + s\mathfrak{f}^{[1]} + q\mathfrak{f} = 0, \quad r\mathfrak{f} = 0 \quad \Rightarrow \quad \mathfrak{f} = 0.$$

In particular, the Sturm–Liouville equation (7.10.2) is definite if the weight function r is positive on an open interval.

The matrix function Δ in (7.10.1) induces the spaces $\mathcal{L}_\Delta^2(\imath)$ and $L_\Delta^2(\imath)$. With the weight r it is natural to introduce the space $\mathcal{L}_r^2(\imath)$ of all complex measurable functions φ for which

$$\int_\imath |\varphi(s)|^2 r(s)\, ds = \int_\imath \varphi(s)^* r(s) \varphi(s)\, ds < \infty.$$

The corresponding semi-inner product is denoted by $(\cdot, \cdot)_r$ and the corresponding Hilbert space of equivalence classes of elements from $\mathcal{L}_r^2(\imath)$ is denoted by $L_r^2(\imath)$. It is clear that the mapping R defined by

$$f = \begin{pmatrix} f_1 \\ f_2 \end{pmatrix} \in \mathcal{L}_\Delta^2(\imath) \mapsto f_1 \in \mathcal{L}_r^2(\imath)$$

is an isometry with respect to the semi-inner products, thanks to the identity

$$(f, f)_\Delta = \int_\imath (f_1(s)^* \ f_2(s)^*) \begin{pmatrix} r(s) & 0 \\ 0 & 0 \end{pmatrix} \begin{pmatrix} f_1(s) \\ f_2(s) \end{pmatrix} ds = (f_1, f_1)_r.$$

Furthermore, this mapping is onto, since each function in $\mathcal{L}_r^2(\imath)$ can be regarded as the first component of an element in $\mathcal{L}_\Delta^2(\imath)$ with the understanding that the second component can be any measurable function. Therefore, the mapping R induces a unitary operator, again denoted by R, from $L_\Delta^2(\imath)$ onto $L_r^2(\imath)$.

Assume now that the system or, equivalently, the Sturm–Liouville equation is definite. In the Hilbert space $L_\Delta^2(\imath)$ there are the preminimal relation T_0, minimal relation $T_{\min}$, and maximal relation $T_{\max}$ associated with the canonical system $Jf' - Hf = \Delta g$:

$$T_0 \subset \overline{T}_0 = T_{\min} \subset T_{\max} = (T_{\min})^*.$$

Likewise, one can define corresponding relations in the Hilbert space $L_r^2(\imath)$. The maximal relation $\mathcal{T}_{\max}$ is defined as follows:

$$\mathcal{T}_{\max} = \left\{ \{\mathfrak{f}, \mathfrak{g}\} \in L_r^2(\imath) \times L_r^2(\imath) : -(p\mathfrak{f}^{[1]})' + s\mathfrak{f}^{[1]} + q\mathfrak{f} = r\mathfrak{g} \right\},$$

in the sense that there exist representatives $\mathfrak{f}$ and $\mathfrak{g} \in \mathcal{L}_r^2(\imath)$ of $\mathfrak{f}$ and $\mathfrak{g}$, respectively, such that $\mathfrak{f} \in AC(\imath)$, $p\mathfrak{f}^{[1]} \in AC(\imath)$, and (7.10.2) holds. It is clear that the definiteness of the canonical system or, equivalently, of the equation (7.10.2) implies that each element $\mathfrak{f} \in \operatorname{dom} \mathcal{T}_{\max}$ has a unique representative $\mathfrak{f}$ such that $\mathfrak{f} \in AC(\imath)$, $p\mathfrak{f}^{[1]} \in AC(\imath)$; cf. Lemma 7.6.1. The preminimal relation $\mathcal{T}_0$ and the minimal relation $\mathcal{T}_{\min}$ are defined by

$$\mathcal{T}_0 = \left\{ \{\mathfrak{f}, \mathfrak{g}\} \in \mathcal{T}_{\max} : \mathfrak{f} \text{ has compact support} \right\} \quad \text{and} \quad \mathcal{T}_{\min} = \overline{\mathcal{T}_0}.$$

It is not difficult to see that the mapping $\widehat{R}$ defined by

$$\widehat{R}\{f, g\} = \{Rf, Rg\}, \quad \{f, g\} \in L_\Delta^2(\imath) \times L_\Delta^2(\imath),$$

takes $T_{\max}$ one-to-one onto $\mathcal{T}_{\max}$, including absolutely continuous representatives, and that with $\{\mathfrak{f}, \mathfrak{g}\} = \widehat{R}\{f, g\}$ and $\{\mathfrak{h}, \mathfrak{k}\} = \widehat{R}\{h, k\}$:

$$(g, h)_\Delta - (f, k)_\Delta = (\mathfrak{g}, \mathfrak{h})_r - (\mathfrak{f}, \mathfrak{k})_r, \quad \{f, g\}, \{h, k\} \in T_{\max}. \tag{7.10.3}$$

Similarly, $\widehat{R}$ takes T_0 one-to-one onto $\mathcal{T}_0$ and hence $\widehat{R}$ takes $T_{\min}$ one-to-one onto $\mathcal{T}_{\min}$.

In the Hilbert space $L_r^2(\imath)$ the relations $\mathcal{T}_0$, $\mathcal{T}_{\min}$, and $\mathcal{T}_{\max}$ associated with the Sturm–Liouville equation (7.10.2) satisfy

$$\mathcal{T}_0 \subset \overline{\mathcal{T}_0} = \mathcal{T}_{\min} \subset \mathcal{T}_{\max} = (\mathcal{T}_{\min})^*.$$

Furthermore, R maps $\ker(T_{\max} - \lambda)$ one-to-one onto $\ker(\mathcal{T}_{\max} - \lambda)$. Since the functions p, q, s, and r are real, it follows that the defect numbers of $T_{\min}$ and $\mathcal{T}_{\min}$ are equal. Let $\{\mathcal{G}, \Gamma_0, \Gamma_1\}$ be a boundary triplet for $T_{\max}$ and let $\mathcal{T}_{\max}$ be the corresponding maximal relation for the Sturm–Liouville operator. Then the mappings Γ_0' and Γ_1' from $\mathcal{T}_{\max}$ to $\mathcal{G}$ given by

$$\Gamma_0'\{\mathfrak{f}, \mathfrak{g}\} = \Gamma_0\{f, g\} \quad \text{and} \quad \Gamma_1'\{\mathfrak{f}, \mathfrak{g}\} = \Gamma_1\{f, g\}, \quad \{\mathfrak{f}, \mathfrak{g}\} = \widehat{R}\{f, g\}, \tag{7.10.4}$$

form a boundary triplet for $\mathcal{T}_{\max}$; cf. (7.10.3). The boundary triplet $\{\mathcal{G}, \Gamma_0, \Gamma_1\}$ and the one in (7.10.4) have the same Weyl function.

Via the above identification, the discussion and the results for canonical systems with regular, quasiregular, and singular endpoints in Section 7.7 and Section 7.8 remain valid for weighted Sturm–Liouville equations of the form (7.10.2). Note that in the special case where $s(t) = 0$ and $r(t) > 0$ for almost all $t \in \imath$ the Sturm–Liouville expression (7.10.2) coincides with the Sturm–Liouville expression studied in Chapter 6.

Special canonical systems

This subsection is devoted to the special class of canonical differential equations which have the form

$$Jf' = \lambda \Delta f + \Delta g \tag{7.10.5}$$

on an open interval $\imath = (a, b)$, i.e., the class of canonical systems of the form (7.2.3) with $H = 0$. It will be assumed that the system is real and definite on $\imath$. Here definiteness means that the identity $\Delta(t)e = 0$ for some $e \in \mathbb{C}^2$ and all $t \in \imath$ implies $e = 0$. Note that Lemma 7.2.5 shows that any real definite canonical system of the form (7.2.3) can be transformed into the form (7.10.5) with a possible real shift of the eigenvalue parameter. For this class of equations the limit-point and limit-circle classification at an endpoint can be characterized in terms of the integrability of the function Δ.

Theorem 7.10.1. *Let the canonical system* (7.10.5) *be real and definite, and let the endpoint a be regular. Then there is the alternative:*

(i) Δ *is integrable and the endpoint b is in the limit-circle case;*

(ii) Δ *is not integrable and the endpoint b is in the limit-point case.*

Proof. By assumption, the canonical system (7.10.5) is real and definite, and the endpoint a is regular. If Δ is integrable on $\imath$, then b is quasiregular, which implies that b is in the limit-circle case; see Corollary 7.4.6. Therefore, it suffices to show that if Δ not integrable, then b is in the limit-point case. Hence, assume that Δ is not integrable at b, so that

$$\infty = \int_a^b |\Delta(s)| \, ds \le \int_a^b \operatorname{tr} \Delta(s) \, ds, \tag{7.10.6}$$

where the estimate $|\Delta(s)| \le \operatorname{tr} \Delta(s)$ follows from (7.1.6). In order to show that the endpoint b is in the limit-point case one must verify that

$$\lim_{t \to b} h(t)^* J f(t) = 0$$

for all $\{f, g\}, \{h, k\} \in T_{\max}$; cf. Lemma 7.6.8. Since the limit on the left-hand side exists due to Lemma 7.6.4, it suffices to verify the weaker statement

$$\liminf_{t \to b} h(t)^* J f(t) = 0 \tag{7.10.7}$$

for all $\{f, g\}, \{h, k\} \in T_{\max}$. According to Corollary 7.6.6 and the von Neumann formula in Theorem 1.7.11, it then suffices to prove (7.10.7) for elements of the form $f = u_f + v_f$ and $h = u_h + v_h$, where

$$\{u_f, \lambda u_f\}, \{u_h, \lambda u_h\} \in T_{\max} \quad \text{and} \quad \{v_f, \mu v_f\}, \{v_h, \mu v_h\} \in T_{\max}$$

with λ and μ in different half-planes. Finally, by polarization, it is clearly sufficient to show that

$$\liminf_{t \to b} f(t)^* J f(t) = 0 \tag{7.10.8}$$

for $f = u + v$, where $\{u, \lambda u\}, \{v, \mu v\} \in T_{\max}$. The proof of (7.10.8) is carried out in five steps.

Step 1. Let u be a solution of the homogeneous equation $Jy' = \lambda \Delta y$ which satisfies $u \in \mathcal{L}^2_\Delta(\imath)$. Then

$$u(t) = u(a) + \int_a^t J^{-1} \lambda \Delta(s) u(s) \, ds. \tag{7.10.9}$$

Hence, it follows from (7.10.9) as in Lemma 7.1.4 that

$$|u(t)| \leq |u(a)| + |\lambda| \int_a^t |\Delta(s) u(s)| \, ds$$

$$\leq |u(a)| + |\lambda| \left(\int_a^t |\Delta(s)| \, ds \right)^{\frac{1}{2}} \left(\int_a^t |\Delta(s)^{\frac{1}{2}} u(s)|^2 \, ds \right)^{\frac{1}{2}}$$

$$\leq |u(a)| + |\lambda| \left(\int_a^t |\Delta(s)| \, ds \right)^{\frac{1}{2}} \|u\|_\Delta.$$

Due to the estimate $|\Delta(s)| \leq \operatorname{tr} \Delta(s)$ one obtains

$$|u(t)| \leq |u(a)| + |\lambda| \left(\int_a^t \operatorname{tr} \Delta(s) \, ds \right)^{\frac{1}{2}} \|u\|_\Delta. \tag{7.10.10}$$

It is clear that for a solution v of the homogeneous equation $Jy' = \mu \Delta y$ which satisfies $v \in \mathcal{L}^2_\Delta(\imath)$ one obtains the similar inequality

$$|v(t)| \leq |v(a)| + |\mu| \left(\int_a^t \operatorname{tr} \Delta(s) \, ds \right)^{\frac{1}{2}} \|v\|_\Delta. \tag{7.10.11}$$

Step 2. Let u and v be as in Step 1 with λ and μ in different half-planes. Due to the assumption $\int_a^b \operatorname{tr} \Delta(s) \, ds = \infty$ in (7.10.6), one can choose $t_0 > a$ so large that $\int_a^t \operatorname{tr} \Delta(s) \, ds \geq 1$ for $t \geq t_0$. Then it follows from the estimates (7.10.10) and (7.10.11) that

$$|u(t)| \leq \left(\int_a^t \operatorname{tr} \Delta(s) \, ds \right)^{\frac{1}{2}} \left(|u(a)| + |\lambda| \|u\|_\Delta \right), \quad t \geq t_0,$$

$$|v(t)| \leq \left(\int_a^t \operatorname{tr} \Delta(s) \, ds \right)^{\frac{1}{2}} \left(|v(a)| + |\mu| \|v\|_\Delta \right), \quad t \geq t_0.$$

Consequently, for $f = u + v$,

$$|f(t)| \le C_f \left(\int_a^t \operatorname{tr} \Delta(s) \, ds \right)^{\frac{1}{2}}, \quad t \ge t_0, \tag{7.10.12}$$

where $C_f = |u(a)| + |\lambda| \, \|u\|_\Delta + |v(a)| + |\mu| \, \|v\|_\Delta$.

Step 3. Define the 2×2 matrix function Δ_0 by

$$\Delta_0(t) = \begin{cases} (\operatorname{tr} \Delta(t))^{-1} \Delta(t), & \operatorname{tr} \Delta(t) \ne 0, \\ \frac{1}{2} I, & \operatorname{tr} \Delta(t) = 0, \end{cases}$$

for almost every $t \in \imath$. Since $\operatorname{tr} \Delta(t) = 0$ implies $\Delta(t) = 0$ (see (7.1.6)), one has $\Delta(t) = (\operatorname{tr} \Delta(t)) \Delta_0(t)$. Then Δ_0 is nonnegative and

$$\Delta_0 = \begin{pmatrix} \alpha & \beta \\ \beta & \delta \end{pmatrix},$$

where the functions α and δ are nonnegative with $\alpha + \delta = 1$, and the function β is real. Define the matrix function Δ_1 by

$$\begin{aligned} \Delta_1 &= \left((\operatorname{sgn} \beta) \alpha^{\frac{1}{2}} \quad \delta^{\frac{1}{2}} \right)^* \left((\operatorname{sgn} \beta) \alpha^{\frac{1}{2}} \quad \delta^{\frac{1}{2}} \right) \\ &= \begin{pmatrix} \alpha & (\operatorname{sgn} \beta) \alpha^{\frac{1}{2}} \delta^{\frac{1}{2}} \\ (\operatorname{sgn} \beta) \alpha^{\frac{1}{2}} \delta^{\frac{1}{2}} & \delta \end{pmatrix}; \end{aligned} \tag{7.10.13}$$

then Δ_1 is nonnegative. Moreover, since $\alpha^{\frac{1}{2}} \delta^{\frac{1}{2}} \ge (\operatorname{sgn} \beta) \beta$ it follows that the matrix

$$2\Delta_0 - \Delta_1 = \begin{pmatrix} \alpha & 2\beta - (\operatorname{sgn} \beta) \alpha^{\frac{1}{2}} \delta^{\frac{1}{2}} \\ 2\beta - (\operatorname{sgn} \beta) \alpha^{\frac{1}{2}} \delta^{\frac{1}{2}} & \delta \end{pmatrix}$$

is nonnegative. Therefore, $\Delta_1(t) \le 2\Delta_0(t)$ and one has the estimate

$$(\operatorname{tr} \Delta(t)) \Delta_1(t) \le 2(\operatorname{tr} \Delta(t)) \Delta_0(t) = 2\Delta(t) \tag{7.10.14}$$

for almost every $t \in \imath$.

Step 4. It will be shown that

$$\liminf_{t \to b} \left[f(t)^* \Delta_1(t) f(t) \int_a^t \operatorname{tr} \Delta(s) \, ds \right] = 0 \tag{7.10.15}$$

for $f \in \mathcal{L}^2_\Delta(\imath)$ and, in particular, for $f = u + v$ as in Step 2. In fact, assume that (7.10.15) does not hold. Then there exist $a < a' < b$ and $\varepsilon > 0$ such that for all $t \ge a'$

$$\varepsilon \le f(t)^* \Delta_1(t) f(t) \int_a^t \operatorname{tr} \Delta(s) \, ds$$

or, equivalently,

$$\varepsilon \, \frac{\operatorname{tr} \Delta(t)}{\int_a^t \operatorname{tr} \Delta(s) \, ds} \leq f(t)^* \Delta_1(t) f(t) \operatorname{tr} \Delta(t). \tag{7.10.16}$$

Integration of the right-hand side of (7.10.16) together with (7.10.14) lead to

$$\int_{a'}^b f(t)^* \Delta_1(t) f(t) \operatorname{tr} \Delta(t) \, dt \leq 2 \int_{a'}^b f(t)^* \Delta(t) f(t) \, dt < \infty,$$

while integration of the left-hand side of (7.10.16) gives

$$\varepsilon \int_{a'}^b \frac{\operatorname{tr} \Delta(t)}{\int_a^t \operatorname{tr} \Delta(s) \, ds} \, dt = \varepsilon \int_{a'}^b \frac{d}{dt} \left(\log \int_a^t \operatorname{tr} \Delta(s) \, ds \right) dt = \infty,$$

due to (7.10.6). This contradiction shows that (7.10.15) is valid.

Step 5. It will be shown that for $f = u + v$ as in Step 2 the limit in (7.10.15) implies the limit in (7.10.8). It is helpful to introduce the notation

$$\varphi = \left((\operatorname{sgn} \beta) \alpha^{\frac{1}{2}} \quad \delta^{\frac{1}{2}} \right) \begin{pmatrix} f_1 \\ f_2 \end{pmatrix},$$

so that $|\varphi|^2 = f^* \Delta_1 f$; cf. (7.10.13). Then the limit result (7.10.15) can be written as

$$\liminf_{t \to b} \left[|\varphi(t)|^2 \int_a^t \operatorname{tr} \Delta(s) \, ds \right] = 0. \tag{7.10.17}$$

Observe that the term $f^* J f$ in (7.10.8) is given by

$$f^* J f = 2i \operatorname{Im} (\bar{f}_2 f_1). \tag{7.10.18}$$

To estimate the term $|\operatorname{Im} (\bar{f}_2 f_1)|$ note that, by the definition of the function φ,

$$\bar{f}_1 \varphi = (\operatorname{sgn} \beta) \alpha^{\frac{1}{2}} \bar{f}_1 f_1 + \delta^{\frac{1}{2}} \bar{f}_1 f_2 \quad \text{and} \quad \bar{f}_2 \varphi = (\operatorname{sgn} \beta) \alpha^{\frac{1}{2}} \bar{f}_2 f_1 + \delta^{\frac{1}{2}} \bar{f}_2 f_2.$$

This yields the identities

$$\operatorname{Im} (\bar{f}_1 \varphi) = \delta^{\frac{1}{2}} \operatorname{Im} (\bar{f}_1 f_2) \quad \text{and} \quad \operatorname{Im} (\bar{f}_2 \varphi) = (\operatorname{sgn} \beta) \alpha^{\frac{1}{2}} \operatorname{Im} (\bar{f}_2 f_1).$$

Therefore, it is clear that

$$\delta^{\frac{1}{2}} |\operatorname{Im} (\bar{f}_1 f_2)| = |\operatorname{Im} (\bar{f}_1 \varphi)| \leq |f_1| \, |\varphi| \tag{7.10.19}$$

and

$$\alpha^{\frac{1}{2}} |\operatorname{Im} (\bar{f}_2 f_1)| = |\operatorname{Im} (\bar{f}_2 \varphi)| \leq |f_2| \, |\varphi|. \tag{7.10.20}$$

Since $\alpha + \delta = 1$, one has $\alpha < \frac{1}{2}$ if and only if $\delta \geq \frac{1}{2}$. Note that $x \geq \frac{1}{2}$ if and only if $1/\sqrt{x} \leq \sqrt{2}$, so it follows from (7.10.18) and (7.10.19)–(7.10.20) that

$$|f^* J f| \leq \begin{cases} 2\sqrt{2} \, |\varphi| |f_2|, & \alpha \geq \frac{1}{2}, \\ 2\sqrt{2} \, |\varphi| |f_1|, & \delta \geq \frac{1}{2}. \end{cases}$$

Therefore, if $t \geq t_0$, then (7.10.12) implies that

$$|f(t)^* J f(t)| \leq 2\sqrt{2} \, C_f |\varphi(t)| \left(\int_a^t \operatorname{tr} \Delta(s) \, ds \right)^{\frac{1}{2}}, \quad t \geq t_0.$$

Combined with (7.10.17) this shows that (7.10.8) is satisfied. □

The next corollary follows from Theorem 7.10.1 and (7.1.6).

Corollary 7.10.2. *Let the canonical system* (7.10.5) *be real and definite, and let the endpoint a be regular. Assume that Δ is trace-normed in the sense that* $\operatorname{tr} \Delta = 1$. *Then the following alternative holds:*

(i) *if $b \in \mathbb{R}$, then the limit-circle case prevails;*

(ii) *if $b = \infty$, then the limit-point case prevails.*

Next two simple examples for trace-normed canonical systems are discussed.

Example 7.10.3. Let $\imath = (-1, 1)$ and define the matrix function Δ by

$$\Delta(t) = \begin{pmatrix} 1 & 0 \\ 0 & 0 \end{pmatrix}, \quad t \in (-1, 0), \qquad \Delta(t) = \begin{pmatrix} 0 & 0 \\ 0 & 1 \end{pmatrix}, \quad t \in (0, 1).$$

A measurable function $f = (f_1, f_2)^\top$ belongs to $\mathcal{L}_\Delta^2(\imath)$ if and only if

$$\int_{-1}^0 |f_1(t)|^2 \, dt < \infty \quad \text{and} \quad \int_0^1 |f_2(t)|^2 \, dt < \infty,$$

and for $f, g \in \mathcal{L}_\Delta^2(\imath)$ the semi-inner product is given by

$$(f, g)_\Delta = \int_{-1}^0 \bar{g}_1(t) f_1(t) \, dt + \int_0^1 \bar{g}_2(t) f_2(t) \, dt.$$

Hence, an element $f \in \mathcal{L}_\Delta^2(\imath)$ has Δ-norm 0 if and only if

$$\begin{pmatrix} f_1(t) \\ f_2(t) \end{pmatrix} = \begin{pmatrix} 0 \\ f_2(t) \end{pmatrix} \quad \text{for a.e. } t \in (-1, 0)$$

and

$$\begin{pmatrix} f_1(t) \\ f_2(t) \end{pmatrix} = \begin{pmatrix} f_1(t) \\ 0 \end{pmatrix} \quad \text{for a.e. } t \in (0, 1),$$

where f_2 on $(-1, 0)$ and f_1 on $(0, 1)$ are completely arbitrary complex measurable functions.

It is straightforward to see that the regular canonical system $J f' = \Delta g$ is definite. Hence, in the Hilbert space $L_\Delta^2(\imath)$ the maximal relation

$$T_{\max} = \{ \{f, g\} \in L_\Delta^2(\imath) \times L_\Delta^2(\imath) : J f' = \Delta g \}$$

is well defined and for each $\{f, g\} \in T_{\max}$ the equivalence class f contains a unique absolutely continuous representative such that $Jf' = \Delta g$; cf. Lemma 7.6.1. In fact, an absolutely continuous function $f = (f_1, f_2)^\top$ satisfies $Jf' = \Delta g$ with $g \in L^2_\Delta(\imath)$ if and only if

$$\begin{pmatrix} f_1(t) \\ f_2(t) \end{pmatrix} = \begin{pmatrix} \gamma_1 \\ \gamma_2 + \int_t^0 g_1(s)\, ds \end{pmatrix} \qquad \text{for a.e. } t \in (-1, 0)$$

and

$$\begin{pmatrix} f_1(t) \\ f_2(t) \end{pmatrix} = \begin{pmatrix} \gamma_1 + \int_0^t g_2(s)\, ds \\ \gamma_2 \end{pmatrix} \qquad \text{for a.e. } t \in (0, 1)$$

for some constants $\gamma_1, \gamma_2 \in \mathbb{C}$. From the equality

$$\int_{-1}^1 \left(f(t) - \begin{pmatrix} \gamma_1 \\ \gamma_2 \end{pmatrix} \right)^* \Delta(t) \left(f(t) - \begin{pmatrix} \gamma_1 \\ \gamma_2 \end{pmatrix} \right) dt = 0$$

it follows that $f = (f_1, f_2)^\top$ and $(\gamma_1, \gamma_2)^\top$ are in the same equivalence class in $L^2_\Delta(\imath)$. Therefore,

$$\dim \left(\operatorname{dom} T_{\max} \right) = 2,$$

and the functions $\varphi = (1, 0)^\top$ and $\psi = (0, 1)^\top$ form an orthonormal system in $\operatorname{dom} T_{\max}$. Furthermore, it follows from the representation of $T_{\min}$ in Lemma 7.7.1 that $\{f, g\} \in T_{\min}$ if and only if

$$\gamma_1 = 0, \quad \gamma_2 = 0, \quad \int_{-1}^0 g_1(t)\, dt = 0, \quad \int_0^1 g_2(t)\, dt = 0.$$

Hence,

$$\operatorname{dom} T_{\min} = \{0\} \quad \text{and} \quad \operatorname{mul} T_{\min} = (\operatorname{dom} T_{\max})^\perp,$$

and

$$\operatorname{mul} T_{\max} = (\operatorname{dom} T_{\min})^\perp = L^2_\Delta(\imath).$$

The boundary mappings in Theorem 7.7.2 are given by

$$\Gamma_0\{f, g\} = \frac{1}{\sqrt{2}} \begin{pmatrix} 2\gamma_1 + \int_0^1 g_2(t)\, dt \\ 2\gamma_2 + \int_{-1}^0 g_1(t)\, dt \end{pmatrix} \quad \text{and} \quad \Gamma_1\{f, g\} = \frac{1}{\sqrt{2}} \begin{pmatrix} \int_{-1}^0 g_1(t)\, dt \\ \int_0^1 g_2(t)\, dt \end{pmatrix}.$$

In order to compute the γ-field and Weyl function corresponding to the boundary triplet $\{\mathbb{C}^2, \Gamma_0, \Gamma_1\}$ fix a fundamental system by $Y(-1, \lambda) = I$. Then

$$Y(t, \lambda) = \begin{pmatrix} 1 & 0 \\ -\lambda t - \lambda & 1 \end{pmatrix} \qquad \text{for a.e. } t \in (-1, 0),$$

$$Y(t, \lambda) = \begin{pmatrix} -\lambda^2 t + 1 & \lambda t \\ -\lambda & 1 \end{pmatrix} \qquad \text{for a.e. } t \in (0, 1).$$

Hence, it follows from Theorem 7.7.2 that the γ-field γ and the Weyl function M are given by

$$\gamma(\cdot,\lambda) = Y(\cdot,\lambda)\frac{\sqrt{2}}{4-\lambda^2}\begin{pmatrix} 2 & -\lambda \\ \lambda & 2-\lambda^2 \end{pmatrix} \quad \text{and } M(\lambda) = \frac{1}{4-\lambda^2}\begin{pmatrix} 2\lambda & -\lambda^2 \\ -\lambda^2 & 2\lambda \end{pmatrix}.$$

In particular, the poles of M are $\{-2,2\}$ and hence the spectrum of A_0 consists of the eigenvalues 2 and -2 which both have multiplicity 1.

The next example is a variant of Example 7.10.3 in the limit-point case.

Example 7.10.4. Let $\imath = (-1,\infty)$ and define the matrix function Δ by

$$\Delta(t) = \begin{pmatrix} 1 & 0 \\ 0 & 0 \end{pmatrix}, \quad t \in (-1,0), \qquad \Delta(t) = \begin{pmatrix} 0 & 0 \\ 0 & 1 \end{pmatrix}, \quad t \in (0,\infty).$$

As in Example 7.10.3, a measurable function $f = (f_1,f_2)^\top$ belongs to $\mathcal{L}_\Delta^2(\imath)$ if and only if

$$\int_{-1}^0 |f_1(t)|^2\, dt < \infty \quad \text{and} \quad \int_0^\infty |f_2(t)|^2\, dt < \infty.$$

The semi-inner product and the elements with Δ-norm 0 are as in Example 7.10.3, except that the interval $(0,1)$ has to be replaced by $(0,\infty)$. Furthermore, the canonical system $Jf' = \Delta g$ is definite and in the limit-point case; cf. Corollary 7.10.2. Hence, the maximal relation

$$T_{\max} = \{\{f,g\} \in L_\Delta^2(\imath) \times L_\Delta^2(\imath) : Jf' = \Delta g\}$$

is well defined in $L_\Delta^2(\imath)$. In a similar way as in Example 7.10.3 it follows that an absolutely continuous function $f = (f_1,f_2)^\top \in L_\Delta^2(\imath)$ satisfies $Jf' = \Delta g$ with $g \in L_\Delta^2(\imath)$ if and only if

$$\begin{pmatrix} f_1(t) \\ f_2(t) \end{pmatrix} = \begin{pmatrix} \gamma_1 \\ \int_t^0 g_1(s)\, ds \end{pmatrix} \qquad \text{for a.e. } t \in (-1,0)$$

and

$$\begin{pmatrix} f_1(t) \\ f_2(t) \end{pmatrix} = \begin{pmatrix} \gamma_1 + \int_0^t g_2(s)\, ds \\ 0 \end{pmatrix} \qquad \text{for a.e. } t \in (0,\infty)$$

hold for some constant $\gamma_1 \in \mathbb{C}$. The functions $f = (f_1,f_2)^\top$ and $(\gamma_1,0)^\top$ are in the same equivalence class in $L_\Delta^2(\imath)$ and therefore

$$\dim\left(\operatorname{dom} T_{\max}\right) = 1,$$

and $\operatorname{dom} T_{\max}$ is spanned by the function $\varphi = (1,0)^\top$. It follows from Lemma 7.8.1 that $\{f,g\} \in T_{\min}$ if and only if

$$\gamma_1 = 0 \quad \text{and} \quad \int_{-1}^0 g_1(t)\, dt = 0.$$

Hence, $\operatorname{dom} T_{\min} = \{0\}$, and $\operatorname{mul} T_{\min}$ and $\operatorname{mul} T_{\max}$ are related as in Example 7.10.3.

The boundary mappings in Theorem 7.8.2 are given by

$$\Gamma_0\{f,g\} = \gamma_1 \quad \text{and} \quad \Gamma_1\{f,g\} = \int_{-1}^{0} g_1(t)\, dt.$$

To compute the γ-field and Weyl function corresponding to the boundary triplet $\{\mathbb{C}, \Gamma_0, \Gamma_1\}$ use the fundamental system

$$Y(t,\lambda) = \begin{pmatrix} 1 & 0 \\ -\lambda t - \lambda & 1 \end{pmatrix} \quad \text{for a.e. } t \in (-1,0),$$

$$Y(t,\lambda) = \begin{pmatrix} -\lambda^2 t + 1 & \lambda t \\ -\lambda & 1 \end{pmatrix} \quad \text{for a.e. } t \in (0,\infty).$$

Clearly, not both colums of $Y(\cdot,\lambda)$ belong to $L^2_\Delta(\imath)$, but the function $\chi(\cdot,\lambda)$ given by

$$\begin{pmatrix} 1 \\ -\lambda t \end{pmatrix} \quad \text{for a.e. } t \in (-1,0), \qquad \begin{pmatrix} 1 \\ 0 \end{pmatrix} \quad \text{for a.e. } t \in (0,\infty),$$

belongs to $L^2_\Delta(\imath)$ and satisfies $J\chi'(\cdot,\lambda) = \lambda \Delta \chi(\cdot,\lambda)$. Hence, by Theorem 7.8.2, the γ-field γ and the Weyl function M are given by

$$\gamma(\cdot,\lambda) = Y(\cdot,\lambda) \begin{pmatrix} 1 \\ \lambda \end{pmatrix} \quad \text{and} \quad M(\lambda) = \lambda.$$

The Domains of Schrödinger Operators

For the multi-dimensional Schrödinger operator $-\Delta + V$ with a bounded real potential V on a bounded domain $\Omega \subset \mathbb{R}^n$ with a C^2-smooth boundary a boundary triplet and a Weyl function will be constructed. The self-adjoint realizations of $-\Delta + V$ in $L^2(\Omega)$ and their spectral properties will be investigated. One of the main difficulties here is to provide trace mappings on the domain of the maximal realization, in such a way that the second Green identity remains valid in an appropriate form. It is necessary to introduce and study Sobolev spaces on the domain Ω and its boundary $\partial\Omega$, which will be done in Section 8.2; in this context also the rigged Hilbert spaces from Section 8.1 arise as Sobolev spaces and their duals. The minimal and maximal operators, and the Dirichlet and Neumann trace maps on the maximal domain will be discussed in Section 8.3, and in Section 8.4 a boundary triplet and Weyl function for the maximal operator associated with $-\Delta + V$ is provided. The self-adjoint realizations, their spectral properties, and some natural boundary conditions are also discussed in Section 8.4. The class of semibounded self-adjoint realizations of $-\Delta + V$ in $L^2(\Omega)$ and the corresponding semibounded forms are studied in Section 8.5. For this purpose a boundary pair which is compatible with the boundary triplet in Section 8.4 is provided. Orthogonal couplings of Schrödinger operators are treated in Section 8.6 for the model problem in which $\mathbb{R}^n$ decomposes into a bounded C^2-domain Ω_+ and an unbounded component $\Omega_- = \mathbb{R}^n \setminus \overline{\Omega_+}$. Finally, in Section 8.7 the more general setting of Schrödinger operators on bounded Lipschitz domains is briefly discussed.

8.1 Rigged Hilbert spaces

In this preparatory section the notion of rigged Hilbert spaces or Gelfand triples is briefly recalled. For this, let $\mathfrak{G}$ and $\mathfrak{H}$ be Hilbert spaces and assume that $\mathfrak{G}$ is densely and continuously embedded in $\mathfrak{H}$, that is, one has $\mathfrak{G} \subset \mathfrak{H}$ and the embedding operator $\iota : \mathfrak{G} \hookrightarrow \mathfrak{H}$ is continuous with dense range and $\ker \iota = \{0\}$. In the following the dual space $\mathfrak{H}'$ is identified with $\mathfrak{H}$, but the dual space $\mathfrak{G}'$ of antilinear continuous functionals is not identified with $\mathfrak{G}$. Instead, the isometric isomorphism

$$\mathfrak{I} : \mathfrak{G}' \to \mathfrak{G}, \quad g' \mapsto \mathfrak{I}g', \quad \text{where} \quad (\mathfrak{I}g', g)_{\mathfrak{G}} = g'(g), \quad g \in \mathfrak{G}, \tag{8.1.1}$$

is written explicitly whenever used. In fact, in this section the continuous (antilinear) functionals on $\mathfrak{G}$ will be identified via the scalar product in $\mathfrak{H}$. In the following usually the notation

$$\langle g', g \rangle_{\mathfrak{G}' \times \mathfrak{G}} := g'(g) \tag{8.1.2}$$

is employed for the (antilinear) dual pairing in (8.1.1), and when no confusion can arise the index is suppressed, that is, one writes $\langle g', g \rangle = g'(g)$ for (8.1.2).

In the above setting the dual operator of the embedding operator $\iota : \mathfrak{G} \hookrightarrow \mathfrak{H}$ is given by

$$\iota' : \mathfrak{H} \hookrightarrow \mathfrak{G}', \quad (\iota'h)(g) = (h, \iota g)_{\mathfrak{H}}, \quad g \in \mathfrak{G}, \tag{8.1.3}$$

and in terms of the pairing $\langle \cdot, \cdot \rangle$ this means

$$\langle \iota'h, g \rangle = (h, \iota g)_{\mathfrak{H}}, \quad h \in \mathfrak{H}, \ g \in \mathfrak{G}. \tag{8.1.4}$$

Since the scalar product $(\cdot, \cdot)_{\mathfrak{H}}$ is antilinear in the second argument one has $\langle \iota'h, \lambda g \rangle = \bar{\lambda} \langle \iota'h, g \rangle$ for $\lambda \in \mathbb{C}$, and hence $\iota'h$ is indeed antilinear. Observe that the dual operator ι' in (8.1.3) is continuous since ι is continuous. Moreover, from the identity $\ker \iota' = (\operatorname{ran} \iota)^{\perp_{\mathfrak{H}}}$ it follows that ι' is injective, and the range of ι' is dense in $\mathfrak{G}'$ since $\ker \iota'' = (\operatorname{ran} \iota')^{\perp_{\mathfrak{G}'}}$ and $\iota = \iota''$ as $\mathfrak{G}$ is reflexive. Thus,

$$\mathfrak{G} \overset{\iota}{\hookrightarrow} \mathfrak{H} \overset{\iota'}{\hookrightarrow} \mathfrak{G}' \quad \text{with } \operatorname{ran} \iota \subset \mathfrak{H} \text{ dense and } \operatorname{ran} \iota' \subset \mathfrak{G}' \text{ dense,}$$

and since $\mathfrak{G}$ can be viewed as a subspace of $\mathfrak{H}$, and $\mathfrak{H}$ can be viewed as a subspace of $\mathfrak{G}'$, instead of (8.1.4) also the notation

$$\langle h, g \rangle = (h, g)_{\mathfrak{H}}, \quad h \in \mathfrak{H}, \ g \in \mathfrak{G}, \tag{8.1.5}$$

will be used. The present situation will appear naturally in the context of Sobolev spaces later in this chapter. First the terminology will be fixed in the next definition.

Definition 8.1.1. Let $\mathfrak{G}$ and $\mathfrak{H}$ be Hilbert spaces such that $\mathfrak{G}$ is densely and continuously embedded in $\mathfrak{H}$. Then the triple $\{\mathfrak{G}, \mathfrak{H}, \mathfrak{G}'\}$ is a called a *Gelfand triple* or a *rigged Hilbert space*.

Assume now that $\{\mathfrak{G}, \mathfrak{H}, \mathfrak{G}'\}$ is a Gelfand triple. Since the embedding operator $\iota : \mathfrak{G} \hookrightarrow \mathfrak{H}$ is continuous, one has $\|g\|_{\mathfrak{H}} \leq C\|g\|_{\mathfrak{G}}$ for all $g \in \mathfrak{G}$ with the constant

$C = \|\iota\| > 0$. Moreover, as $\mathfrak{G}$ is a Hilbert space it follows from Lemma 5.1.9 that the symmetric form

$$\mathfrak{t}[g_1, g_2] := (g_1, g_2)_{\mathfrak{G}}, \quad \mathrm{dom}\,\mathfrak{t} = \mathfrak{G},$$

is densely defined and closed in $\mathfrak{H}$ with a positive lower bound. Hence, by the first representation theorem (Theorem 5.1.18) there exists a unique self-adjoint operator T with the same positive lower bound in $\mathfrak{H}$, such that $\mathrm{dom}\,T \subset \mathrm{dom}\,\mathfrak{t}$ and

$$(g_1, g_2)_{\mathfrak{G}} = \mathfrak{t}[g_1, g_2] = (Tg_1, g_2)_{\mathfrak{H}}, \quad g_1 \in \mathrm{dom}\,T,\ g_2 \in \mathfrak{G}.$$

Moreover, if $R := T^{\frac{1}{2}}$, then the second representation theorem (Theorem 5.1.23) implies $\mathrm{dom}\,R = \mathrm{dom}\,\mathfrak{t}$ and

$$(g_1, g_2)_{\mathfrak{G}} = \mathfrak{t}[g_1, g_2] = (Rg_1, Rg_2)_{\mathfrak{H}}, \quad g_1, g_2 \in \mathrm{dom}\,R = \mathfrak{G}. \tag{8.1.6}$$

Note that R is a uniformly positive self-adjoint operator in $\mathfrak{H}$.

In the next lemma some more properties of the Gelfand triple $\{\mathfrak{G}, \mathfrak{H}, \mathfrak{G}'\}$ and the operator R are collected.

Lemma 8.1.2. *Let $\{\mathfrak{G}, \mathfrak{H}, \mathfrak{G}'\}$ be a Gelfand triple, let $\mathfrak{J} : \mathfrak{G}' \to \mathfrak{G}$ be the isometric isomorphism in* (8.1.1), *and let R be the uniformly positive self-adjoint operator in $\mathfrak{H}$ such that* (8.1.6) *holds. Then the following statements hold:*

(i) *The Hilbert space $\mathfrak{G}'$ coincides with the completion of $\mathfrak{H}$ equipped with the inner product $(R^{-1}\cdot, R^{-1}\cdot)_{\mathfrak{H}}$.*

(ii) *The operators $\iota_+ = R : \mathfrak{G} \to \mathfrak{H}$ and $\iota_- = R\mathfrak{J} : \mathfrak{G}' \to \mathfrak{H}$ are isometric isomorphisms such that*

$$(\iota_- g', \iota_+ g)_{\mathfrak{H}} = \langle g', g \rangle, \quad g \in \mathfrak{G},\ g' \in \mathfrak{G}'. \tag{8.1.7}$$

(iii) *For all $h \in \mathfrak{H}$ one has $\iota_- h = R^{-1}h$.*

(iv) *For all $h \in \mathfrak{H}$ and $g \in \mathfrak{G}$ one has $\iota_+ \iota_- h = h$ and $\iota_- \iota_+ g = g$.*

(v) *The operator R^{-2} can be extended by continuity to an isometric operator $\widetilde{R}^{-2} : \mathfrak{G}' \to \mathfrak{G}$ which coincides with the isometric isomorphism $\mathfrak{J} : \mathfrak{G}' \to \mathfrak{G}$.*

Proof. (i) Consider an element $g' \in \mathfrak{G}'$ and assume, in addition, that $g' \in \mathfrak{H}$. Then one has

$$\|g'\|_{\mathfrak{G}'} = \sup_{g \in \mathfrak{G} \setminus \{0\}} \frac{|g'(g)|}{\|g\|_{\mathfrak{G}}} = \sup_{g \in \mathfrak{G} \setminus \{0\}} \frac{|\langle g', g \rangle|}{\|g\|_{\mathfrak{G}}} = \sup_{g \in \mathfrak{G} \setminus \{0\}} \frac{|(g', g)_{\mathfrak{H}}|}{\|g\|_{\mathfrak{G}}},$$

where (8.1.2) was used in the second equality, and $g' \in \mathfrak{H}$ and (8.1.5) were used in the last step. Since R is uniformly positive, one has $R^{-1} \in \mathbf{B}(\mathfrak{H})$, and using (8.1.6) one obtains

$$\|g'\|_{\mathfrak{G}'} = \sup_{g \in \mathfrak{G} \setminus \{0\}} \frac{|(R^{-1}g', Rg)_{\mathfrak{H}}|}{\|Rg\|_{\mathfrak{H}}} = \sup_{h \in \mathfrak{H} \setminus \{0\}} \frac{|(R^{-1}g', h)_{\mathfrak{H}}|}{\|h\|_{\mathfrak{H}}} = \|R^{-1}g'\|_{\mathfrak{H}}.$$

Therefore, $\|g'\|_{\mathfrak{G}'} = \|R^{-1}g'\|_{\mathfrak{H}}$ for all $g' \in \mathfrak{H} \subset \mathfrak{G}'$ and as $\mathfrak{H}$ is dense in $\mathfrak{G}'$ with respect to the norm $\|\cdot\|_{\mathfrak{G}'}$, one concludes that $\mathfrak{G}'$ coincides with the completion of $\mathfrak{H}$ with respect to the norm $\|R^{-1}\cdot\|_{\mathfrak{H}}$.

(ii) Observe that by the definition of ι_+ and (8.1.6) one has

$$\|\iota_+ g\|_{\mathfrak{H}} = \|Rg\|_{\mathfrak{H}} = \|g\|_{\mathfrak{G}}, \quad g \in \mathfrak{G} = \operatorname{dom} \iota_+ = \operatorname{dom} R,$$

and hence $\iota_+ : \mathfrak{G} \to \mathfrak{H}$ is isometric. Moreover, since R is bijective, it follows that ι_+ is an isometric isomorphism. Similarly, for $g' \in \mathfrak{G}'$ one has

$$\|\iota_- g'\|_{\mathfrak{H}} = \|R\mathfrak{J}g'\|_{\mathfrak{H}} = \|\mathfrak{J}g'\|_{\mathfrak{G}} = \|g'\|_{\mathfrak{G}'},$$

where in the last step it was used that $\mathfrak{J} : \mathfrak{G}' \to \mathfrak{G}$ is an isometric isomorphism. In order to check the identity (8.1.7), let $g' \in \mathfrak{G}'$ and $g \in \mathfrak{G}$. Then (8.1.6) and (8.1.1) imply

$$(\iota_- g', \iota_+ g)_{\mathfrak{H}} = (R\mathfrak{J}g', Rg)_{\mathfrak{H}} = (\mathfrak{J}g', g)_{\mathfrak{G}} = \langle g', g \rangle. \tag{8.1.8}$$

(iii) Let $g' \in \mathfrak{H} \subset \mathfrak{G}'$ and $g \in \mathfrak{G}$. By (8.1.5), one has

$$\langle g', g \rangle = (g', g)_{\mathfrak{H}} = (R^{-1}g', Rg)_{\mathfrak{H}} = (R^{-1}g', \iota_+ g)_{\mathfrak{H}}$$

and comparing this with (8.1.8) it follows that $R^{-1}g' = \iota_- g'$ for all $g' \in \mathfrak{H}$.

(iv) By the definition of ι_+ and (iii) it is clear that $\iota_+ \iota_- h = RR^{-1}h = h$. Similarly, $\iota_- \iota_+ g = \iota_- Rg = R^{-1}Rg = g$ for $g \in \mathfrak{G}$ by (iii).

(v) For $h \in \mathfrak{H}$ one has $\|R^{-2}h\|_{\mathfrak{G}} = \|R^{-1}h\|_{\mathfrak{H}} = \|h\|_{\mathfrak{G}'}$ by (8.1.6) and (i), and since $\mathfrak{H}$ is dense in $\mathfrak{G}'$, it follows that R^{-2} admits an extension to an isometric operator $\widetilde{R}^{-2} : \mathfrak{G}' \to \mathfrak{G}$. Moreover, for $h \in \mathfrak{H}$ it follows from the definition of ι_- in (ii) and (iii) that

$$R\mathfrak{J}h = \iota_- h = R^{-1}h, \quad \text{and hence} \quad \mathfrak{J}h = R^{-2}h.$$

Thus $\mathfrak{J}$ and the restriction R^{-2} of $\widetilde{R}^{-2}$ coincide on the dense subspace $\mathfrak{H} \subset \mathfrak{G}'$. This implies $\mathfrak{J} = \widetilde{R}^{-2}$. $\qquad\square$

Now a different point of view is taken on Gelfand triples. In the next lemma it is shown that the powers R^s for $s \geq 0$ of a uniformly positive self-adjoint operator R in $\mathfrak{H}$ give rise to Gelfand triples with certain compatibility properties.

Lemma 8.1.3. *Let $\mathfrak{H}$ be a Hilbert space and let R be a uniformly positive self-adjoint operator R in $\mathfrak{H}$. Let $s \geq 0$ and equip $\mathfrak{G}_s := \operatorname{dom} R^s$ with the inner product*

$$(h, k)_{\mathfrak{G}_s} := (R^s h, R^s k)_{\mathfrak{H}}, \quad h, k \in \operatorname{dom} R^s. \tag{8.1.9}$$

Then $\mathfrak{G}_t \subset \mathfrak{G}_s$ for all $t \geq s \geq 0$ and the following statements hold:

(i) *$\{\mathfrak{G}_s, \mathfrak{H}, \mathfrak{G}'_s\}$ is a Gelfand triple and the assertions in Lemma 8.1.2 hold with R, $\mathfrak{G}$, and $\mathfrak{G}'$ replaced by R^s, $\mathfrak{G}_s$, and $\mathfrak{G}'_s$, respectively.*

(ii) *If* $\iota_+ : \mathfrak{G}_1 \to \mathfrak{H}$ *and* $\iota_- : \mathfrak{G}_1' \to \mathfrak{H}$ *denote the isometric isomorphisms corresponding to the Gelfand triple* $\{\mathfrak{G}_1, \mathfrak{H}, \mathfrak{G}_1'\}$ *such that*

$$(\iota_- g', \iota_+ g)_{\mathfrak{H}} = \langle g', g \rangle_{\mathfrak{G}_1' \times \mathfrak{G}_1}, \quad g' \in \mathfrak{G}_1', g \in \mathfrak{G}_1,$$

then their restrictions

$$\iota_+ = R : \mathfrak{G}_{s+1} \to \mathfrak{G}_s \quad \text{and} \quad \iota_- = R^{-1} : \mathfrak{G}_s \to \mathfrak{G}_{s+1}, \quad s \geq 0, \qquad (8.1.10)$$

are isometric isomorphisms such that $\iota_+ \iota_- g = g$ *for* $g \in \mathfrak{G}_s$ *and* $\iota_- \iota_+ l = l$ *for* $l \in \mathfrak{G}_{s+1}$.

Proof. (i) For $s \geq 0$ the self-adjoint operator R^s is uniformly positive in $\mathfrak{H}$ and hence $\mathfrak{G}_s = \operatorname{dom} R^s$ equipped with the inner product (8.1.9) is a Hilbert space which is dense in $\mathfrak{H}$. Moreover, from $R^{-s} \in \mathbf{B}(\mathfrak{H})$ and (8.1.9) one obtains that

$$\|g\|_{\mathfrak{H}} = \|R^{-s} R^s g\|_{\mathfrak{H}} \leq \|R^{-s}\| \|R^s g\|_{\mathfrak{H}} = \|R^{-s}\| \|g\|_{\mathfrak{G}_s}, \quad g \in \mathfrak{G}_s,$$

which shows that the embedding $\mathfrak{G}_s \hookrightarrow \mathfrak{H}$ is continuous. Therefore, if $\mathfrak{G}_s'$ denotes the dual of $\mathfrak{G}_s$, then $\{\mathfrak{G}_s, \mathfrak{H}, \mathfrak{G}_s'\}$ is a Gelfand triple. Comparing (8.1.9) with (8.1.6) shows that the operator R^s plays the same role as the representing operator of the inner product in (8.1.6). Hence, the assertions of Lemma 8.1.2 are valid with R, $\mathfrak{G}$, and $\mathfrak{G}'$ replaced by R^s, $\mathfrak{G}_s$, and $\mathfrak{G}_s'$, respectively.

(ii) Let $s \geq 0$ and consider $l \in \mathfrak{G}_{s+1} = \operatorname{dom} R^{s+1}$. It follows from (8.1.9) that

$$\|Rl\|_{\mathfrak{G}_s} = \|R^s Rl\|_{\mathfrak{H}} = \|R^{s+1} l\|_{\mathfrak{H}} = \|l\|_{\mathfrak{G}_{s+1}}$$

and hence $\iota_+ = R : \mathfrak{G}_{s+1} \to \mathfrak{G}_s$ is isometric. In order to verify that this mapping is onto let $k \in \mathfrak{G}_s$. Then $k \in \mathfrak{H}$, and as R is bijective, there exists $l \in \operatorname{dom} R$ such that $Rl = k$. Therefore, $l = R^{-1} k$ and as $k \in \mathfrak{G}_s = \operatorname{dom} R^s$ one concludes $l \in \operatorname{dom} R^{s+1} = \mathfrak{G}_{s+1}$. This shows that $\iota_+ = R : \mathfrak{G}_{s+1} \to \mathfrak{G}_s$ is an isometric isomorphism for $s \geq 0$. A similar reasoning shows that $\iota_- = R^{-1} : \mathfrak{G}_s \to \mathfrak{G}_{s+1}$ is an isometric isomorphism for $s \geq 0$. The remaining assertions $\iota_+ \iota_- g = g$ for $g \in \mathfrak{G}_s$ and $\iota_- \iota_+ l = l$ for $l \in \mathfrak{G}_{s+1}$ follow immediately from (8.1.10). $\qquad \square$

8.2 Sobolev spaces, C^2-domains, and trace operators

In this section Sobolev spaces on $\mathbb{R}^n$, open subsets $\Omega \subset \mathbb{R}^n$, and on the boundaries $\partial\Omega$ of C^2-domains are defined and some of their features are briefly recalled. Furthermore, the mapping properties of the Dirichlet and Neumann trace map on a C^2-domain Ω are recalled and the first Green identity is established.

For $s \geq 0$ the scale of L^2-based Sobolev spaces $H^s(\mathbb{R}^n)$ is defined with the help of the (classical) Fourier transform $\mathcal{F} \in \mathbf{B}(L^2(\mathbb{R}^n))$ by

$$H^s(\mathbb{R}^n) := \left\{ f \in L^2(\mathbb{R}^n) : (1 + |\cdot|^2)^{s/2} \mathcal{F} f \in L^2(\mathbb{R}^n) \right\}$$

and $H^s(\mathbb{R}^n)$ is equipped with the natural norm

$$\|f\|_{H^s(\mathbb{R}^n)} := \left\| (1 + |\cdot|^2)^{s/2} \mathcal{F} f \right\|_{L^2(\mathbb{R}^n)}, \qquad f \in H^s(\mathbb{R}^n),$$

and corresponding scalar product

$$(f, g)_{H^s(\mathbb{R}^n)} := \left((1 + |\cdot|^2)^{s/2} \mathcal{F} f, (1 + |\cdot|^2)^{s/2} \mathcal{F} g \right)_{L^2(\mathbb{R}^n)}, \qquad f, g \in H^s(\mathbb{R}^n).$$

Then the space $H^s(\mathbb{R}^n)$ is a separable Hilbert space for every $s \geq 0$ and one has $H^0(\mathbb{R}^n) = L^2(\mathbb{R}^n)$. It is also useful to note that the space $C_0^\infty(\mathbb{R}^n)$ is dense in $H^s(\mathbb{R}^n)$ for all $s \geq 0$. Since the Fourier transform is a unitary operator in $L^2(\mathbb{R}^n)$, it is clear that

$$\mathcal{R} = \mathcal{F}^{-1}(1 + |\cdot|^2)^{1/2} \mathcal{F}$$

is a uniformly positive self-adjoint operator in $L^2(\mathbb{R}^n)$ such that $\operatorname{dom} \mathcal{R} = H^1(\mathbb{R}^n)$. Furthermore, for each $s \geq 0$ one has

$$\mathcal{R}^s = \mathcal{F}^{-1}(1 + |\cdot|^2)^{s/2} \mathcal{F}$$

and hence $\mathcal{R}^s$ for $s \geq 0$ is also a uniformly positive self-adjoint operator in $L^2(\mathbb{R}^n)$ such that $\operatorname{dom} \mathcal{R}^s = H^s(\mathbb{R}^n)$. Note that the scalar product in $H^s(\mathbb{R}^n)$ satisfies

$$(f, g)_{H^s(\mathbb{R}^n)} = (\mathcal{R}^s f, \mathcal{R}^s g)_{L^2(\mathbb{R}^n)}, \qquad f, g \in H^s(\mathbb{R}^n),$$

for all $s \geq 0$. In particular, $\mathcal{R}$ plays the same role as the operator R in (8.1.6) and $\mathcal{R}^s$ plays the same role as the operator R^s in (8.1.9). Hence, $\mathcal{R}^s$, $s \geq 0$, gives rise to a Gelfand triple $\{H^s(\mathbb{R}^n), L^2(\mathbb{R}^n), H^{-s}(\mathbb{R}^n)\}$, where $H^{-s}(\mathbb{R}^n)$ denotes the dual space consisting of continuous antilinear functionals on $H^s(\mathbb{R}^n)$. From Lemma 8.1.3 it is now clear that the restrictions $\mathcal{R} : H^{s+1}(\mathbb{R}^n) \to H^s(\mathbb{R}^n)$ and $\mathcal{R}^{-1} : H^s(\mathbb{R}^n) \to H^{s+1}(\mathbb{R}^n)$ are isometric isomorphisms for $s \geq 0$.

For a nonempty open subset $\Omega \subset \mathbb{R}^n$ and $s \geq 0$ define

$$H^s(\Omega) := \left\{ f \in L^2(\Omega) : \text{there exists } g \in H^s(\mathbb{R}^n) \text{ such that } f = g|_\Omega \right\}$$

and endow this space with the norm

$$\|f\|_{H^s(\Omega)} := \inf_{\substack{g \in H^s(\mathbb{R}^n) \\ f = g|_\Omega}} \|g\|_{H^s(\mathbb{R}^n)}, \qquad f \in H^s(\Omega). \tag{8.2.1}$$

The space $H^s(\Omega)$ is a separable Hilbert space; the corresponding scalar product will be denoted by $(\cdot, \cdot)_{H^s(\Omega)}$. For $s \geq 0$ the space $C^\infty(\overline{\Omega}) := \{\varphi|_\Omega : \varphi \in C_0^\infty(\mathbb{R}^n)\}$ is dense in $H^s(\Omega)$. The closure of $C_0^\infty(\Omega)$ in $H^s(\Omega)$ is a closed subspace of $H^s(\Omega)$; it is denoted by

$$H_0^s(\Omega) := \overline{C_0^\infty(\Omega)}^{\|\cdot\|_{H^s(\Omega)}}. \tag{8.2.2}$$

In order to define Sobolev spaces on the boundary $\partial\Omega$ of some domain $\Omega \subset \mathbb{R}^n$ assume first that $\phi : \mathbb{R}^{n-1} \to \mathbb{R}$ is a C^2-function. The vectors in $\mathbb{R}^{n-1}$ will be denoted by $x' = (x_1, \ldots, x_{n-1})^\top \in \mathbb{R}^{n-1}$ and the notation $(x', x_n)^\top$ is used for $(x_1, \ldots, x_n)^\top \in \mathbb{R}^n$. Then the domain

$$\Omega_\phi := \{(x', x_n)^\top \in \mathbb{R}^n : x_n < \phi(x')\} \tag{8.2.3}$$

is called a C^2-*hypograph* and its boundary is given by

$$\partial\Omega_\phi = \{(x', \phi(x'))^\top \in \mathbb{R}^n : x' \in \mathbb{R}^{n-1}\}.$$

For a measurable function $h : \partial\Omega_\phi \to \mathbb{C}$ the surface integral on $\partial\Omega_\phi$ is defined as

$$\int_{\partial\Omega_\phi} h \, d\sigma := \int_{\mathbb{R}^{n-1}} h(x', \phi(x'))\sqrt{1 + |\nabla\phi(x')|^2} \, dx'. \tag{8.2.4}$$

If $\mathbf{1}_B$ denotes the characteristic function of a Borel set $B \subset \partial\Omega_\phi$, then the surface integral in (8.2.4) induces a surface measure

$$\sigma(B) = \int_{\partial\Omega_\phi} \mathbf{1}_B \, d\sigma. \tag{8.2.5}$$

This surface measure also gives rise to the usual L^2-space on $\partial\Omega_\phi$, which will be denoted by $L^2(\partial\Omega_\phi)$. Furthermore, for $s \in [0, 2]$ define the Sobolev space of order s on $\partial\Omega_\phi$ by

$$H^s(\partial\Omega_\phi) := \{h \in L^2(\partial\Omega_\phi) : x' \mapsto h(x', \phi(x')) \in H^s(\mathbb{R}^{n-1})\}$$

and equip $H^s(\partial\Omega_\phi)$ with the corresponding Hilbert space scalar product

$$(h, k)_{H^s(\partial\Omega_\phi)} := \big(h(\cdot, \phi(\cdot)), k(\cdot, \phi(\cdot))\big)_{H^s(\mathbb{R}^{n-1})}, \quad h, k \in H^s(\partial\Omega_\phi). \tag{8.2.6}$$

Note that the operator $V_\phi : H^s(\partial\Omega_\phi) \to H^s(\mathbb{R}^{n-1})$ that maps $h \in H^s(\partial\Omega_\phi)$ to the function $x' \mapsto h(x', \phi(x')) \in H^s(\mathbb{R}^{n-1})$ is an isometric isomorphism.

In the next step the notion of C^2-hypograph is replaced by a bounded domain with a C^2-smooth boundary, that is, the boundary is locally the boundary of a C^2-hypograph.

Definition 8.2.1. A bounded nonempty open subset $\Omega \subset \mathbb{R}^n$ is called a C^2-*domain* if there exist open sets $U_1, \ldots, U_l \subset \mathbb{R}^n$ and (possibly up to rotations of coordinates) C^2-hypographs $\Omega_1, \ldots, \Omega_l \subset \mathbb{R}^n$ such that

$$\partial\Omega \subset \bigcup_{j=1}^l U_j \quad \text{and} \quad \Omega \cap U_j = \Omega_j \cap U_j, \quad j = 1, \ldots, l.$$

Let $\Omega \subset \mathbb{R}^n$ be a bounded C^2-domain as in Definition 8.2.1. Then the boundary $\partial\Omega \subset \mathbb{R}^n$ is compact and there exists a partition of unity subordinate to the open cover $\{U_j\}$ of $\partial\Omega$, that is, there exist functions $\eta_j \in C_0^\infty(\mathbb{R}^n)$, $j = 1, \ldots, l$, with $\operatorname{supp} \eta_j \subset U_j$ such that $0 \leq \eta_j(x) \leq 1$ for all $x \in \mathbb{R}^n$ and $\sum_{j=1}^l \eta_j(x) = 1$ for all $x \in \partial\Omega$. For a measurable function $h : \partial\Omega \to \mathbb{C}$ the surface integral on $\partial\Omega$ is defined as

$$\int_{\partial\Omega} h \, d\sigma := \sum_{j=1}^l \int_{\mathbb{R}^{n-1}} \eta_j(x', \phi_j(x')) h(x', \phi_j(x')) \sqrt{1 + |\nabla\phi_j(x')|^2} \, dx',$$

where the C^2-functions $\phi_j : \mathbb{R}^{n-1} \to \mathbb{R}$ define the C^2-hypographs Ω_j as in (8.2.3) and the possible rotation of coordinates is suppressed. This surface integral induces a surface measure and the notion of an L^2-space $L^2(\partial\Omega)$ in the same way as in (8.2.4) and (8.2.5). In the present setting the Sobolev space $H^s(\partial\Omega)$ for $s \in [0,2]$ is now defined by

$$H^s(\partial\Omega) := \{h \in L^2(\partial\Omega) : \eta_j h \in H^s(\partial\Omega_j), \, j = 1, \dots, l\}$$

and is equipped with the corresponding Hilbert space scalar product

$$(h, k)_{H^s(\partial\Omega)} = \sum_{j=1}^l (\eta_j h, \eta_j k)_{H^s(\partial\Omega_j)}, \quad h, k \in H^s(\partial\Omega). \tag{8.2.7}$$

It follows from the construction that $H^s(\partial\Omega)$ is densely and continuously embedded in $L^2(\partial\Omega)$ for $s \in [0,2]$. Furthermore, since $\partial\Omega$ is a compact subset of $\mathbb{R}^n$, the embedding

$$H^t(\partial\Omega) \hookrightarrow H^s(\partial\Omega), \quad 0 \leq s < t \leq 2, \tag{8.2.8}$$

is compact; see, e.g., [774, Theorem 7.10].

For later purposes it is convenient to use an equivalent characterization of the spaces $H^s(\partial\Omega)$ via interpolation; cf. [573, Theorem B.11]. More precisely, as in (8.1.6) it follows that there exists a unique uniformly positive self-adjoint operator Q in $L^2(\partial\Omega)$ such that

$$\operatorname{dom} Q = H^2(\partial\Omega) \quad \text{and} \quad (h, k)_{H^2(\partial\Omega)} = (Qh, Qk)_{L^2(\partial\Omega)} \tag{8.2.9}$$

for all $h, k \in H^2(\partial\Omega)$. It can be shown that the spaces $H^s(\partial\Omega)$ coincide with the domains $\operatorname{dom} Q^{s/2}$ for $s \in [0,2]$ and that $(Q^{s/2}\cdot, Q^{s/2}\cdot)_{L^2(\partial\Omega)}$ defines a scalar product and equivalent norm in $H^s(\partial\Omega)$. The dual space of the antilinear continuous functionals on $H^s(\partial\Omega)$ is denoted by $H^{-s}(\partial\Omega)$, $s \in [0,2]$. Then one obtains the following statement from Lemma 8.1.2 and Lemma 8.1.3.

Corollary 8.2.2. *Let $\Omega \subset \mathbb{R}^n$ be a bounded C^2-domain and let $s \in [0,2]$. Then the following statements hold:*

(i) *$\{H^s(\partial\Omega), L^2(\partial\Omega), H^{-s}(\partial\Omega)\}$ is a Gelfand triple and the assertions in Lemma 8.1.2 hold with R, $\mathfrak{G}$, and $\mathfrak{G}'$ replaced by $Q^{s/2}$, $H^s(\partial\Omega)$, and $H^{-s}(\partial\Omega)$, respectively.*

(ii) *If $\iota_+ : H^{1/2}(\partial\Omega) \to L^2(\partial\Omega)$ and $\iota_- : H^{-1/2}(\partial\Omega) \to L^2(\partial\Omega)$ denote the isometric isomorphisms from Lemma 8.1.2 (ii) corresponding to the Gelfand triple $\{H^{1/2}(\partial\Omega), L^2(\partial\Omega), H^{-1/2}(\partial\Omega)\}$ such that*

$$(\iota_-\varphi, \iota_+\psi)_{L^2(\partial\Omega)} = \langle \varphi, \psi \rangle_{H^{-1/2}(\partial\Omega) \times H^{1/2}(\partial\Omega)}$$

holds for $\varphi \in H^{-1/2}(\partial\Omega)$ and $\psi \in H^{1/2}(\partial\Omega)$, then for $s \in [0, 3/2]$ their restrictions

$$\iota_+ = Q^{1/4} : H^{s+1/2}(\partial\Omega) \to H^s(\partial\Omega)$$

and

$$\iota_- = Q^{-1/4} : H^s(\partial\Omega) \to H^{s+1/2}(\partial\Omega)$$

are isometric isomorphisms such that $\iota_+\iota_-\phi = \phi$ for $\phi \in H^s(\partial\Omega)$ and $\iota_-\iota_+\chi = \chi$ for $\chi \in H^{s+1/2}(\partial\Omega)$; here Q is the uniformly positive self-adjoint operator in (8.2.9).

Assume now that $\Omega \subset \mathbb{R}^n$ is a bounded C^2-domain as in Definition 8.2.1. The weak derivative of order $|\alpha|$ of an L^2-function f is denoted by $D^\alpha f$ in the following; as usual, here $\alpha \in \mathbb{N}_0^n$ stands for a multiindex and $|\alpha| = \alpha_1 + \cdots + \alpha_n$. Then for $k \in \mathbb{N}_0$ one has

$$H^k(\Omega) = \{f \in L^2(\Omega) : D^\alpha f \in L^2(\Omega) \text{ for all } \alpha \in \mathbb{N}_0^n \text{ with } |\alpha| \leq k\}$$

and

$$\|f\|_k := \sum_{|\alpha| \leq k} \|D^\alpha f\|_{L^2(\Omega)}, \quad f \in H^k(\Omega), \tag{8.2.10}$$

is equivalent to the norm on $H^k(\Omega)$ in (8.2.1); cf. [573, Theorem 3.30]. Recall also that for $k \in \mathbb{N}$ there exists $C_k > 0$ such that the *Poincaré inequality*

$$\|f\|_k \leq C_k \sum_{|\alpha| = k} \|D^\alpha f\|_{L^2(\Omega)}, \quad f \in H_0^k(\Omega), \tag{8.2.11}$$

is valid. In particular, for $f \in C_0^\infty(\Omega)$ and $k = 2$, integration by parts and the Schwarz theorem give

$$\sum_{|\alpha|=2} \|D^\alpha f\|_{L^2(\Omega)}^2 = \sum_{|\alpha|=2} (D^\alpha f, D^\alpha f)_{L^2(\Omega)}$$

$$= \sum_{j,k=1}^n (\partial_j \partial_k f, \partial_j \partial_k f)_{L^2(\Omega)}$$

$$= \sum_{j,k=1}^n (\partial_j^2 f, \partial_k^2 f)_{L^2(\Omega)}$$

$$= \|\Delta f\|_{L^2(\Omega)}^2,$$

and this equality extends to all $f \in H_0^2(\Omega)$ by (8.2.2). As a consequence one obtains the following useful fact.

Lemma 8.2.3. *The mapping $f \mapsto \|\Delta f\|_{L^2(\Omega)}$ is a norm on $H_0^2(\Omega)$ which is equivalent to the norms $\|\cdot\|_2$ and $\|\cdot\|_{H^2(\Omega)}$ in (8.2.10) and (8.2.1), respectively.*

Let again $\Omega \subset \mathbb{R}^n$ be a bounded C^2-domain and denote the unit normal vector field pointing outwards of $\partial\Omega$ by ν. The notion of *trace operator* or *trace map* and some of their properties are discussed next. Recall first that the mapping

$$C^\infty(\overline{\Omega}) \ni f \mapsto \left\{ f|_{\partial\Omega}, \frac{\partial f}{\partial\nu}\Big|_{\partial\Omega} \right\} \in H^{3/2}(\partial\Omega) \times H^{1/2}(\partial\Omega)$$

extends by continuity to a continuous operator

$$H^2(\Omega) \ni f \mapsto \{\tau_D f, \tau_N f\} \in H^{3/2}(\partial\Omega) \times H^{1/2}(\partial\Omega), \tag{8.2.12}$$

which is surjective; here

$$\tau_D : H^2(\Omega) \to H^{3/2}(\partial\Omega) \tag{8.2.13}$$

denotes the *Dirichlet trace operator* and

$$\tau_N : H^2(\Omega) \to H^{1/2}(\partial\Omega) \tag{8.2.14}$$

denotes the *Neumann trace operator*. In particular, for all $f \in C^\infty(\overline{\Omega})$ one has

$$\tau_D f = f|_\Omega \quad \text{and} \quad \tau_N f = \frac{\partial f}{\partial\nu}\Big|_{\partial\Omega},$$

respectively. With the help of the trace operators one has another useful characterization of the space $H_0^2(\Omega)$ in (8.2.2), namely,

$$H_0^2(\Omega) = \{f \in H^2(\Omega) : \tau_D f = \tau_N f = 0\}. \tag{8.2.15}$$

It will also be used that the Dirichlet trace operator $\tau_D : H^2(\Omega) \to H^{3/2}(\partial\Omega)$ admits a continuous surjective extension

$$\tau_D^{(1)} : H^1(\Omega) \to H^{1/2}(\partial\Omega), \tag{8.2.16}$$

which, in analogy to (8.2.15), leads to the characterization

$$H_0^1(\Omega) = \{f \in H^1(\Omega) : \tau_D^{(1)} f = 0\}. \tag{8.2.17}$$

Recall next that for $f \in H^2(\Omega)$ and $g \in H^1(\Omega)$ the *first Green identity*

$$(-\Delta f, g)_{L^2(\Omega)} = (\nabla f, \nabla g)_{L^2(\Omega;\mathbb{C}^n)} - \left(\tau_N f, \tau_D^{(1)} g\right)_{L^2(\partial\Omega)} \tag{8.2.18}$$

holds. Note that $\tau_N f, \tau_D g \in H^{1/2}(\partial\Omega)$ by (8.2.14) and (8.2.16). If, in addition, also $g \in H^2(\Omega)$, then one concludes from (8.2.18) the second Green identity

$$(-\Delta f, g)_{L^2(\Omega)} - (f, -\Delta g)_{L^2(\Omega)} = (\tau_D f, \tau_N g)_{L^2(\partial\Omega)} - (\tau_N f, \tau_D g)_{L^2(\partial\Omega)}, \quad (8.2.19)$$

which is valid for all $f, g \in H^2(\Omega)$.

In the next lemma it will be shown that the Neumann trace operator τ_N in (8.2.14) admits an extension to the subspace of $H^1(\Omega)$ consisting of all those functions $f \in H^1(\Omega)$ such that $\Delta f \in L^2(\Omega)$, and it turns out that the first Green identity (8.2.18) remains valid in an extended form. Here, and in the following, the expression Δf is understood in the sense of distributions. If, in addition, one has that $\Delta f \in L^2(\Omega)$, then Δf is a regular distribution generated by the function $\Delta f \in L^2(\Omega)$ via

$$(\Delta f)(\varphi) = \int_\Omega (\Delta f)(x)\,\overline{\varphi(x)}\,dx, \qquad \varphi \in C_0^\infty(\Omega). \quad (8.2.20)$$

Lemma 8.2.4. *For $f \in H^1(\Omega)$ with $\Delta f \in L^2(\Omega)$ there exists a unique element $\varphi \in H^{-1/2}(\partial\Omega)$ such that*

$$(-\Delta f, g)_{L^2(\Omega)} = (\nabla f, \nabla g)_{L^2(\Omega;\mathbb{C}^n)} - \left\langle \varphi, \tau_D^{(1)} g \right\rangle_{H^{-1/2}(\partial\Omega)\times H^{1/2}(\partial\Omega)} \quad (8.2.21)$$

holds for all $g \in H^1(\Omega)$. In the following the notation $\tau_N^{(1)} f := \varphi$ will be used.

Proof. Notice that there exists a bounded right inverse of $\tau_D^{(1)}$ in (8.2.16), that is, there is a bounded operator $\eta : H^{1/2}(\partial\Omega) \to H^1(\Omega)$ with the property

$$\tau_D^{(1)} \eta \psi = \psi, \qquad \psi \in H^{1/2}(\partial\Omega).$$

For a fixed $f \in H^1(\Omega)$ such that $\Delta f \in L^2(\Omega)$ define the antilinear functional $\varphi : H^{1/2}(\partial\Omega) \to \mathbb{C}$ by

$$\varphi(\psi) := (\nabla f, \nabla\eta\psi)_{L^2(\Omega;\mathbb{C}^n)} + (\Delta f, \eta\psi)_{L^2(\Omega)}, \quad \psi \in H^{1/2}(\partial\Omega). \quad (8.2.22)$$

Then one has

$$|\varphi(\psi)| \leq \|\nabla f\|_{L^2(\Omega;\mathbb{C}^n)} \|\nabla\eta\psi\|_{L^2(\Omega;\mathbb{C}^n)} + \|\Delta f\|_{L^2(\Omega)} \|\eta\psi\|_{L^2(\Omega)}$$
$$\leq C\|\eta\psi\|_{H^1(\Omega)}$$
$$\leq C'\|\psi\|_{H^{1/2}(\partial\Omega)}$$

with some constants $C, C' > 0$, and hence $\varphi \in H^{-1/2}(\partial\Omega)$. Thus, (8.2.22) can also be written in the form

$$\langle \varphi, \psi \rangle_{H^{-1/2}(\partial\Omega)\times H^{1/2}(\partial\Omega)} = (\nabla f, \nabla\eta\psi)_{L^2(\Omega;\mathbb{C}^n)} + (\Delta f, \eta\psi)_{L^2(\Omega)}, \quad (8.2.23)$$

where $\psi \in H^{1/2}(\partial\Omega)$. Now let $g \in H^1(\Omega)$ and set $g_0 := g - \eta\tau_D^{(1)} g$. Then it follows from the characterization of the space $H_0^1(\Omega)$ in (8.2.17) that $g_0 \in H_0^1(\Omega)$, and hence (8.2.2) shows that there is a sequence $(g_m) \subset C_0^\infty(\Omega)$ such that $g_m \to g_0$ in $H^1(\Omega)$. It follows that

$$
\begin{aligned}
(\nabla f, \nabla g_0)_{L^2(\Omega;\mathbb{C}^n)} &= \lim_{m \to \infty} (\nabla f, \nabla g_m)_{L^2(\Omega;\mathbb{C}^n)} \\
&= - \lim_{m \to \infty} (\Delta f, g_m)_{L^2(\Omega)} \\
&= -(\Delta f, g_0)_{L^2(\Omega)}
\end{aligned}
$$

and one obtains, together with (8.2.23) (and with $\psi = \tau_{\mathrm{D}}^{(1)} g$), that

$$
\begin{aligned}
(\nabla f, \nabla g)_{L^2(\Omega;\mathbb{C}^n)} &= \left(\nabla f, \nabla\left(g_0 + \eta \tau_{\mathrm{D}}^{(1)} g\right)\right)_{L^2(\Omega;\mathbb{C}^n)} \\
&= -(\Delta f, g_0)_{L^2(\Omega)} + \left(\nabla f, \nabla\left(\eta \tau_{\mathrm{D}}^{(1)} g\right)\right)_{L^2(\Omega;\mathbb{C}^n)} \\
&= -(\Delta f, g_0)_{L^2(\Omega)} + \left\langle \varphi, \tau_{\mathrm{D}}^{(1)} g \right\rangle_{H^{-1/2}(\partial\Omega) \times H^{1/2}(\partial\Omega)} - \left(\Delta f, \eta \tau_{\mathrm{D}}^{(1)} g\right)_{L^2(\Omega)} \\
&= (-\Delta f, g)_{L^2(\Omega)} + \left\langle \varphi, \tau_{\mathrm{D}}^{(1)} g \right\rangle_{H^{-1/2}(\partial\Omega) \times H^{1/2}(\partial\Omega)}.
\end{aligned}
$$

This shows that $\varphi \in H^{-1/2}(\partial\Omega)$ in (8.2.22)–(8.2.23) satisfies (8.2.21).

It remains to check that $\varphi \in H^{-1/2}(\partial\Omega)$ in (8.2.21) is unique. Suppose that $\varphi_1, \varphi_2 \in H^{-1/2}(\partial\Omega)$ satisfy (8.2.21). Then

$$
\left\langle \varphi_1 - \varphi_2, \tau_{\mathrm{D}}^{(1)} g \right\rangle_{H^{-1/2}(\partial\Omega) \times H^{1/2}(\partial\Omega)} = 0
$$

for all $g \in H^1(\Omega)$. As $\tau_{\mathrm{D}}^{(1)} : H^1(\Omega) \to H^{1/2}(\partial\Omega)$ is surjective it follows that $\varphi_1 - \varphi_2 = 0$ and hence φ in (8.2.21) is unique. $\qquad\square$

Remark 8.2.5. The assertion in Lemma 8.2.4 and its proof extend in a natural manner to all $f \in H^1(\Omega)$ such that $-\Delta f \in H^1(\Omega)^*$. In this situation there still exists a unique element $\varphi \in H^{-1/2}(\partial\Omega)$ such that (instead of (8.2.21)) one has the slightly more general first Green identity

$$
\langle -\Delta f, g \rangle_{H^1(\Omega)^* \times H^1(\Omega)} = (\nabla f, \nabla g)_{L^2(\Omega;\mathbb{C}^n)} - \left\langle \varphi, \tau_{\mathrm{D}}^{(1)} g \right\rangle_{H^{-1/2}(\partial\Omega) \times H^{1/2}(\partial\Omega)}
$$

for all $g \in H^1(\Omega)$; cf. [573, Lemma 4.3].

8.3 Trace maps for the maximal Schrödinger operator

The differential expression $-\Delta + V$ is considered on a bounded domain Ω, where the function $V \in L^\infty(\Omega)$ is assumed to be real. One then associates with $-\Delta + V$ a preminimal, minimal and maximal operator in $L^2(\Omega)$, which are adjoints of each other. Furthermore, the Dirichlet and Neumann operators are defined via the corresponding sesquilinear form and the first representation theorem, and some of their properties are collected. In the case where Ω is a bounded C^2-domain it is shown in Theorem 8.3.9 and Theorem 8.3.10 that the Dirichlet and Neumann trace operators in the previous section admit continuous extensions to the maximal domain; this is a key ingredient in the construction of a boundary triplet in the next section.

Let $n \geq 2$, let $\Omega \subset \mathbb{R}^n$ be a bounded domain, and assume that the function $V \in L^\infty(\Omega)$ is real. The *preminimal operator* associated to the differential expression $-\Delta + V$ is defined as

$$T_0 = -\Delta + V, \qquad \mathrm{dom}\, T_0 = C_0^\infty(\Omega).$$

It follows immediately from

$$(T_0 f, f)_{L^2(\Omega)} = (\nabla f, \nabla f)_{L^2(\Omega;\mathbb{C}^n)} + (V f, f)_{L^2(\Omega)}, \quad f \in \mathrm{dom}\, T_0,$$

that T_0 is a densely defined symmetric operator in $L^2(\Omega)$ which is bounded from below with $v_- := \mathrm{essinf}\, V$ as a lower bound, so that $T_0 - v_-$ is nonnegative. Actually, for $f \in \mathrm{dom}\, T_0$ one has

$$\big((T_0 - v_-)f, f\big)_{L^2(\Omega)} = (\nabla f, \nabla f)_{L^2(\Omega;\mathbb{C}^n)} + \big((V - v_-)f, f\big)_{L^2(\Omega)} \geq \|\nabla f\|_{L^2(\Omega;\mathbb{C}^n)}^2$$

and hence, by the Poincaré inequality (8.2.11),

$$\big((T_0 - v_-)f, f\big)_{L^2(\Omega)} \geq C\|f\|_1^2 \geq C\|f\|_{L^2(\Omega)}^2 \tag{8.3.1}$$

with some constant $C > 0$. This shows that $T_0 - v_-$ is uniformly positive.

The closure of T_0 in $L^2(\Omega)$ is the *minimal operator*

$$T_{\min} = -\Delta + V, \qquad \mathrm{dom}\, T_{\min} = H_0^2(\Omega). \tag{8.3.2}$$

In fact, using Lemma 8.2.3 and the fact that $V \in L^\infty(\Omega)$ one obtains that the graph norm

$$\| \cdot \|_{L^2(\Omega)} + \|T_{\min} \cdot \|_{L^2(\Omega)}$$

is equivalent to the H^2-norm on the closed subspace $H_0^2(\Omega)$ of $H^2(\Omega)$. Hence, $T_{\min}$ is a closed operator in $L^2(\Omega)$ and it follows from (8.2.2) that $\overline{T}_0 = T_{\min}$. Therefore, $T_{\min}$ is a densely defined closed symmetric operator in $L^2(\Omega)$ and $T_{\min} - v_-$ is uniformly positive.

Besides the preminimal and minimal operator, also the *maximal operator* $T_{\max}$ associated with $-\Delta + V$ in $L^2(\Omega)$ will be important in the sequel; it is defined by

$$T_{\max} = -\Delta + V,$$
$$\mathrm{dom}\, T_{\max} = \{f \in L^2(\Omega) : -\Delta f + V f \in L^2(\Omega)\}. \tag{8.3.3}$$

Here the expression Δf for $f \in L^2(\Omega)$ is understood in the distributional sense. Since $V \in L^\infty(\Omega)$, it is clear that $f \in L^2(\Omega)$ belongs to $\mathrm{dom}\, T_{\max}$ if and only if $\Delta f \in L^2(\Omega)$, that is, the (regular) distribution Δf is generated by an L^2-function; cf. (8.2.20). Observe that $H^2(\Omega) \subset \mathrm{dom}\, T_{\max}$, and it will also turn out that $H^2(\Omega) \neq \mathrm{dom}\, T_{\max}$.

Proposition 8.3.1. *Let T_0, $T_{\min}$, and $T_{\max}$ be the preminimal, minimal, and maximal operator associated with $-\Delta + V$ in $L^2(\Omega)$, respectively. Then $\overline{T}_0 = T_{\min}$, and*

$$(T_{\min})^* = T_{\max} \quad and \quad T_{\min} = (T_{\max})^*. \tag{8.3.4}$$

Proof. It has already been shown above that $\overline{T}_0 = T_{\min}$ holds. In particular, this implies $T_0^* = (T_{\min})^*$ and thus for the first identity in (8.3.4) it suffices to show $T_0^* = T_{\max}$. Furthermore, since multiplication by $V \in L^\infty(\Omega)$ is a bounded operator in $L^2(\Omega)$, it is no restriction to assume $V = 0$ in the following. Let $f \in \operatorname{dom} T_0^*$ and consider $T_0^* f \in L^2(\Omega)$ as a distribution. Then one has for all $\varphi \in C_0^\infty(\Omega) = \operatorname{dom} T_0$

$$(T_0^* f)(\varphi) = (T_0^* f, \overline{\varphi})_{L^2(\Omega)} = (f, T_0 \overline{\varphi})_{L^2(\Omega)} = (f, -\Delta \overline{\varphi})_{L^2(\Omega)} = (-\Delta f)(\varphi),$$

and hence $-\Delta f = T_0^* f \in L^2(\Omega)$. Thus, $f \in \operatorname{dom} T_{\max}$ and $T_{\max} f = T_0^* f$. Conversely, for $f \in \operatorname{dom} T_{\max}$ and all $\varphi \in C_0^\infty(\Omega) = \operatorname{dom} T_0$ one has

$$(T_0 \varphi, f)_{L^2(\Omega)} = (-\Delta \varphi, f)_{L^2(\Omega)} = (\varphi, -\Delta f)_{L^2(\Omega)},$$

that is, $f \in \operatorname{dom} T_0^*$ and $T_0^* f = -\Delta f = T_{\max} f$. Thus, the first identity in (8.3.4) has been shown. The second identity in (8.3.4) follows by taking adjoints. $\square$

In the following the self-adjoint *Dirichlet realization* A_D and the self-adjoint *Neumann realization* A_N of $-\Delta + V$ in $L^2(\Omega)$ will play an important role. The operators A_D and A_N will be introduced via the corresponding sesquilinear forms using the first representation theorem. More precisely, consider the densely defined forms

$$\mathfrak{t}_D[f, g] = (\nabla f, \nabla g)_{L^2(\Omega; \mathbb{C}^n)} + (V f, g)_{L^2(\Omega)}, \quad \operatorname{dom} \mathfrak{t}_D = H_0^1(\Omega),$$

and

$$\mathfrak{t}_N[f, g] = (\nabla f, \nabla g)_{L^2(\Omega; \mathbb{C}^n)} + (V f, g)_{L^2(\Omega)}, \quad \operatorname{dom} \mathfrak{t}_N = H^1(\Omega),$$

in $L^2(\Omega)$. It is easy to see that both forms are semibounded from below and that $v_- = \operatorname{essinf} V$ is a lower bound. The same argument as in (8.3.1) using the Poincaré inequality (8.2.11) on $\operatorname{dom} \mathfrak{t}_D = H_0^1(\Omega)$ implies the stronger statement that the form $\mathfrak{t}_D - v_-$ is uniformly positive. Furthermore, it follows from the definitions that the form $(\nabla \cdot, \nabla \cdot)_{L^2(\Omega; \mathbb{C}^n)}$ defined on $H_0^1(\Omega)$ or $H^1(\Omega)$ is closed in $L^2(\Omega)$; cf. Lemma 5.1.9. Since $V \in L^\infty(\Omega)$, it is clear that the form $(V \cdot, \cdot)_{L^2(\Omega)}$ is bounded on $L^2(\Omega)$ and hence it follows from Theorem 5.1.16 that also the forms $\mathfrak{t}_D$ and $\mathfrak{t}_N$ are closed in $L^2(\Omega)$. Therefore, by the first representation theorem (Theorem 5.1.18), there exist unique semibounded self-adjoint operators A_D and A_N in $L^2(\Omega)$ associated with $\mathfrak{t}_D$ and $\mathfrak{t}_N$, respectively, such that

$$(A_D f, g)_{L^2(\Omega)} = \mathfrak{t}_D[f, g] \quad \text{for } f \in \operatorname{dom} A_D, \ g \in \operatorname{dom} \mathfrak{t}_D,$$

and

$$(A_N f, g)_{L^2(\Omega)} = \mathfrak{t}_N[f, g] \quad \text{for } f \in \operatorname{dom} A_N, \ g \in \operatorname{dom} \mathfrak{t}_N.$$

The self-adjoint operators A_D and A_N are called the *Dirichlet operator* and *Neumann operator*, respectively. In the next propositions some properties of these operators are discussed.

Proposition 8.3.2. *Assume that $\Omega \subset \mathbb{R}^n$ is a bounded domain. Then the Dirichlet operator A_D is given by*

$$A_D f = -\Delta f + V f,$$
$$\text{dom}\, A_D = \{ f \in H_0^1(\Omega) : -\Delta f + V f \in L^2(\Omega) \}, \tag{8.3.5}$$

and for all $\lambda \in \rho(A_D)$ the resolvent $(A_D - \lambda)^{-1}$ is a compact operator in $L^2(\Omega)$. The Dirichlet operator A_D coincides with the Friedrichs extension S_F of the minimal operator $T_{\min}$ in (8.3.2). In particular, $A_D - v_-$ is uniformly positive. Furthermore, if $\Omega \subset \mathbb{R}^n$ is a bounded C^2-domain, then the Dirichlet operator A_D is given by

$$A_D f = -\Delta f + V f,$$
$$\text{dom}\, A_D = \{ f \in H^1(\Omega) : -\Delta f + V f \in L^2(\Omega), \tau_D^{(1)} f = 0 \}. \tag{8.3.6}$$

Proof. Observe that for $f \in \text{dom}\, A_D$ and $g \in C_0^\infty(\Omega) \subset \text{dom}\, \mathfrak{t}_D$ one has

$$(A_D f, g)_{L^2(\Omega)} = \mathfrak{t}_D[f, g] = (\nabla f, \nabla g)_{L^2(\Omega; \mathbb{C}^n)} + (V f, g)_{L^2(\Omega)} = \left(-\Delta f + V f \right)(\bar{g}),$$

where $-\Delta f + V f$ is viewed as a distribution. Since this identity holds for all $g \in C_0^\infty(\Omega)$ and A_D is an operator in $L^2(\Omega)$ it follows that

$$-\Delta f + V f = A_D f \in L^2(\Omega).$$

Therefore, A_D is given by (8.3.5). In the case that $\Omega \subset \mathbb{R}^n$ has a C^2-smooth boundary the form of the domain of A_D in (8.3.6) follows from (8.3.5) and (8.2.17).

Next it will be shown that for $\lambda \in \rho(A_D)$ the resolvent $(A_D - \lambda)^{-1}$ is a compact operator in $L^2(\Omega)$. For this observe first that

$$(A_D - \lambda)^{-1} : L^2(\Omega) \to H_0^1(\Omega), \quad \lambda \in \rho(A_D), \tag{8.3.7}$$

is everywhere defined and closed as an operator from $L^2(\Omega)$ into $H_0^1(\Omega)$. In fact, if $f_n \to f$ in $L^2(\Omega)$ and $(A_D - \lambda)^{-1} f_n \to h$ in $H_0^1(\Omega)$, then $(A_D - \lambda)^{-1} f_n \to h$ in $L^2(\Omega)$, and since the operator $(A_D - \lambda)^{-1}$ is everywhere defined and continuous in $L^2(\Omega)$, it is clear that $(A_D - \lambda)^{-1} f = h$. Hence, the operator in (8.3.7) is bounded by the closed graph theorem. By Rellich's theorem the embedding $H_0^1(\Omega) \hookrightarrow L^2(\Omega)$ is compact, and it follows that $(A_D - \lambda)^{-1}$, $\lambda \in \rho(A_D)$, is a compact operator in $L^2(\Omega)$.

It remains to verify that A_D is the Friedrichs extension S_F of $T_{\min} = \overline{T}_0$ or, equivalently, the Friedrichs extension of T_0; cf. Lemma 5.3.1 and Definition 5.3.2. For this, consider the form $\mathfrak{t}_{T_0}[f, g] = (T_0 f, g)_{L^2(\Omega)}$, defined for $f, g \in \text{dom}\, T_0$, and note that

$$\mathfrak{t}_{T_0}[f, g] = (\nabla f, \nabla g)_{L^2(\Omega; \mathbb{C}^n)} + (V f, g)_{L^2(\Omega)}, \quad \text{dom}\, \mathfrak{t}_{T_0} = C_0^\infty(\Omega).$$

Observe that $f_m \to_{t_{T_0}} f$ if and only if $f_m \to f$ in $L^2(\Omega)$ and (∇f_m) is a Cauchy sequence in $L^2(\Omega; \mathbb{C}^n)$. Hence, $f_m \to_{t_{T_0}} f$ implies $f_m \to f$ in the norm of $H^1(\Omega)$, and so $f \in H_0^1(\Omega)$ by (8.2.2). Therefore, by (5.1.16), the closure of the form t_{T_0} is given by

$$\widetilde{t}_{T_0}[f, g] = \lim_{m \to \infty} t_{T_0}[f_m, g_m] = (\nabla f, \nabla g)_{L^2(\Omega; \mathbb{C}^n)} + (Vf, g)_{L^2(\Omega)},$$

where $f, g \in H_0^1(\Omega)$ and $f_m \to_{t_{T_0}} f$, $g_m \to_{t_{T_0}} g$. Hence, $\widetilde{t}_{T_0} = t_D$, and since by Definition 5.3.2 the Friedrichs extension of T_0 is the unique self-adjoint operator corresponding to the closed form $\widetilde{t}_{T_0}$, the assertion follows. $\qquad \square$

In order to specify the Neumann operator A_N, the first Green identity and the trace operators $\tau_D^{(1)} : H^1(\Omega) \to H^{1/2}(\partial\Omega)$ and $\tau_N^{(1)} : H^1(\Omega) \to H^{-1/2}(\partial\Omega)$ will be used; cf. Lemma 8.2.4. For this reason in the next proposition it is assumed that $\Omega \subset \mathbb{R}^n$ is a bounded C^2-domain. It is also important to note that the Neumann operator A_N below differs from the Kreĭn–von Neumann extension and the Kreĭn type extensions in Definition 5.4.2; cf. Section 8.5 for more details.

Proposition 8.3.3. *Assume that $\Omega \subset \mathbb{R}^n$ is a bounded C^2-domain. Then the Neumann operator A_N is given by*

$$\begin{aligned} A_N f &= -\Delta f + Vf, \\ \mathrm{dom}\, A_N &= \{f \in H^1(\Omega) : -\Delta f + Vf \in L^2(\Omega),\, \tau_N^{(1)} f = 0\}, \end{aligned} \tag{8.3.8}$$

and for all $\lambda \in \rho(A_N)$ the resolvent $(A_N - \lambda)^{-1}$ is a compact operator in $L^2(\Omega)$.

Proof. In a first step it follows for $f \in \mathrm{dom}\, A_N \subset H^1(\Omega)$ and all $g \in C_0^\infty(\Omega)$ in the same way as in the proof of Proposition 8.3.2 that

$$(A_N f, g)_{L^2(\Omega)} = t_N[f, g] = (\nabla f, \nabla g)_{L^2(\Omega; \mathbb{C}^n)} + (Vf, g)_{L^2(\Omega)} = (-\Delta f + Vf)(\bar{g}),$$

and hence $A_N f = (-\Delta + V)f \in L^2(\Omega)$. In particular, for $f \in \mathrm{dom}\, A_N$ one has $f \in H^1(\Omega)$ and $-\Delta f \in L^2(\Omega)$, so that Lemma 8.2.4 applies and yields

$$\begin{aligned} (A_N f, g)_{L^2(\Omega)} &= t_N[f, g] \\ &= (\nabla f, \nabla g)_{L^2(\Omega; \mathbb{C}^n)} + (Vf, g)_{L^2(\Omega)} \\ &= \big((-\Delta + V)f, g\big)_{L^2(\Omega)} + \big\langle \tau_N^{(1)} f, \tau_D^{(1)} g \big\rangle_{H^{-1/2}(\partial\Omega) \times H^{1/2}(\partial\Omega)} \end{aligned}$$

for all $g \in \mathrm{dom}\, t_N = H^1(\Omega)$. As $A_N f = (-\Delta + V)f$, one concludes that

$$\big\langle \tau_N^{(1)} f, \tau_D^{(1)} g \big\rangle_{H^{-1/2}(\partial\Omega) \times H^{1/2}(\partial\Omega)} = 0 \quad \text{for all } g \in H^1(\Omega).$$

Since $\tau_D^{(1)} : H^1(\Omega) \to H^{1/2}(\partial\Omega)$ is surjective, it follows that $\tau_N^{(1)} f = 0$. This implies the representation (8.3.8).

To show that the resolvent $(A_N - \lambda)^{-1}$ is a compact operator in $L^2(\Omega)$ one argues in the same way as in the proof of Proposition 8.3.2. In fact, the operator

$$(A_N - \lambda)^{-1} : L^2(\Omega) \to H^1(\Omega), \quad \lambda \in \rho(A_N),$$

is everywhere defined and closed, and hence bounded by the closed graph theorem. Since $\Omega \subset \mathbb{R}^n$ is a bounded C^2-domain, the embedding $H^1(\Omega) \hookrightarrow L^2(\Omega)$ is compact and this implies that $(A_N - \lambda)^{-1}$, $\lambda \in \rho(A_N)$, is a compact operator in $L^2(\Omega)$. $\square$

It is known that functions f in dom A_D or dom A_N are locally H^2-regular, that is, for every compact subset $K \subset \Omega$ the restriction of f to K is in $H^2(K)$. The next theorem is an important elliptic regularity result which ensures H^2-regularity of the functions in dom A_D or dom A_N in (8.3.6) and (8.3.8), respectively, up to the boundary if the bounded domain Ω is C^2-smooth in the sense of Definition 8.2.1.

Theorem 8.3.4. *Assume that $\Omega \subset \mathbb{R}^n$ is a bounded C^2-domain. Then one has*

$$A_D f = -\Delta f + V f, \quad \mathrm{dom}\, A_D = \{f \in H^2(\Omega) : \tau_D f = 0\},$$

and

$$A_N f = -\Delta f + V f, \quad \mathrm{dom}\, A_N = \{f \in H^2(\Omega) : \tau_N f = 0\}.$$

Note that under the assumptions in Theorem 8.3.4 the domain of the Dirichlet operator A_D is $H^2(\Omega) \cap H_0^1(\Omega)$; cf. (8.2.17). The direct sum decompositions in the next corollary follow immediately from Theorem 1.7.1 when considering the operator $T = -\Delta + V$, dom $T = H^2(\Omega)$, and taking into account that $A_D \subset T$ and $A_N \subset T$.

Corollary 8.3.5. *Assume that $\Omega \subset \mathbb{R}^n$ is a bounded C^2-domain and denote by $\tau_D : H^2(\Omega) \to H^{3/2}(\partial\Omega)$ and $\tau_N : H^2(\Omega) \to H^{1/2}(\partial\Omega)$ the Dirichlet and Neumann trace operator in (8.2.13) and (8.2.14), respectively. Then for $\lambda \in \rho(A_D)$ one has the direct sum decomposition*

$$\begin{aligned} H^2(\Omega) &= \mathrm{dom}\, A_D \dotplus \{f_\lambda \in H^2(\Omega) : (-\Delta + V)f_\lambda = \lambda f_\lambda\} \\ &= \ker \tau_D \dotplus \{f_\lambda \in H^2(\Omega) : (-\Delta + V)f_\lambda = \lambda f_\lambda\}, \end{aligned} \tag{8.3.9}$$

and for $\lambda \in \rho(A_N)$ one has the direct sum decomposition

$$\begin{aligned} H^2(\Omega) &= \mathrm{dom}\, A_N \dotplus \{f_\lambda \in H^2(\Omega) : (-\Delta + V)f_\lambda = \lambda f_\lambda\} \\ &= \ker \tau_N \dotplus \{f_\lambda \in H^2(\Omega) : (-\Delta + V)f_\lambda = \lambda f_\lambda\}. \end{aligned} \tag{8.3.10}$$

As a consequence of the decomposition (8.3.9) in Corollary 8.3.5 and (8.2.12) one concludes that the so-called Dirichlet-to-Neumann map in the next definition is a well-defined operator from $H^{3/2}(\partial\Omega)$ into $H^{1/2}(\partial\Omega)$.

Definition 8.3.6. Let $\Omega \subset \mathbb{R}^n$ be a bounded C^2-domain, let A_D be the self-adjoint Dirichlet operator, and let $\tau_D : H^2(\Omega) \to H^{3/2}(\partial\Omega)$ and $\tau_N : H^2(\Omega) \to H^{1/2}(\partial\Omega)$

be the Dirichlet and Neumann trace operator in (8.2.13) and (8.2.14), respectively. For $\lambda \in \rho(A_D)$ the *Dirichlet-to-Neumann map* is defined as

$$D(\lambda) : H^{3/2}(\partial\Omega) \to H^{1/2}(\partial\Omega), \qquad \tau_D f_\lambda \mapsto \tau_N f_\lambda,$$

where $f_\lambda \in H^2(\Omega)$ is such that $(-\Delta + V)f_\lambda = \lambda f_\lambda$.

Note that for $\lambda \in \rho(A_D) \cap \rho(A_N)$ both decompositions (8.3.9) and (8.3.10) in Corollary 8.3.5 hold and together with (8.2.12) this implies that the Dirichlet-to-Neumann map $D(\lambda)$ is a bijective operator from $H^{3/2}(\partial\Omega)$ onto $H^{1/2}(\partial\Omega)$.

A further useful consequence of Theorem 8.3.4 is given by the following a priori estimates.

Corollary 8.3.7. *Assume that $\Omega \subset \mathbb{R}^n$ is a bounded C^2-domain and let A_D and A_N be the Dirichlet and Neumann operator, respectively. Then there exist constants $C_D > 0$ and $C_N > 0$ such that*

$$\|f\|_{H^2(\Omega)} \leq C_D\big(\|f\|_{L^2(\Omega)} + \|A_D f\|_{L^2(\Omega)}\big), \quad f \in \operatorname{dom} A_D,$$

and

$$\|g\|_{H^2(\Omega)} \leq C_N\big(\|g\|_{L^2(\Omega)} + \|A_N g\|_{L^2(\Omega)}\big), \quad g \in \operatorname{dom} A_N.$$

Proof. One verifies in the same way as in the proof of Proposition 8.3.2 that for $\lambda \in \rho(A_D)$ the operator $(A_D - \lambda)^{-1} : L^2(\Omega) \to H^2(\Omega)$ is everywhere defined and closed, and hence bounded. For $f \in \operatorname{dom} A_D$ choose $h \in L^2(\Omega)$ such that $f = (A_D - \lambda)^{-1}h$. Then

$$\|f\|_{H^2(\Omega)} = \|(A_D - \lambda)^{-1}h\|_{H^2(\Omega)} \leq C\|h\|_{L^2(\Omega)} = C_D\|(A_D - \lambda)f\|_{L^2(\Omega)}$$

for some $C > 0$ and $C_D > 0$, and the first estimate follows. The second estimate is proved in the same way. $\qquad\square$

The next lemma is an important ingredient in the following.

Lemma 8.3.8. *Let $T_{\max}$ be the maximal operator associated to $-\Delta + V$ in (8.3.3). Then the space $C^\infty(\overline{\Omega})$ is dense in $\operatorname{dom} T_{\max}$ with respect to the graph norm.*

Proof. Since $V \in L^\infty(\Omega)$ is bounded, the graph norms

$$\big(\|\cdot\|_{L^2(\Omega)}^2 + \|T_{\max}\cdot\|_{L^2(\Omega)}^2\big)^{1/2} \quad \text{and} \quad \big(\|\cdot\|_{L^2(\Omega)}^2 + \|\Delta\cdot\|_{L^2(\Omega)}^2\big)^{1/2}$$

are equivalent on $\operatorname{dom} T_{\max}$, and hence it is no restriction to assume that $V = 0$. Now suppose that $f \in \operatorname{dom} T_{\max}$ is such that for all $g \in C^\infty(\overline{\Omega})$

$$0 = (f, g)_{L^2(\Omega)} + (\Delta f, \Delta g)_{L^2(\Omega)}. \tag{8.3.11}$$

Then (8.3.11) holds for all $g \in C_0^\infty(\Omega)$, so that $0 = (f + \Delta^2 f)(\bar{g})$, where $f + \Delta^2 f$ is viewed as a distribution. As $f \in L^2(\Omega)$, one concludes that

$$\Delta^2 f = -f \in L^2(\Omega). \tag{8.3.12}$$

Next it will be shown that
$$\Delta f \in H_0^2(\Omega). \qquad (8.3.13)$$
In fact, choose an open ball B such that $\overline{\Omega} \subset B$ and let $h \in C_0^\infty(B)$. Let
$$\widetilde{A}_{\mathrm{D}} = -\Delta, \qquad \operatorname{dom} \widetilde{A}_{\mathrm{D}} = H^2(B) \cap H_0^1(B),$$
be the self-adjoint Dirichlet Laplacian in $L^2(B)$; cf. Theorem 8.3.4. Since B is bounded, one has $0 \in \rho(\widetilde{A}_{\mathrm{D}})$ by Proposition 8.3.2. As $h \in C_0^\infty(B)$, elliptic regularity yields $\widetilde{A}_{\mathrm{D}}^{-1} h \in C^\infty(B)$ and hence $(\widetilde{A}_{\mathrm{D}}^{-1} h)|_\Omega \in C^\infty(\overline{\Omega})$ for the restriction onto Ω. Denote by $\widetilde{f}$ and $\widetilde{\Delta f}$ the extension of f and Δf by zero to B. Then it follows with the help of (8.3.11) that

$$
\begin{aligned}
(\widetilde{A}_{\mathrm{D}}^{-1}\widetilde{f}, h)_{L^2(B)} &= (\widetilde{f}, \widetilde{A}_{\mathrm{D}}^{-1} h)_{L^2(B)} \\
&= \big(f, (\widetilde{A}_{\mathrm{D}}^{-1} h)|_\Omega\big)_{L^2(\Omega)} \\
&= -\big(\Delta f, \Delta(\widetilde{A}_{\mathrm{D}}^{-1} h)|_\Omega\big)_{L^2(\Omega)} \\
&= (-\widetilde{\Delta f}, h)_{L^2(B)}
\end{aligned}
$$

holds for $h \in C_0^\infty(B)$. This yields $-\widetilde{\Delta f} = \widetilde{A}_{\mathrm{D}}^{-1}\widetilde{f} \in H^2(B)$. Moreover, as $\widetilde{\Delta f}$ vanishes outside of Ω it follows that $\Delta f \in H_0^2(\Omega)$, that is, (8.3.13) holds.

Now choose a sequence $(\psi_k) \subset C_0^\infty(\Omega)$ such that $\psi_k \to \Delta f$ in $H^2(\Omega)$. Then, by (8.3.12),

$$
\begin{aligned}
0 \le (\Delta f, \Delta f)_{L^2(\Omega)} &= \lim_{k\to\infty} (\psi_k, \Delta f)_{L^2(\Omega)} = \lim_{k\to\infty} (\Delta \psi_k, f)_{L^2(\Omega)} \\
&= (\Delta^2 f, f)_{L^2(\Omega)} = -(f, f)_{L^2(\Omega)} \le 0,
\end{aligned}
$$

that is, $f = 0$ in (8.3.11). Hence, $C^\infty(\overline{\Omega})$ is dense in $\operatorname{dom} T_{\max}$ with respect to the graph norm. $\qquad\square$

The following result on the extension of the Dirichlet trace operator onto $\operatorname{dom} T_{\max}$ is essential for the construction of a boundary triplet for $T_{\max}$.

Theorem 8.3.9. *Assume that $\Omega \subset \mathbb{R}^n$ is a bounded C^2-domain. Then the Dirichlet trace operator $\tau_{\mathrm{D}} : H^2(\Omega) \to H^{3/2}(\partial\Omega)$ in (8.2.13) admits a unique extension to a continuous surjective operator*

$$\widetilde{\tau}_{\mathrm{D}} : \operatorname{dom} T_{\max} \to H^{-1/2}(\partial\Omega),$$

where $\operatorname{dom} T_{\max}$ is equipped with the graph norm. Furthermore,

$$\ker \widetilde{\tau}_{\mathrm{D}} = \ker \tau_{\mathrm{D}} = \operatorname{dom} A_{\mathrm{D}}.$$

Proof. In the following fix $\lambda \in \rho(A_{\mathrm{D}})$ and consider the operator

$$\Upsilon := -\tau_{\mathrm{N}}(A_{\mathrm{D}} - \overline{\lambda})^{-1} : L^2(\Omega) \to H^{1/2}(\partial\Omega). \qquad (8.3.14)$$

Since $(A_D - \bar\lambda)^{-1} : L^2(\Omega) \to H^2(\Omega)$ is everywhere defined and closed, it is clear that $(A_D - \bar\lambda)^{-1} : L^2(\Omega) \to H^2(\Omega)$ is continuous and maps onto $\operatorname{dom} A_D$. Hence, it follows from Theorem 8.3.4 and (8.2.12) that $\Upsilon \in \mathbf{B}(L^2(\Omega), H^{1/2}(\partial\Omega))$ in (8.3.14) is a surjective operator.

Next it will be shown that

$$\ker \Upsilon = \mathfrak{N}_\lambda(T_{\max})^\perp, \tag{8.3.15}$$

where $\mathfrak{N}_\lambda(T_{\max}) = \ker(T_{\max} - \lambda)$. In fact, for the inclusion ($\subset$) in (8.3.15), assume that

$$\Upsilon h = -\tau_N (A_D - \bar\lambda)^{-1} h = 0$$

for some $h \in L^2(\Omega)$. Then it follows from Theorem 8.3.4 that

$$(A_D - \bar\lambda)^{-1} h \in \operatorname{dom} A_D \cap \operatorname{dom} A_N$$

and hence $(A_D - \bar\lambda)^{-1} h \in \operatorname{dom} T_{\min}$ by (8.2.15) and (8.3.2). For $f_\lambda \in \mathfrak{N}_\lambda(T_{\max})$ one concludes, together with Proposition 8.3.1, that

$$
\begin{aligned}
(f_\lambda, h)_{L^2(\Omega)} &= \big(f_\lambda, (T_{\min} - \bar\lambda)(A_D - \bar\lambda)^{-1} h\big)_{L^2(\Omega)} \\
&= \big((T_{\max} - \lambda)f_\lambda, (A_D - \bar\lambda)^{-1} h\big)_{L^2(\Omega)} \\
&= 0,
\end{aligned}
$$

which shows $h \in \mathfrak{N}_\lambda(T_{\max})^\perp$. For the inclusion ($\supset$) in (8.3.15), let $h \in \mathfrak{N}_\lambda(T_{\max})^\perp$. Then $h \in \operatorname{ran}(T_{\min} - \bar\lambda)$, and hence there exists $k \in \operatorname{dom} T_{\min} = H_0^2(\Omega)$ such that $h = (T_{\min} - \bar\lambda)k$. It follows that

$$\Upsilon h = -\tau_N (A_D - \bar\lambda)^{-1} h = -\tau_N (A_D - \bar\lambda)^{-1}(T_{\min} - \bar\lambda)k = -\tau_N k = 0,$$

which shows that $h \in \ker \Upsilon$. This completes the proof of (8.3.15).

From (8.3.14) and (8.3.15) it follows that the restriction of Υ to $\mathfrak{N}_\lambda(T_{\max})$ is an isomorphism from $\mathfrak{N}_\lambda(T_{\max})$ onto $H^{1/2}(\partial\Omega)$. This implies that the dual operator

$$\Upsilon' : H^{-1/2}(\partial\Omega) \to L^2(\Omega) \tag{8.3.16}$$

is bounded and invertible, and by the closed range theorem (see Theorem 1.3.5 for the Hilbert space adjoint) one has

$$\operatorname{ran} \Upsilon' = (\ker \Upsilon)^\perp = \mathfrak{N}_\lambda(T_{\max}).$$

The inverse $(\Upsilon')^{-1}$ is regarded as an isomorphism from $\mathfrak{N}_\lambda(T_{\max})$ onto $H^{-1/2}(\partial\Omega)$. Now recall the direct sum decomposition

$$\operatorname{dom} T_{\max} = \operatorname{dom} A_D + \mathfrak{N}_\lambda(T_{\max})$$

from Theorem 1.7.1 or Corollary 1.7.5, and write the elements $f \in \operatorname{dom} T_{\max}$ accordingly,

$$f = f_D + f_\lambda, \quad f_D \in \operatorname{dom} A_D, \quad f_\lambda \in \mathfrak{N}_\lambda(T_{\max}).$$

Define the mapping

$$\widetilde{\tau}_D : \operatorname{dom} T_{\max} \to H^{-1/2}(\partial\Omega), \qquad f \mapsto \widetilde{\tau}_D f = (\Upsilon')^{-1} f_\lambda. \tag{8.3.17}$$

Next it will be shown that $\widetilde{\tau}_D$ is an extension of the Dirichlet trace operator $\tau_D : H^2(\Omega) \to H^{3/2}(\partial\Omega)$. For this, consider $\varphi \in \operatorname{ran} \tau_D = H^{3/2}(\partial\Omega) \subset H^{-1/2}(\partial\Omega)$ and note that by (8.3.9) and (8.2.12) there exists a unique $f_\lambda \in H^2(\Omega)$ such that

$$(-\Delta + V)f_\lambda = \lambda f_\lambda \quad \text{and} \quad \tau_D f_\lambda = \varphi. \tag{8.3.18}$$

Let $h \in L^2(\Omega)$ and set $k := (A_D - \bar\lambda)^{-1} h$. Then, by (8.3.14), the fact that $\tau_D k = 0$, and the second Green identity (8.2.19),

$$
\begin{aligned}
(\Upsilon'\varphi, h)_{L^2(\Omega)} &= \langle \varphi, \Upsilon h \rangle_{H^{-1/2}(\partial\Omega) \times H^{1/2}(\partial\Omega)} \\
&= (\varphi, \Upsilon h)_{L^2(\partial\Omega)} \\
&= -\left(\varphi, \tau_N (A_D - \bar\lambda)^{-1} h\right)_{L^2(\partial\Omega)} \\
&= -(\tau_D f_\lambda, \tau_N k)_{L^2(\partial\Omega)} + (\tau_N f_\lambda, \tau_D k)_{L^2(\partial\Omega)} \\
&= -\left((-\Delta + V)f_\lambda, k\right)_{L^2(\Omega)} + \left(f_\lambda, (-\Delta + V)k\right)_{L^2(\Omega)} \\
&= -(\lambda f_\lambda, k)_{L^2(\Omega)} + (f_\lambda, A_D k)_{L^2(\Omega)} \\
&= \left(f_\lambda, (A_D - \bar\lambda)k\right)_{L^2(\Omega)} \\
&= (f_\lambda, h)_{L^2(\Omega)},
\end{aligned}
$$

and thus $\Upsilon'\varphi = f_\lambda$. Hence, the restriction of Υ' to $H^{3/2}(\partial\Omega)$ maps $\varphi \in H^{3/2}(\partial\Omega)$ to the unique $H^2(\Omega)$-solution f_λ of the boundary value problem (8.3.18), that is, to the unique element $f_\lambda \in \mathfrak{N}_\lambda(T_{\max}) \cap H^2(\Omega)$ such that $\tau_D f_\lambda = \varphi$. Therefore, $(\Upsilon')^{-1}$ maps the elements in $\mathfrak{N}_\lambda(T_{\max}) \cap H^2(\Omega)$ onto their Dirichlet boundary values, that is,

$$(\Upsilon')^{-1} f_\lambda = \tau_D f_\lambda \quad \text{for} \quad f_\lambda \in \mathfrak{N}_\lambda(T_{\max}) \cap H^2(\Omega).$$

By definition $\widetilde{\tau}_D f_D = 0 = \tau_D f_D$ for $f_D \in \operatorname{dom} A_D$. Therefore, if $f \in H^2(\Omega)$ is decomposed according to (8.3.9) as

$$f = f_D + f_\lambda, \quad f_D \in \operatorname{dom} A_D, \quad f_\lambda \in \mathfrak{N}_\lambda(T_{\max}) \cap H^2(\Omega),$$

then

$$\widetilde{\tau}_D f = \widetilde{\tau}_D(f_D + f_\lambda) = (\Upsilon')^{-1} f_\lambda = \tau_D f_\lambda = \tau_D f,$$

so that $\widetilde{\tau}_D$ in (8.3.17) is an extension of τ_D. Note that by construction $\widetilde{\tau}_D$ is surjective. Furthermore, the property $\ker \widetilde{\tau}_D = \ker \tau_D$ is clear from the definition.

It remains to show that $\widetilde{\tau}_D$ in (8.3.17) is continuous with respect to the graph norm on $\operatorname{dom} T_{\max}$. For this, consider $f = f_D + f_\lambda \in \operatorname{dom} T_{\max}$ with $f_D \in \operatorname{dom} A_D$ and $f_\lambda \in \mathfrak{N}_\lambda(T_{\max})$, and note that

$$
\begin{aligned}
f_\lambda &= f - f_D = f - (A_D - \lambda)^{-1}(T_{\max} - \lambda)f_D \\
&= f - (A_D - \lambda)^{-1}(T_{\max} - \lambda)f.
\end{aligned}
$$

Since $(\Upsilon')^{-1} : \mathfrak{N}_\lambda(T_{\max}) \to H^{-1/2}(\partial\Omega)$ is an isomorphism and hence, in particular, bounded, one has

$$
\begin{aligned}
\|\widetilde{\tau}_\mathrm{D} f\|_{H^{-1/2}(\partial\Omega)} &= \|(\Upsilon')^{-1} f_\lambda\|_{H^{-1/2}(\partial\Omega)} \\
&\leq C\|f_\lambda\|_{L^2(\Omega)} \\
&\leq C\big(\|f\|_{L^2(\Omega)} + \|(A_\mathrm{D} - \lambda)^{-1}(T_{\max} - \lambda)f\|_{L^2(\Omega)}\big) \\
&\leq C'\big(\|f\|_{L^2(\Omega)} + \|(T_{\max} - \lambda)f\|_{L^2(\Omega)}\big) \\
&\leq C''\big(\|f\|_{L^2(\Omega)} + \|T_{\max}f\|_{L^2(\Omega)}\big)
\end{aligned}
$$

with some constants $C, C', C'' > 0$. Thus, $\widetilde{\tau}_\mathrm{D}$ is continuous. The proof of Theorem 8.3.9 is complete. □

The following result is parallel to Theorem 8.3.9 and can be proved in a similar way.

Theorem 8.3.10. *Assume that $\Omega \subset \mathbb{R}^n$ is a bounded C^2-domain. Then the Neumann trace operator $\tau_\mathrm{N} : H^2(\Omega) \to H^{1/2}(\partial\Omega)$ in (8.2.14) admits a unique extension to a continuous surjective operator*

$$
\widetilde{\tau}_\mathrm{N} : \operatorname{dom} T_{\max} \to H^{-3/2}(\partial\Omega),
$$

where $\operatorname{dom} T_{\max}$ is equipped with the graph norm. Furthermore,

$$
\ker \widetilde{\tau}_\mathrm{N} = \ker \tau_\mathrm{N} = \operatorname{dom} A_\mathrm{N}.
$$

As a consequence of Theorem 8.3.9 and Theorem 8.3.10 one can also extend the second Green identity in (8.2.19) to elements $f \in \operatorname{dom} T_{\max}$ and $g \in H^2(\Omega)$.

Corollary 8.3.11. *Assume that $\Omega \subset \mathbb{R}^n$ is a bounded C^2-domain, and let*

$$
\widetilde{\tau}_\mathrm{D} : \operatorname{dom} T_{\max} \to H^{-1/2}(\partial\Omega) \quad and \quad \widetilde{\tau}_\mathrm{N} : \operatorname{dom} T_{\max} \to H^{-3/2}(\partial\Omega)
$$

be the unique continuous extensions of the Dirichlet and Neumann trace operators

$$
\tau_\mathrm{D} : H^2(\Omega) \to H^{3/2}(\partial\Omega) \quad and \quad \tau_\mathrm{N} : H^2(\Omega) \to H^{1/2}(\partial\Omega)
$$

from Theorem 8.3.9 and Theorem 8.3.10, respectively. Then the second Green identity in (8.2.19) extends to

$$
\begin{aligned}
(T_{\max}f, g)_{L^2(\Omega)} &- (f, T_{\max}g)_{L^2(\Omega)} \\
&= \langle \widetilde{\tau}_\mathrm{D} f, \tau_\mathrm{N} g \rangle_{H^{-1/2}(\partial\Omega) \times H^{1/2}(\partial\Omega)} - \langle \widetilde{\tau}_\mathrm{N} f, \tau_\mathrm{D} g \rangle_{H^{-3/2}(\partial\Omega) \times H^{3/2}(\partial\Omega)}
\end{aligned}
$$

for $f \in \operatorname{dom} T_{\max}$ and $g \in H^2(\Omega)$.

Proof. Let $f \in \operatorname{dom} T_{\max}$ and $g \in H^2(\Omega)$. Since $C^\infty(\overline{\Omega})$ is dense in $\operatorname{dom} T_{\max}$ with respect to the graph norm by Lemma 8.3.8 and $C^\infty(\overline{\Omega}) \subset H^2(\Omega) \subset \operatorname{dom} T_{\max}$, there exists a sequence $(f_n) \subset H^2(\Omega)$ such that $f_n \to f$ and $T_{\max}f_n \to T_{\max}f$

in $L^2(\Omega)$. Moreover, $\tau_D f_n \to \widetilde{\tau}_D f$ in $H^{-1/2}(\partial\Omega)$ and $\tau_N f_n \to \widetilde{\tau}_N f$ in $H^{-3/2}(\partial\Omega)$, because $\widetilde{\tau}_D$ and $\widetilde{\tau}_N$ are continuous with respect to the graph norm. Therefore, with the help of the second Green identity (8.2.19), one concludes that

$$
\begin{aligned}
&(T_{\max} f, g)_{L^2(\Omega)} - (f, T_{\max} g)_{L^2(\Omega)} \\
&= \lim_{n\to\infty} (T_{\max} f_n, g)_{L^2(\Omega)} - \lim_{n\to\infty} (f_n, T_{\max} g)_{L^2(\Omega)} \\
&= \lim_{n\to\infty} \left[(\tau_D f_n, \tau_N g)_{L^2(\partial\Omega)} - (\tau_N f_n, \tau_D g)_{L^2(\partial\Omega)} \right] \\
&= \lim_{n\to\infty} \left[\langle \tau_D f_n, \tau_N g \rangle_{H^{-1/2}(\partial\Omega) \times H^{1/2}(\partial\Omega)} - \langle \tau_N f_n, \tau_D g \rangle_{H^{-3/2}(\partial\Omega) \times H^{3/2}(\partial\Omega)} \right] \\
&= \langle \widetilde{\tau}_D f, \tau_N g \rangle_{H^{-1/2}(\partial\Omega) \times H^{1/2}(\partial\Omega)} - \langle \widetilde{\tau}_N f, \tau_D g \rangle_{H^{-3/2}(\partial\Omega) \times H^{3/2}(\partial\Omega)},
\end{aligned}
$$

which completes the proof. $\qquad\square$

Note that, by construction, there exists a bounded right inverse for the extended Dirichlet trace operator $\widetilde{\tau}_D$ (see (8.3.16)–(8.3.17)) and similarly there exists a bounded right inverse for the extended Neumann trace operator $\widetilde{\tau}_N$. This also implies that the Dirichlet-to-Neumann map in Definition 8.3.6 admits a natural extension to a bounded mapping from from $H^{-1/2}(\partial\Omega)$ into $H^{-3/2}(\partial\Omega)$.

Corollary 8.3.12. *Assume that $\Omega \subset \mathbb{R}^n$ is a bounded C^2-domain and let $\widetilde{\tau}_D$ and $\widetilde{\tau}_N$ be the unique continuous extensions of the Dirichlet and Neumann trace operators from Theorem 8.3.9 and Theorem 8.3.10, respectively. Then for $\lambda \in \rho(A_D)$ the Dirichlet-to-Neumann map in Definition 8.3.6 admits an extension to a bounded operator*

$$
\widetilde{D}(\lambda) : H^{-1/2}(\partial\Omega) \to H^{-3/2}(\partial\Omega), \qquad \widetilde{\tau}_D f_\lambda \mapsto \widetilde{\tau}_N f_\lambda,
$$

where $f_\lambda \in \mathfrak{N}_\lambda(T_{\max})$.

For later purposes the following fact is provided.

Proposition 8.3.13. *The minimal operator $T_{\min}$ in (8.3.2) is simple.*

Proof. Since A_D is a self-adjoint extension of $T_{\min}$ with discrete spectrum, it suffices to check that $T_{\min}$ has no eigenvalues; cf. Proposition 3.4.8. For this, assume that $T_{\min} f = \lambda f$ for some $\lambda \in \mathbb{R}$ and some $f \in \mathrm{dom}\, T_{\min}$. Since $\mathrm{dom}\, T_{\min} = H_0^2(\Omega)$, there exist $(f_k) \in C_0^\infty(\Omega)$ such that $f_k \to f$ in $H^2(\Omega)$. Denote the zero extensions of f and f_k to all of $\mathbb{R}^n$ by $\widetilde{f}$ and $\widetilde{f}_k$, respectively. Then $\widetilde{f}_k \to \widetilde{f}$ in $L^2(\mathbb{R}^n)$ and for all $h \in C_0^\infty(\mathbb{R}^n)$ and $\alpha \in \mathbb{N}_0^n$ such that $|\alpha| \leq 2$ one computes

$$
\begin{aligned}
\int_{\mathbb{R}^n} \widetilde{f}(x) D^\alpha h(x)\, dx &= \lim_{k\to\infty} \int_{\mathbb{R}^n} \widetilde{f}_k(x) D^\alpha h(x)\, dx \\
&= (-1)^{|\alpha|} \lim_{k\to\infty} \int_{\mathbb{R}^n} (D^\alpha \widetilde{f}_k)(x) h(x)\, dx \\
&= (-1)^{|\alpha|} \lim_{k\to\infty} \int_\Omega (D^\alpha f_k)(x) h(x)\, dx
\end{aligned}
$$

$$= (-1)^{|\alpha|} \int_\Omega (D^\alpha f)(x)h(x)dx$$

$$= (-1)^{|\alpha|} \int_{\mathbb{R}^n} \widetilde{(D^\alpha f)}(x)h(x)dx,$$

where $\widetilde{(D^\alpha f)}$ denotes the zero extension of $D^\alpha f$ to all of $\mathbb{R}^n$. It follows from this computation that

$$D^\alpha \widetilde{f} = \widetilde{(D^\alpha f)} \in L^2(\mathbb{R}^n), \qquad |\alpha| \leq 2,$$

and hence $\widetilde{f} \in H^2(\mathbb{R}^n)$. Furthermore, if $\widetilde{V} \in L^\infty(\mathbb{R}^n)$ denotes some real extension of V, then $(-\Delta + \widetilde{V})\widetilde{f} = \lambda \widetilde{f}$ and since $\widetilde{f}$ vanishes on an open subset of $\mathbb{R}^n$, the unique continuation principle (see, e.g., [652, Theorem XIII.63]) implies $\widetilde{f} = 0$, so that $f = 0$. Therefore, $T_{\min}$ has no eigenvalues and now Proposition 3.4.8 shows that $T_{\min}$ is simple. □

8.4 A boundary triplet for the maximal Schrödinger operator

In this section a boundary triplet $\{L^2(\partial\Omega), \Gamma_0, \Gamma_1\}$ for the maximal operator $T_{\max}$ in (8.3.3) is provided under the assumption that $\Omega \subset \mathbb{R}^n$ is a bounded C^2-domain. The corresponding Weyl function is closely connected to the extended Dirichlet-to-Neumann map in Corollary 8.3.12. As examples, Neumann and Robin type boundary conditions are discussed, and it is also explained that there exist self-adjoint realizations of $-\Delta + V$ in $L^2(\Omega)$ which are not semibounded and which may have essential spectrum of rather arbitrary form.

Recall from Corollary 8.2.2 that

$$\{H^{1/2}(\partial\Omega),\ L^2(\partial\Omega), H^{-1/2}(\partial\Omega)\}$$

is a Gelfand triple and there exist isometric isomorphisms $\iota_\pm : H^{\pm\frac{1}{2}}(\partial\Omega) \to L^2(\partial\Omega)$ such that

$$\langle \varphi, \psi \rangle_{H^{-1/2}(\partial\Omega) \times H^{1/2}(\partial\Omega)} = (\iota_-\varphi, \iota_+\psi)_{L^2(\partial\Omega)}$$

holds for all $\varphi \in H^{-1/2}(\partial\Omega)$ and $\psi \in H^{1/2}(\partial\Omega)$. For the definition of the boundary mappings in the next proposition recall also the definition and the properties of the Dirichlet operator A_D (see Theorem 8.3.4), as well as the direct sum decomposition

$$\mathrm{dom}\, T_{\max} = \mathrm{dom}\, A_D \dotplus \mathfrak{N}_\eta(T_{\max}), \qquad (8.4.1)$$

which holds for all $\eta \in \rho(A_D)$. In particular, since A_D is semibounded from below, one may choose $\eta \in \rho(A_D) \cap \mathbb{R}$ in (8.4.1). Further, let $\tau_N : H^2(\Omega) \to H^{1/2}(\partial\Omega)$ be the Neumann trace operator in (8.2.12) and let $\widetilde{\tau}_D : \mathrm{dom}\, T_{\max} \to H^{-1/2}(\partial\Omega)$ be the extension of the Dirichlet trace operator in Theorem 8.3.9.

Theorem 8.4.1. *Let $\Omega \subset \mathbb{R}^n$ be a bounded C^2-domain, let A_D be the self-adjoint Dirichlet realization of $-\Delta + V$ in $L^2(\Omega)$ in Theorem 8.3.4, fix $\eta \in \rho(A_D) \cap \mathbb{R}$, and decompose $f \in \operatorname{dom} T_{\max}$ according to (8.4.1) in the form $f = f_D + f_\eta$, where $f_D \in \operatorname{dom} A_D$ and $f_\eta \in \mathfrak{N}_\eta(T_{\max})$. Then $\{L^2(\partial\Omega), \Gamma_0, \Gamma_1\}$, where*

$$\Gamma_0 f = \iota_- \widetilde{\tau}_D f \quad and \quad \Gamma_1 f = -\iota_+ \tau_N f_D, \qquad f = f_D + f_\eta \in \operatorname{dom} T_{\max},$$

is a boundary triplet for $(T_{\min})^* = T_{\max}$ such that

$$A_0 = A_D \qquad and \qquad A_1 = T_{\min} \,\widehat{+}\, \mathfrak{N}_\eta(T_{\max}). \tag{8.4.2}$$

Proof. Let $f, g \in \operatorname{dom} T_{\max}$ and decompose f and g in the form $f = f_D + f_\eta$ and $g = g_D + g_\eta$ with $f_D, g_D \in \operatorname{dom} A_D \subset H^2(\Omega)$ and $f_\eta, g_\eta \in \mathfrak{N}_\eta(T_{\max})$. Since A_D is self-adjoint,

$$(T_{\max} f_D, g_D)_{L^2(\Omega)} = (A_D f_D, g_D)_{L^2(\Omega)} = (f_D, A_D g_D)_{L^2(\Omega)} = (f_D, T_{\max} g_D)_{L^2(\Omega)}$$

and since η is real, one also has

$$(T_{\max} f_\eta, g_\eta)_{L^2(\Omega)} = (\eta f_\eta, g_\eta)_{L^2(\Omega)} = (f_\eta, \eta g_\eta)_{L^2(\Omega)} = (f_\eta, T_{\max} g_\eta)_{L^2(\Omega)}.$$

Therefore, one obtains

$$
\begin{aligned}
(T_{\max} f, g)_{L^2(\Omega)} &- (f, T_{\max} g)_{L^2(\Omega)} \\
&= \big(T_{\max}(f_D + f_\eta), g_D + g_\eta\big)_{L^2(\Omega)} - \big(f_D + f_\eta, T_{\max}(g_D + g_\eta)\big)_{L^2(\Omega)} \\
&= (T_{\max} f_\eta, g_D)_{L^2(\Omega)} + (T_{\max} f_D, g_\eta)_{L^2(\Omega)} \\
&\quad - (f_\eta, T_{\max} g_D)_{L^2(\Omega)} - (f_D, T_{\max} g_\eta)_{L^2(\Omega)}.
\end{aligned}
$$

Let $\widetilde{\tau}_N$ be the extension of the Neumann trace to $\operatorname{dom} T_{\max}$ from Theorem 8.3.10. Then it follows together with Corollary 8.3.11 and $\tau_D f_D = \tau_D g_D = 0$ that

$$
\begin{aligned}
(T_{\max} f_\eta, g_D)_{L^2(\Omega)} &- (f_\eta, T_{\max} g_D)_{L^2(\Omega)} \\
&= \langle \widetilde{\tau}_D f_\eta, \tau_N g_D \rangle_{H^{-1/2}(\partial\Omega) \times H^{1/2}(\partial\Omega)} - \langle \widetilde{\tau}_N f_\eta, \tau_D g_D \rangle_{H^{-3/2}(\partial\Omega) \times H^{3/2}(\partial\Omega)} \\
&= \langle \widetilde{\tau}_D f_\eta, \tau_N g_D \rangle_{H^{-1/2}(\partial\Omega) \times H^{1/2}(\partial\Omega)}
\end{aligned}
$$

and

$$
\begin{aligned}
(T_{\max} f_D, g_\eta)_{L^2(\Omega)} &- (f_D, T_{\max} g_\eta)_{L^2(\Omega)} \\
&= \langle \tau_D f_D, \widetilde{\tau}_N g_\eta \rangle_{H^{3/2}(\partial\Omega) \times H^{-3/2}(\partial\Omega)} - \langle \tau_N f_D, \widetilde{\tau}_D g_\eta \rangle_{H^{1/2}(\partial\Omega) \times H^{-1/2}(\partial\Omega)} \\
&= -\langle \tau_N f_D, \widetilde{\tau}_D g_\eta \rangle_{H^{1/2}(\partial\Omega) \times H^{-1/2}(\partial\Omega)}.
\end{aligned}
$$

Hence,

$$
\begin{aligned}
(T_{\max} f, g)_{L^2(\Omega)} &- (f, T_{\max} g)_{L^2(\Omega)} \\
&= \langle \widetilde{\tau}_D f_\eta, \tau_N g_D \rangle_{H^{-1/2}(\partial\Omega) \times H^{1/2}(\partial\Omega)} - \langle \tau_N f_D, \widetilde{\tau}_D g_\eta \rangle_{H^{1/2}(\partial\Omega) \times H^{-1/2}(\partial\Omega)} \\
&= \big(\iota_- \widetilde{\tau}_D f_\eta, \iota_+ \tau_N g_D\big)_{L^2(\partial\Omega)} - \big(\iota_+ \tau_N f_D, \iota_- \widetilde{\tau}_D g_\eta\big)_{L^2(\partial\Omega)}
\end{aligned}
$$

and, since $f_D, g_D \in \ker \tau_D = \ker \tilde{\tau}_D$ according to Theorem 8.3.9, one sees that

$$
\begin{aligned}
(T_{\max}f, g)_{L^2(\Omega)} &- (f, T_{\max}g)_{L^2(\Omega)} \\
&= \left(\iota_- \tilde{\tau}_D f, \iota_+ \tau_N g_D \right)_{L^2(\partial\Omega)} - \left(\iota_+ \tau_N f_D, \iota_- \tilde{\tau}_D g \right)_{L^2(\partial\Omega)} \\
&= \left(-\iota_+ \tau_N f_D, \iota_- \tilde{\tau}_D g \right)_{L^2(\partial\Omega)} - \left(\iota_- \tilde{\tau}_D f, -\iota_+ \tau_N g_D \right)_{L^2(\partial\Omega)} \\
&= (\Gamma_1 f, \Gamma_0 g)_{L^2(\partial\Omega)} - (\Gamma_0 f, \Gamma_1 g)_{L^2(\partial\Omega)}
\end{aligned}
$$

for all $f, g \in \operatorname{dom} T_{\max}$, that is, the abstract Green identity is satisfied. To verify the surjectivity of the mapping

$$
\begin{pmatrix} \Gamma_0 \\ \Gamma_1 \end{pmatrix} : \operatorname{dom} T_{\max} \to L^2(\partial\Omega) \times L^2(\partial\Omega), \tag{8.4.3}
$$

let $\varphi, \psi \in L^2(\partial\Omega)$ and consider $\iota_-^{-1}\varphi \in H^{-1/2}(\partial\Omega)$ and $-\iota_-^{-1}\psi \in H^{1/2}(\partial\Omega)$. Observe that by (8.2.12) the Neumann trace operator τ_N is a surjective mapping from $\{h \in H^2(\Omega) : \tau_D h = 0\}$ onto $H^{1/2}(\partial\Omega)$, that is, $\tau_N : \operatorname{dom} A_D \to H^{1/2}(\partial\Omega)$ is onto, and hence there exists $f_D \in \operatorname{dom} A_D$ such that $\tau_N f_D = -\iota_+^{-1}\psi$. Next recall from Theorem 8.3.9 that the extended Dirichlet trace operator $\tilde{\tau}_D$ maps $\operatorname{dom} T_{\max}$ onto $H^{-1/2}(\partial\Omega)$ and that $\ker \tilde{\tau}_D = \ker \tau_D = \operatorname{dom} A_D$. Hence, it follows from the direct sum decomposition $\operatorname{dom} T_{\max} = \operatorname{dom} A_D + \mathfrak{N}_\eta(T_{\max})$ that the restriction $\tilde{\tau}_D : \mathfrak{N}_\eta(T_{\max}) \to H^{-1/2}(\partial\Omega)$ is bijective, in particular, there exists $f_\eta \in \mathfrak{N}_\eta(T_{\max})$ such that $\tilde{\tau}_D f_\eta = \iota_-^{-1}\varphi$. Now it follows that $f := f_D + f_\eta \in \operatorname{dom} T_{\max}$ satisfies

$$
\Gamma_0 f = \iota_- \tilde{\tau}_D f = \iota_- \tilde{\tau}_D f_\eta = \iota_- \iota_-^{-1} \varphi = \varphi
$$

and

$$
\Gamma_1 f = -\iota_+ \tau_N f_D = \iota_+ \iota_+^{-1} \psi = \psi,
$$

and hence the mapping in (8.4.3) is onto. Thus, $\{L^2(\partial\Omega), \Gamma_0, \Gamma_1\}$ is a boundary triplet for $(T_{\min})^* = T_{\max}$, as claimed.

From the definition of Γ_0 and $\ker \tilde{\tau}_D = \ker \tau_D = \operatorname{dom} A_D$ it is clear that $\operatorname{dom} A_D = \ker \Gamma_0$, and hence the self-adjoint extension corresponding to Γ_0 coincides with the Dirichlet operator A_D, that is, the first identity in (8.4.2) holds. It remains to check the second identity in (8.4.2). For this let $f = f_D + f_\eta \in \ker \Gamma_1$, which means $\tau_N f_D = 0$. Thus, $f_D \in \operatorname{dom} T_{\min}$ by (8.2.15) and it follows that $A_1 \subset T_{\min} \hat{+} \mathfrak{N}_\eta(T_{\max})$. The inclusion $T_{\min} \hat{+} \mathfrak{N}_\eta(T_{\max}) \subset A_1$ is clear from the definition of Γ_1. This leads to the second identity in (8.4.2). $\qquad \square$

Remark 8.4.2. The boundary triplet $\{L^2(\partial\Omega), \Gamma_0, \Gamma_1\}$ in Theorem 8.4.1 is closely related to the boundary triplet $\{\mathfrak{N}_\eta(T_{\max}), \Gamma_0', \Gamma_1'\}$ in Corollary 5.5.12, where

$$
\Gamma_0' f = f_\eta \quad \text{and} \quad \Gamma_1' f = P_{\mathfrak{N}_\eta(T_{\max})}(A_D - \eta)f_D, \quad f = f_D + f_\eta \in \operatorname{dom} T_{\max}.
$$

In fact, one has $\ker \Gamma_0 = \ker \Gamma_0'$ and $\ker \Gamma_1 = \ker \Gamma_1'$, and hence

$$
\begin{pmatrix} \Gamma_0' \\ \Gamma_1' \end{pmatrix} = \begin{pmatrix} W_{11} & 0 \\ 0 & W_{22} \end{pmatrix} \begin{pmatrix} \Gamma_0 \\ \Gamma_1 \end{pmatrix}
$$

with some 2×2 operator matrix $\mathcal{W} = (W_{ij})_{i,j=1}^2$ as in Theorem 2.5.1, see also Corollary 2.5.5. In the present situation it follows from Theorem 8.3.9 and (8.4.1) that the restriction $\iota_-\widetilde{\tau}_D : \mathfrak{N}_\eta(T_{\max}) \to L^2(\partial\Omega)$ is bijective and one concludes $W_{11} = (\iota_-\widetilde{\tau}_D)^{-1}$. Now the properties of $\mathcal{W}$ imply that $W_{22} = (\iota_-\widetilde{\tau}_D)^*$.

With the help of the extended Dirichlet-to-Neumann map in Corollary 8.3.12 one obtains a more explicit description of the domain of the self-adjoint operator A_1 in (8.4.2).

Proposition 8.4.3. *Let $\Omega \subset \mathbb{R}^n$ be a bounded C^2-domain, let A_D be the self-adjoint Dirichlet realization of $-\Delta + V$ in $L^2(\Omega)$, and fix $\eta \in \rho(A_D) \cap \mathbb{R}$. Moreover, let $\widetilde{D}(\eta)$ be the extended Dirichlet-to-Neumann map in Corollary 8.3.12. Then the self-adjoint extension A_1 of $T_{\min}$ in (8.4.2) is defined on*

$$\operatorname{dom} A_1 = \big\{ f \in \operatorname{dom} T_{\max} : \widetilde{\tau}_N f = \widetilde{D}(\eta)\widetilde{\tau}_D f \big\}. \tag{8.4.4}$$

In the case that $\eta < m(A_D)$, where $m(A_D)$ denotes the lower bound of A_D, the operator A_1 coincides with the Kreĭn type extension $S_{K,\eta}$ of $T_{\min}$ in Definition 5.4.2. In particular, if $m(A_D) > 0$ and $\eta = 0$, then $A_1 = S_{K,0}$ is the Kreĭn–von Neumann extension of $T_{\min}$.

Proof. It is clear from Theorem 8.4.1 that

$$\operatorname{dom} A_1 = \ker \Gamma_1 = \big\{ f = f_D + f_\eta \in \operatorname{dom} T_{\max} : \tau_N f_D = 0 \big\}.$$

Let $\widetilde{\tau}_N$ be the extension of the Neumann trace τ_N to the maximal domain in Theorem 8.3.10. Then the boundary condition $\tau_N f_D = 0$ can be rewritten as $\widetilde{\tau}_N f = \widetilde{\tau}_N f_\eta$, where $f = f_D + f_\eta \in \operatorname{dom} T_{\max}$. With the help of the extended Dirichlet-to-Neumann map

$$\widetilde{D}(\eta) : H^{-1/2}(\partial\Omega) \to H^{-3/2}(\partial\Omega), \quad \widetilde{\tau}_D f_\eta \mapsto \widetilde{\tau}_N f_\eta, \qquad f_\eta \in \mathfrak{N}_\eta(T_{\max}),$$

one obtains $\widetilde{\tau}_N f_\eta = \widetilde{D}(\eta)\widetilde{\tau}_D f_\eta = \widetilde{D}(\eta)\widetilde{\tau}_D f$, which implies (8.4.4).

If $\eta \in \mathbb{R}$ is chosen smaller than the lower bound $m(A_D)$ of A_D, then it follows from the second identity in (8.4.2), Lemma 5.4.1, and Definition 5.4.2 that the Kreĭn type extension $S_{K,\eta} = T_{\min} \widehat{+} \mathfrak{N}_\eta(T_{\max})$ of $T_{\min}$ and A_1 coincide. In the special case $m(A_D) > 0$ and $\eta = 0$ one has $A_1 = S_{K,0}$, which is the Kreĭn–von Neumann extension of $T_{\min}$; cf. Definition 5.4.2. $\qquad\square$

In the next proposition the γ-field and the Weyl function corresponding to the boundary triplet $\{L^2(\partial\Omega), \Gamma_0, \Gamma_1\}$ in Theorem 8.4.1 are provided. Note that for $f = f_D + f_\eta$ decomposed as in (8.4.1) one has

$$\Gamma_0 f = \iota_-\widetilde{\tau}_D f = \iota_-\widetilde{\tau}_D f_\eta,$$

as $\ker \widetilde{\tau}_D = \ker \tau_D = \operatorname{dom} A_D$ by Theorem 8.3.9. It is also clear from (8.4.1) that Γ_0 is a bijective mapping from $\mathfrak{N}_\eta(T_{\max})$ onto $L^2(\partial\Omega)$.

Proposition 8.4.4. *Let $\{L^2(\partial\Omega), \Gamma_0, \Gamma_1\}$ be the boundary triplet for $(T_{\min})^* = T_{\max}$ in Theorem 8.4.1 and let $f_\eta(\varphi)$ be the unique element in $\mathfrak{N}_\eta(T_{\max})$ such that $\Gamma_0 f_\eta(\varphi) = \varphi$. Then for all $\lambda \in \rho(A_D)$ the γ-field corresponding to the boundary triplet $\{L^2(\partial\Omega), \Gamma_0, \Gamma_1\}$ is given by*

$$\gamma(\lambda)\varphi = \big(I + (\lambda - \eta)(A_D - \lambda)^{-1}\big)f_\eta(\varphi), \quad \varphi \in L^2(\partial\Omega), \qquad (8.4.5)$$

and $f_\lambda(\varphi) := \gamma(\lambda)\varphi$ is the unique element in $\mathfrak{N}_\lambda(T_{\max})$ such that $\Gamma_0 f_\lambda(\varphi) = \varphi$. Furthermore, one has

$$\gamma(\lambda)^* = -\iota_+ \tau_N (A_D - \bar\lambda)^{-1}, \quad \lambda \in \rho(A_D). \qquad (8.4.6)$$

The Weyl function M corresponding to the boundary triplet $\{L^2(\partial\Omega), \Gamma_0, \Gamma_1\}$ is given by

$$M(\lambda)\varphi = (\eta - \lambda)\iota_+ \tau_N (A_D - \lambda)^{-1} f_\eta(\varphi), \quad \varphi \in L^2(\partial\Omega).$$

In particular, $\gamma(\eta)\varphi = f_\eta(\varphi)$ and $M(\eta)\varphi = 0$ for all $\varphi \in L^2(\partial\Omega)$.

Proof. Since by definition $\gamma(\eta)$ is the inverse of the restriction of Γ_0 to $\mathfrak{N}_\eta(T_{\max})$, it is clear that $\gamma(\eta)\varphi = f_\eta(\varphi)$, where $f_\eta(\varphi)$ is the unique element in $\mathfrak{N}_\eta(T_{\max})$ such that $\Gamma_0 f_\eta(\varphi) = \varphi$. Both (8.4.5) and (8.4.6) are consequences of Proposition 2.3.2. In order to compute the Weyl function note that

$$\gamma(\lambda)\varphi = f_\eta(\varphi) + (\lambda - \eta)(A_D - \lambda)^{-1} f_\eta(\varphi)$$

is decomposed in $(\lambda - \eta)(A_D - \lambda)^{-1} f_\eta(\varphi) \in \mathrm{dom}\, A_D$ and $f_\eta(\varphi) \in \mathfrak{N}_\eta(T_{\max})$, and hence by the definition of Γ_1 it follows that

$$\begin{aligned} M(\lambda)\varphi = \Gamma_1\gamma(\lambda)\varphi &= -\iota_+ \tau_N\big[(\lambda - \eta)(A_D - \lambda)^{-1} f_\eta(\varphi)\big] \\ &= (\eta - \lambda)\iota_+ \tau_N (A_D - \lambda)^{-1} f_\eta(\varphi). \end{aligned}$$

The assertion $M(\eta)\varphi = 0$ for all $\varphi \in L^2(\partial\Omega)$ is clear from the above. □

The Weyl function M in Proposition 8.4.4 is closely connected with the Dirichlet-to-Neumann map $D(\lambda)$ and its extension $\widetilde{D}(\lambda)$, $\lambda \in \rho(A_D)$, in Definition 8.3.6 and Corollary 8.3.12. This connection will be made explicit in the next lemma. First, consider $f = f_D + f_\eta \in \mathrm{dom}\, T_{\max}$ as in (8.4.1). In the present situation one has

$$\tau_N f_D = \widetilde{\tau}_N f_D = \widetilde{\tau}_N f - \widetilde{\tau}_N f_\eta.$$

Hence, making use of $\widetilde{D}(\eta)\widetilde{\tau}_D f_\eta = \widetilde{\tau}_N f_\eta$ (see Corollary 8.3.12) and the identity $\ker \widetilde{\tau}_D = \mathrm{dom}\, A_D$, it follows that

$$\tau_N f_D = \widetilde{\tau}_N f - \widetilde{D}(\eta)\widetilde{\tau}_D f_\eta = \widetilde{\tau}_N f - \widetilde{D}(\eta)\widetilde{\tau}_D f. \qquad (8.4.7)$$

Lemma 8.4.5. *Let M be the Weyl function corresponding to the boundary triplet in Theorem 8.4.1 and let $\widetilde{D}(\lambda)$, $\lambda \in \rho(A_{\mathrm{D}})$, be the extended Dirichlet-to-Neumann map in Corollary 8.3.12. Then the regularization property*

$$\operatorname{ran}\big(\widetilde{D}(\eta) - \widetilde{D}(\lambda)\big) \subset H^{1/2}(\partial\Omega) \tag{8.4.8}$$

holds and one has

$$M(\lambda)\varphi = \iota_+\big(\widetilde{D}(\eta) - \widetilde{D}(\lambda)\big)\iota_-^{-1}\varphi, \qquad \varphi \in L^2(\partial\Omega), \tag{8.4.9}$$

and

$$M(\lambda)\varphi = \iota_+\big(D(\eta) - D(\lambda)\big)\iota_-^{-1}\varphi, \qquad \varphi \in H^2(\partial\Omega), \tag{8.4.10}$$

Proof. For $\psi \in H^{-1/2}(\partial\Omega)$ choose $f_\lambda \in \mathfrak{N}_\lambda(T_{\max})$ such that $\widetilde{\tau}_{\mathrm{D}} f_\lambda = \psi$ or, equivalently, $\Gamma_0 f_\lambda = \iota_- \psi$. Decompose f_λ in the form $f_\lambda = f_{\mathrm{D}}^\lambda + f_{\lambda,\eta}$ with $f_{\mathrm{D}}^\lambda \in \operatorname{dom} A_{\mathrm{D}}$ and $f_{\lambda,\eta} \in \mathfrak{N}_\eta(T_{\max})$. Then one computes

$$\big(\widetilde{D}(\eta) - \widetilde{D}(\lambda)\big)\psi = \widetilde{D}(\eta)\widetilde{\tau}_{\mathrm{D}} f_\lambda - \widetilde{\tau}_{\mathrm{N}} f_\lambda = -\tau_{\mathrm{N}} f_{\mathrm{D}}^\lambda, \tag{8.4.11}$$

where (8.4.7) was used in the last step for $f = f_\lambda$. Since $f_{\mathrm{D}}^\lambda \in \operatorname{dom} A_{\mathrm{D}} \subset H^2(\Omega)$, the regularization property (8.4.8) follows from (8.2.12). From (8.4.11) one also concludes that

$$\iota_+\big(\widetilde{D}(\eta) - \widetilde{D}(\lambda)\big)\iota_-^{-1}\Gamma_0 f_\lambda = -\iota_+\tau_{\mathrm{N}} f_{\mathrm{D}}^\lambda = \Gamma_1 f_\lambda,$$

and since $M(\lambda)\Gamma_0 f_\lambda = \Gamma_1 f_\lambda$ by the definition of the Weyl function, this shows (8.4.9).

It remains to prove the second assertion (8.4.10). For this note that the restriction of $\iota_-^{-1} : L^2(\partial\Omega) \to H^{-1/2}(\partial\Omega)$ to $H^2(\partial\Omega)$ is an isometric isomorphism from $H^2(\partial\Omega)$ onto $H^{3/2}(\partial\Omega)$ by Corollary 8.2.2. Furthermore, it follows from the definition that the extended Dirichlet-to-Neumann map $\widetilde{D}(\lambda)$ coincides with the Dirichlet-to-Neumann map $D(\lambda)$ on $H^{3/2}(\partial\Omega)$. With these observations it is clear that (8.4.10) follows when restricting (8.4.9) to $H^2(\partial\Omega)$. $\qquad\square$

Remark 8.4.6. The boundary mappings in Theorem 8.4.1 and the corresponding γ-field and Weyl function depend on the choice of $\eta \in \rho(A_{\mathrm{D}}) \cap \mathbb{R}$ and the decomposition of $f \in \operatorname{dom} T_{\max}$ as $f = f_{\mathrm{D}}^\eta + f_\eta$; observe that also $f_{\mathrm{D}} = f_{\mathrm{D}}^\eta \in \operatorname{dom} A_{\mathrm{D}}$ depends on η. Suppose now that the boundary mappings are defined with respect to some other $\eta' \in \rho(A_{\mathrm{D}}) \cap \mathbb{R}$ and decompose f accordingly as $f = f_{\mathrm{D}}^{\eta'} + f_{\eta'}$. If $\Gamma_0^\eta, \Gamma_1^\eta$ denote the boundary mappings in Theorem 8.4.1 with respect to η, and $\Gamma_0^{\eta'}, \Gamma_1^{\eta'}$ denote the boundary mappings in Theorem 8.4.1 with respect to η', then one has

$$\begin{pmatrix} \Gamma_0^{\eta'} \\ \Gamma_1^{\eta'} \end{pmatrix} = \begin{pmatrix} I & 0 \\ -M(\eta') & I \end{pmatrix} \begin{pmatrix} \Gamma_0^\eta \\ \Gamma_1^\eta \end{pmatrix}. \tag{8.4.12}$$

In fact, that $\Gamma_0^{\eta'} f = \Gamma_0^{\eta} f$ for $f \in \operatorname{dom} T_{\max}$ is clear from Theorem 8.4.1, and for the remaining identity in (8.4.12) it follows from Lemma 8.4.5 that

$$
\begin{aligned}
-M(\eta')\Gamma_0^{\eta} f + \Gamma_1^{\eta} f &= \iota_+ \big(\widetilde{D}(\eta') - \widetilde{D}(\eta)\big)\iota_-^{-1}\Gamma_0^{\eta} f + \Gamma_1^{\eta} f \\
&= \iota_+ \big(\widetilde{D}(\eta')\widetilde{\tau}_{\mathrm{D}} f - \widetilde{D}(\eta)\widetilde{\tau}_{\mathrm{D}} f\big) - \iota_+ \tau_{\mathrm{N}} f_{\mathrm{D}}^{\eta} \\
&= \iota_+ \big(\widetilde{D}(\eta')\widetilde{\tau}_{\mathrm{D}} f_{\eta'} - \widetilde{D}(\eta)\widetilde{\tau}_{\mathrm{D}} f_{\eta}\big) - \iota_+ \tau_{\mathrm{N}} f_{\mathrm{D}}^{\eta} \\
&= \iota_+ \big(\widetilde{\tau}_{\mathrm{N}} f_{\eta'} - \widetilde{\tau}_{\mathrm{N}} f_{\eta}\big) - \iota_+ \tau_{\mathrm{N}} f_{\mathrm{D}}^{\eta} \\
&= \iota_+ \tau_{\mathrm{N}} (f_{\mathrm{D}}^{\eta} - f_{\mathrm{D}}^{\eta'}) - \iota_+ \tau_{\mathrm{N}} f_{\mathrm{D}}^{\eta} \\
&= \Gamma_1^{\eta'} f.
\end{aligned}
$$

Finally, note that the γ-fields and Weyl functions of the boundary triplets in Theorem 8.4.1 for different η and η' transform accordingly; cf. Proposition 2.5.3.

Next some classes of extensions of $T_{\min}$ and their spectral properties are briefly discussed. Let $\{L^2(\partial\Omega), \Gamma_0, \Gamma_1\}$ be the boundary triplet in Theorem 8.4.1 with corresponding γ-field γ and Weyl function M in Proposition 8.4.4. According to Corollary 2.1.4, the self-adjoint (maximal dissipative, maximal accumulative) extensions $A_\Theta \subset T_{\max}$ of $T_{\min}$ are in a one-to-one correspondence to the self-adjoint (maximal dissipative, maximal accumulative) relations Θ in $L^2(\partial\Omega)$ via

$$
\begin{aligned}
\operatorname{dom} A_\Theta &= \big\{ f \in \operatorname{dom} T_{\max} : \{\Gamma_0 f, \Gamma_1 f\} \in \Theta \big\} \\
&= \big\{ f \in \operatorname{dom} T_{\max} : \{\iota_- \widetilde{\tau}_{\mathrm{D}} f, -\iota_+ \tau_{\mathrm{N}} f_{\mathrm{D}}\} \in \Theta \big\}.
\end{aligned}
\tag{8.4.13}
$$

If Θ is an operator in $L^2(\partial\Omega)$, then the domain of A_Θ is given by

$$
\operatorname{dom} A_\Theta = \big\{ f \in \operatorname{dom} T_{\max} : \Theta \iota_- \widetilde{\tau}_{\mathrm{D}} f = -\iota_+ \tau_{\mathrm{N}} f_{\mathrm{D}} \big\}.
\tag{8.4.14}
$$

Let Θ be a self-adjoint relation in $L^2(\partial\Omega)$ and let A_Θ be the corresponding self-adjoint realization of $-\Delta + V$ in $L^2(\Omega)$. By Corollary 1.10.9, Θ can be represented in terms of bounded operators $\mathcal{A}, \mathcal{B} \in \mathbf{B}(L^2(\partial\Omega))$ satisfying the conditions $\mathcal{A}^* \mathcal{B} = \mathcal{B}^* \mathcal{A}$, $\mathcal{A}\mathcal{B}^* = \mathcal{B}\mathcal{A}^*$, and $\mathcal{A}^* \mathcal{A} + \mathcal{B}^* \mathcal{B} = I = \mathcal{A}\mathcal{A}^* + \mathcal{B}\mathcal{B}^*$ such that

$$
\Theta = \big\{ \{\mathcal{A}\varphi, \mathcal{B}\varphi\} : \varphi \in L^2(\partial\Omega) \big\} = \big\{ \{\psi, \psi'\} : \mathcal{A}^* \psi' = \mathcal{B}^* \psi \big\}.
$$

In this case one has

$$
\operatorname{dom} A_\Theta = \big\{ f \in \operatorname{dom} T_{\max} : -\mathcal{A}^* \iota_+ \tau_{\mathrm{N}} f_{\mathrm{D}} = \mathcal{B}^* \iota_- \widetilde{\tau}_{\mathrm{D}} f \big\},
$$

and for $\lambda \in \rho(A_\Theta) \cap \rho(A_{\mathrm{D}})$ the Kreĭn formula for the corresponding resolvents

$$
\begin{aligned}
(A_\Theta - \lambda)^{-1} &= (A_{\mathrm{D}} - \lambda)^{-1} + \gamma(\lambda)\big(\Theta - M(\lambda)\big)^{-1}\gamma(\bar{\lambda})^* \\
&= (A_{\mathrm{D}} - \lambda)^{-1} + \gamma(\lambda)\mathcal{A}\big(\mathcal{B} - M(\lambda)\mathcal{A}\big)^{-1}\gamma(\bar{\lambda})^*
\end{aligned}
\tag{8.4.15}
$$

holds by Theorem 2.6.1 and Corollary 2.6.3. Recall that in the present situation the spectrum of $A_D = A_0$ is discrete by Proposition 8.3.2. According to Theorem 2.6.2, $\lambda \in \rho(A_D)$ is an eigenvalue of A_Θ if and only if $\ker(\Theta - M(\lambda))$ or, equivalently, $\ker(\mathcal{B} - M(\lambda)\mathcal{A})$ is nontrivial, and that

$$\ker(A_\Theta - \lambda) = \gamma(\lambda)\ker(\Theta - M(\lambda)) = \gamma(\lambda)\mathcal{A}\ker(\mathcal{B} - M(\lambda)\mathcal{A}).$$

Although Ω is a bounded C^2-domain, it will turn out in Example 8.4.9 that the spectrum of A_Θ is in general not discrete, and thus continuous spectrum may be present. It then follows from Theorem 2.6.2 and Theorem 2.6.5 that $\lambda \in \rho(A_D)$ belongs to the continuous spectrum $\sigma_c(A_\Theta)$ (essential spectrum $\sigma_{\text{ess}}(A_\Theta)$ or discrete spectrum $\sigma_d(A_\Theta)$) of A_Θ if and only if 0 belongs to $\sigma_c(\Theta - M(\lambda))$ ($\sigma_{\text{ess}}(\Theta - M(\lambda))$ or $\sigma_d(\Theta - M(\lambda))$).

For a complete description of the spectrum of A_Θ recall that the symmetric operator $T_{\min}$ is simple according to Proposition 8.3.13 and make use of a transform of the boundary triplet $\{L^2(\partial\Omega), \Gamma_0, \Gamma_1\}$ as in Chapter 3.8. This reasoning implies that λ is an eigenvalue of A_Θ if and only if λ is a pole of the function

$$\lambda \mapsto M_\Theta(\lambda) = (\mathcal{A}^* + \mathcal{B}^* M(\lambda))(\mathcal{B}^* - \mathcal{A}^* M(\lambda))^{-1}.$$

It is important to note in this context that the multiplicity of the eigenvalues of A_Θ is not necessarily finite and that the dimension of the eigenspace $\ker(A_\Theta - \lambda)$ of an isolated eigenvalue λ of A_Θ coincides with the dimension of the range of the residue of M_Θ at λ. Furthermore, the continuous and absolutely continuous spectrum of A_Θ can be characterized as in Section 3.8, e.g., one has

$$\sigma_{\text{ac}}(A_\Theta) = \overline{\bigcup_{\varphi \in L^2(\partial\Omega)} \text{clos}_{\text{ac}}\left(\left\{x \in \mathbb{R} : 0 < \text{Im}\left(M_\Theta(x + i0)\varphi, \varphi\right)_{L^2(\partial\Omega)} < \infty\right\}\right)}.$$

In the special case that the self-adjoint relation Θ in $L^2(\partial\Omega)$ is a bounded operator the boundary condition reads as in (8.4.14) and according to Section 3.8 the spectral properties of the self-adjoint operator A_Θ can also be described with the help of the function

$$\lambda \mapsto (\Theta - M(\lambda))^{-1}.$$

The general boundary conditions in (8.4.13) and (8.4.14) contain also typical classes of boundary conditions that are treated in spectral problems for partial differential operators, as, e.g., Neumann or Robin type boundary conditions. In the following example the standard Neumann boundary conditions are discussed. Note that the Neumann operator does not coincide with the Kreĭn type extension $S_{K,\eta}$ or the Kreĭn–von Neumann extension $S_{K,0}$ of $T_{\min}$ in Proposition 8.4.3.

Example 8.4.7. Let $\{L^2(\partial\Omega), \Gamma_0, \Gamma_1\}$ be the boundary triplet in Theorem 8.4.1 and choose $\eta \in \rho(A_D) \cap \mathbb{R}$ in (8.4.1) in such a way that also $\eta \in \rho(A_N)$, where

A_N denotes the Neumann realization of $-\Delta + V$ in Proposition 8.3.3 and Theorem 8.3.4. Since both self-adjoint operators A_D and A_N are semibounded from below (or both have discrete spectrum), such an η exists. In this situation it follows that the Dirichlet-to-Neumann map

$$D(\eta) : H^{3/2}(\partial\Omega) \to H^{1/2}(\partial\Omega)$$

in Definition 8.3.6 is a bijective mapping. Furthermore, $\iota_+ : H^{1/2}(\partial\Omega) \to L^2(\partial\Omega)$ is bijective and the restriction of $\iota_-^{-1} : L^2(\partial\Omega) \to H^{-1/2}(\partial\Omega)$ to $H^2(\partial\Omega)$ is an isometric isomorphism from $H^2(\partial\Omega)$ onto $H^{3/2}(\partial\Omega)$ according to Corollary 8.2.2. Hence, it is clear that

$$\Theta_N := \iota_+ D(\eta)\iota_-^{-1}, \qquad \operatorname{dom}\Theta_N := H^2(\partial\Omega), \tag{8.4.16}$$

is a densely defined bijective operator in $L^2(\partial\Omega)$. Furthermore, for $\varphi \in H^2(\partial\Omega)$ and $\psi = \iota_-^{-1}\varphi \in H^{3/2}(\partial\Omega)$ it follows from Corollary 8.2.2 that

$$\left(\iota_+ D(\eta)\iota_-^{-1}\varphi, \varphi\right)_{L^2(\partial\Omega)} = \left(\iota_+ D(\eta)\psi, \iota_-\psi\right)_{L^2(\partial\Omega)} = (D(\eta)\psi, \psi)_{L^2(\partial\Omega)}. \tag{8.4.17}$$

Now choose $f_\eta \in H^2(\Omega)$ such that $(-\Delta + V)f_\eta = \eta f_\eta$ and $\tau_D f_\eta = \psi$, which is possible by (8.3.9) and (8.2.12). Then it follows from Definition 8.3.6 and the first Green identity in (8.2.18) that

$$\begin{aligned}
(D(\eta)\psi, \psi)_{L^2(\partial\Omega)} &= \left(D(\eta)\tau_D f_\eta, \tau_D f_\eta\right)_{L^2(\partial\Omega)} \\
&= (\tau_N f_\eta, \tau_D f_\eta)_{L^2(\partial\Omega)} \\
&= \|\nabla f_\eta\|_{L^2(\Omega;\mathbb{C}^n)}^2 + (\Delta f_\eta, f_\eta)_{L^2(\Omega)} \\
&= \|\nabla f_\eta\|_{L^2(\Omega;\mathbb{C}^n)}^2 + ((V - \eta)f_\eta, f_\eta)_{L^2(\Omega)},
\end{aligned} \tag{8.4.18}$$

so that $(D(\eta)\psi, \psi)_{L^2(\partial\Omega)} \in \mathbb{R}$ and hence $(\Theta_N\varphi, \varphi)_{L^2(\partial\Omega)} \in \mathbb{R}$ by (8.4.16)–(8.4.17) for all $\varphi \in H^2(\partial\Omega)$. It follows that the bijective operator Θ_N is symmetric in $L^2(\partial\Omega)$, and hence Θ_N is an unbounded self-adjoint operator in $L^2(\partial\Omega)$ such that $0 \in \rho(\Theta_N)$.

The self-adjoint realization of $-\Delta + V$ in $L^2(\Omega)$ corresponding to the self-adjoint operator Θ_N in (8.4.16) is denoted by A_{Θ_N}. A function $f \in \operatorname{dom}T_{\max}$ belongs to $\operatorname{dom}A_{\Theta_N}$ if and only if

$$\Gamma_0 f = \iota_-\tilde{\tau}_D f \in \operatorname{dom}\Theta_N \quad \text{and} \quad \Gamma_1 f = \Theta_N\Gamma_0 f.$$

Note that $\iota_-\tilde{\tau}_D f \in \operatorname{dom}\Theta_N$ forces $\tilde{\tau}_D f \in H^{3/2}(\partial\Omega)$ and hence $f \in H^2(\Omega)$ and $\tilde{\tau}_D f = \tau_D f$ by (8.2.12) and Theorem 8.3.9. It then follows from (8.4.7) and (8.4.16) that the boundary condition $\Gamma_1 f = \Theta_N\Gamma_0 f$ takes on the form

$$\iota_+ D(\eta)\tau_D f - \iota_+\tau_N f = -\iota_+\tau_N f_D = \Gamma_1 f = \Theta_N\Gamma_0 f = \iota_+ D(\eta)\tau_D f,$$

that is, $\tau_N f = 0$. Hence, it has been shown that $\operatorname{dom} A_{\Theta_N} \subset H^2(\Omega)$ and that $\tau_N f = 0$ for all $f \in \operatorname{dom} A_{\Theta_N}$. Therefore, $A_{\Theta_N} \subset A_N$ and since both operators are self-adjoint one concludes that $A_{\Theta_N} = A_N$.

Note also that by (8.4.16) and Lemma 8.4.5 one has

$$\big(\Theta_N - M(\lambda)\big)\varphi = \iota_+ D(\eta)\iota_-^{-1}\varphi - \iota_+\big(D(\eta) - D(\lambda)\big)\iota_-^{-1}\varphi = \iota_+ D(\lambda)\iota_-^{-1}\varphi$$

for $\varphi \in H^2(\partial\Omega)$ and $\lambda \in \rho(A_D)$. Hence, it follows that $\Theta_N - M(\lambda)$ is a bijective operator in $L^2(\partial\Omega)$ for all $\lambda \in \rho(A_D) \cap \rho(A_N)$ which is defined on $H^2(\partial\Omega)$. Therefore, (8.4.15) implies that the resolvents of A_D and A_N are related via

$$(A_N - \lambda)^{-1} = (A_D - \lambda)^{-1} + \gamma(\lambda)\iota_- D(\lambda)^{-1}\iota_+^{-1}\gamma(\bar\lambda)^*,$$

where γ is the γ-field corresponding to the boundary triplet $\{L^2(\partial\Omega), \Gamma_0, \Gamma_1\}$ in Proposition 8.4.4.

The next example is a generalization of the previous example from Neumann to local and nonlocal Robin boundary conditions.

Example 8.4.8. Let $\{L^2(\partial\Omega), \Gamma_0, \Gamma_1\}$ be as in the previous example and fix some $\eta \in \rho(A_D) \cap \rho(A_N) \cap \mathbb{R}$. Then the operator $\Theta_N = \iota_+ D(\eta)\iota_-^{-1}$ in (8.4.16) is an unbounded self-adjoint operator in $L^2(\partial\Omega)$ with domain $H^2(\partial\Omega)$, and $0 \in \rho(\Theta_N)$. Assume that

$$B : H^{3/2}(\partial\Omega) \to H^{1/2}(\partial\Omega) \tag{8.4.19}$$

is compact as an operator from $H^{3/2}(\partial\Omega)$ into $H^{1/2}(\partial\Omega)$ and that B is symmetric in $L^2(\partial\Omega)$, that is, $(B\psi, \psi)_{L^2(\partial\Omega)} \in \mathbb{R}$ for all $\psi \in \operatorname{dom} B = H^{3/2}(\partial\Omega)$. Then it follows that

$$\iota_+ B\iota_-^{-1} : H^2(\partial\Omega) \to L^2(\partial\Omega)$$

is compact as an operator from $H^2(\partial\Omega)$ into $L^2(\partial\Omega)$ and as in (8.4.17) one sees that $\iota_+ B\iota_-^{-1}$ is symmetric in $L^2(\partial\Omega)$. Consider the operator

$$\Theta_B := \iota_+\big(D(\eta) - B\big)\iota_-^{-1} = \Theta_N - \iota_+ B\iota_-^{-1}, \quad \operatorname{dom}\Theta_B = H^2(\partial\Omega), \tag{8.4.20}$$

and observe that the symmetric operator $\iota_+ B\iota_-^{-1}$ is a relative compact perturbation of the self-adjoint operator Θ_N in $L^2(\partial\Omega)$, that is, the operator

$$\iota_+ B\iota_-^{-1}\Theta_N^{-1}$$

is compact in $L^2(\partial\Omega)$. It is well known from standard perturbation results (see, e.g., [652, Corollary 2 of Theorem XIII.14]) that in this case the perturbed operator Θ_B is self-adjoint in $L^2(\partial\Omega)$.

The self-adjoint realization of $-\Delta + V$ in $L^2(\Omega)$ corresponding to the self-adjoint operator Θ_B in (8.4.20) is denoted by A_{Θ_B}. It is clear that a function $f \in \operatorname{dom} T_{\max}$ belongs to $\operatorname{dom} A_{\Theta_B}$ if and only if

$$\Gamma_0 f = \iota_- \widetilde\tau_D f \in \operatorname{dom}\Theta_B \quad \text{and} \quad \Gamma_1 f = \Theta_B\Gamma_0 f. \tag{8.4.21}$$

In the same way as in Example 8.4.7 the fact that $\iota_- \widetilde{\tau}_D f \in \mathrm{dom}\, \Theta_B$ implies that $f \in H^2(\Omega)$ and $\widetilde{\tau}_D f = \tau_D f$, and the boundary condition $\Gamma_1 f = \Theta_B \Gamma_0 f$ takes the explicit form

$$\iota_+ D(\eta) \tau_D f - \iota_+ \tau_N f = \Gamma_1 f = \Theta_B \Gamma_0 f = \iota_+ \big(D(\eta) - B\big) \tau_D f,$$

that is, $\tau_N f = B \tau_D f$. Conversely, if $f \in H^2(\Omega)$ is such that $\tau_N f = B \tau_D f$, then f satisfies (8.4.21) and hence $f \in \mathrm{dom}\, A_{\Theta_B}$. Thus, it has been shown that the self-adjoint operator A_{Θ_B} is defined on

$$\mathrm{dom}\, A_{\Theta_B} = \big\{ f \in H^2(\Omega) : \tau_N f = B \tau_D f \big\}.$$

In the same way as in the previous example one obtains

$$\Theta_B - M(\lambda) = \iota_+ \big(D(\lambda) - B\big) \iota_-^{-1}$$

and hence for all $\lambda \in \rho(A_D) \cap \rho(A_{\Theta_B})$ one has

$$(A_{\Theta_B} - \lambda)^{-1} = (A_D - \lambda)^{-1} + \gamma(\lambda) \iota_- \big(D(\lambda) - B\big)^{-1} \iota_+^{-1} \gamma(\bar{\lambda})^*.$$

Finally, note that a sufficient condition for the operator B in (8.4.19) to be compact is that $B : H^{3/2}(\partial\Omega) \to H^{1/2+\varepsilon}(\partial\Omega)$ is bounded for some $\varepsilon > 0$, or that $B : H^{3/2-\varepsilon'}(\partial\Omega) \to H^{1/2}(\partial\Omega)$ is bounded for some $\varepsilon' > 0$, since the embeddings $H^{1/2+\varepsilon}(\partial\Omega) \hookrightarrow H^{1/2}(\partial\Omega)$ and $H^{3/2}(\partial\Omega) \hookrightarrow H^{3/2-\varepsilon'}(\partial\Omega)$ are compact by (8.2.8).

In the next example it is shown that the (essential) spectrum of a self-adjoint realization A_Θ of $-\Delta + V$ can be very general, depending on the properties of the parameter Θ. In particular, the self-adjoint realization A_Θ may not be semi-bounded.

Example 8.4.9. Let $\eta \in \rho(A_D) \cap \mathbb{R}$, consider an arbitrary self-adjoint operator Ξ in the Hilbert space $\mathfrak{N}_\eta(T_{\max}) = \ker(T_{\max} - \eta)$, and assume that $\eta \in \rho(\Xi)$. Denote by $P_{\mathfrak{N}_\eta}$ the orthogonal projection in $L^2(\Omega)$ onto $\mathfrak{N}_\eta(T_{\max})$ and let $\iota_{\mathfrak{N}_\eta}$ be the natural embedding of $\mathfrak{N}_\eta(T_{\max})$ into $L^2(\Omega)$.

Let $\{L^2(\partial\Omega), \Gamma_0, \Gamma_1\}$ be the boundary triplet in Theorem 8.4.1 with corresponding γ-field and Weyl function in Proposition 8.4.4. Note that $M(\eta) = 0$ and that both

$$P_{\mathfrak{N}_\eta} \gamma(\eta) : L^2(\partial\Omega) \to \mathfrak{N}_\eta(T_{\max}) \quad \text{and} \quad \gamma(\eta)^* \iota_{\mathfrak{N}_\eta} : \mathfrak{N}_\eta(T_{\max}) \to L^2(\partial\Omega)$$

are isomorphisms. It follows that

$$\Theta := \big(\gamma(\eta)^* \iota_{\mathfrak{N}_\eta}\big) (\Xi - \eta) \big(P_{\mathfrak{N}_\eta} \gamma(\eta)\big)$$

is a self-adjoint operator in $L^2(\partial\Omega)$ with $0 \in \rho(\Theta)$ and

$$\Theta^{-1} = \big(P_{\mathfrak{N}_\eta} \gamma(\eta)\big)^{-1} (\Xi - \eta)^{-1} \big(\gamma(\eta)^* \iota_{\mathfrak{N}_\eta}\big)^{-1}. \tag{8.4.22}$$

Let A_Θ be the corresponding self-adjoint realization of $-\Delta + V$ in (8.4.13)–(8.4.14) defined on

$$\operatorname{dom} A_\Theta = \{ f \in \operatorname{dom} T_{\max} : \Theta \iota_- \widetilde{\tau}_\mathrm{D} f = -\iota_+ \tau_\mathrm{N} f_\mathrm{D} \}.$$

Since $M(\eta) = 0$ and $\eta \in \mathbb{R}$, Kreĭn's formula in (8.4.15) takes the form

$$(A_\Theta - \eta)^{-1} = (A_\mathrm{D} - \eta)^{-1} + \gamma(\eta) \Theta^{-1} \gamma(\eta)^*$$

$$= (A_\mathrm{D} - \lambda)^{-1} + \begin{pmatrix} P_{\mathfrak{N}_\eta} \gamma(\eta) \Theta^{-1} \gamma(\eta)^* \iota_{\mathfrak{N}_\eta} & 0 \\ 0 & 0 \end{pmatrix},$$

where the block operator matrix is acting with respect to the space decomposition $L^2(\Omega) = \mathfrak{N}_\eta(T_{\max}) \oplus (\mathfrak{N}_\eta(T_{\max}))^\perp$. Using (8.4.22) one then concludes that

$$(A_\Theta - \eta)^{-1} = (A_\mathrm{D} - \eta)^{-1} + \begin{pmatrix} (\Xi - \eta)^{-1} & 0 \\ 0 & 0 \end{pmatrix}.$$

In particular, since $(A_\mathrm{D} - \eta)^{-1}$ is compact, well-known perturbation results show that

$$\sigma_\mathrm{ess}\big((A_\Theta - \eta)^{-1}\big) = \sigma_\mathrm{ess}\big((\Xi - \eta)^{-1}\big) \cup \{0\},$$

and hence $\sigma_\mathrm{ess}(A_\Theta) = \sigma_\mathrm{ess}(\Xi)$.

8.5 Semibounded Schrödinger operators

The semibounded self-adjoint realizations of $-\Delta + V$, where $V \in L^\infty(\Omega)$ is real, and the corresponding densely defined closed semibounded forms in $L^2(\Omega)$ are described in this section. For this purpose it is convenient to construct a boundary pair which is compatible with the boundary triplet in Theorem 8.4.1 and to apply the general results from Section 5.6. Under the additional assumption that $V \geq 0$, the nonnegative realizations of $-\Delta + V$ and the corresponding nonnegative forms in $L^2(\Omega)$ are discussed as a special case. In this situation the Kreĭn–von Neumann extension appears as the smallest nonnegative extension.

Let $\Omega \subset \mathbb{R}^n$ be a bounded C^2-domain and let A_D be the self-adjoint Dirichlet realization of $-\Delta + V$. It is clear from Proposition 8.3.2 that A_D coincides with the Friedrichs extension of the minimal operator $T_{\min}$ in (8.3.2) and that A_D is bounded from below with lower bound $m(A_\mathrm{D}) > v_-$, where $v_- = \operatorname{essinf} V$. Furthermore, the resolvent of A_D is compact since the domain Ω is bounded. Therefore, the following description of the semibounded self-adjoint extensions of $T_{\min}$ is an immediate consequence of Proposition 5.5.6 and Proposition 5.5.8.

Proposition 8.5.1. *Let $\Omega \subset \mathbb{R}^n$ be a bounded C^2-domain, let $\{L^2(\partial\Omega), \Gamma_0, \Gamma_1\}$ be the boundary triplet for $(T_{\min})^* = T_{\max}$ from Theorem 8.4.1, and let*

$$A_\Theta = -\Delta + V,$$

$$\operatorname{dom} A_\Theta = \{ f \in \operatorname{dom} T_{\max} : \{\Gamma_0 f, \Gamma_1 f\} \in \Theta \},$$

be a self-adjoint extension of $T_{\min}$ in $L^2(\Omega)$ corresponding to a self-adjoint relation Θ in $L^2(\partial\Omega)$ as in (8.4.13). Then

$$A_\Theta \text{ is semibounded} \quad \Leftrightarrow \quad \Theta \text{ is semibounded.}$$

Recall also from Section 8.3 that the densely defined closed semibounded form t_{A_D} corresponding to A_D is defined on $H_0^1(\Omega)$. Now fix some $\eta < m(A_D)$, use the direct sum decomposition

$$\text{dom}\, T_{\max} = \mathfrak{N}_\eta(T_{\max}) + \text{dom}\, A_D = \mathfrak{N}_\eta(T_{\max}) + \left(H^2(\Omega) \cap H_0^1(\Omega)\right) \qquad (8.5.1)$$

from (8.4.1) and Proposition 8.3.2, and consider the corresponding boundary triplet $\{L^2(\partial\Omega), \Gamma_0, \Gamma_1\}$ for $(T_{\min})^* = T_{\max}$ in Theorem 8.4.1 given by

$$\Gamma_0 f = \iota_- \widetilde{\tau}_D f \quad \text{and} \quad \Gamma_1 f = -\iota_+ \tau_N f_D, \qquad (8.5.2)$$

where $f = f_\eta + f_D \in \text{dom}\, T_{\max}$ with $f_\eta \in \mathfrak{N}_\eta(T_{\max})$ and $f_D \in \text{dom}\, A_D$; cf. (8.5.1). It is clear that $A_0 = A_D$ coincides with the Friedrichs extension of $T_{\min}$ and $A_1 = T_{\min} \widehat{+} \mathfrak{N}_\eta(T_{\max})$ coincides with the Kreĭn type extension $S_{K,\eta}$ of $T_{\min}$; cf. Definition 5.4.2. In order to define a boundary pair for $T_{\min}$ corresponding to $A_1 = S_{K,\eta}$, consider the densely defined closed semibounded form $t_{S_{K,\eta}}$ associated with $S_{K,\eta}$ and recall from Corollary 5.4.16 the direct sum decomposition

$$\text{dom}\, t_{S_{K,\eta}} = \mathfrak{N}_\eta(T_{\max}) + \text{dom}\, t_{A_D} = \mathfrak{N}_\eta(T_{\max}) + H_0^1(\Omega) \qquad (8.5.3)$$

of $\text{dom}\, t_{S_{K,\eta}}$. Comparing (8.5.1) and (8.5.3) one sees that $\text{dom}\, T_{\max} \subset \text{dom}\, t_{S_{K,\eta}}$ and that the domain of the Dirichlet operator A_D in (8.5.1) is replaced by the corresponding form domain in (8.5.3). The functions $f \in \text{dom}\, t_{S_{K,\eta}}$ will be written in the form $f = f_\eta + f_F$, where $f_\eta \in \mathfrak{N}_\eta(T_{\max})$ and $f_F \in \text{dom}\, t_{A_D} = H_0^1(\Omega)$. Now define the mapping

$$\Lambda : \text{dom}\, t_{S_{K,\eta}} \to L^2(\partial\Omega), \quad f \mapsto \Lambda f = \iota_- \widetilde{\tau}_D f_\eta. \qquad (8.5.4)$$

It will be shown next that $\{L^2(\partial\Omega), \Lambda\}$ is a boundary pair that is compatible with the boundary triplet $\{L^2(\partial\Omega), \Gamma_0, \Gamma_1\}$ in the sense of Definition 5.6.4; although the main part of the proof of Lemma 8.5.2 is similar to Example 5.6.9, the details are provided.

Lemma 8.5.2. Let $\Omega \subset \mathbb{R}^n$ be a bounded C^2-domain and let A_D be the self-adjoint Dirichlet realization of $-\Delta + V$ with lower bound $m(A_D)$. Fix $\eta < m(A_D)$, let $\{L^2(\partial\Omega), \Gamma_0, \Gamma_1\}$ be the corresponding boundary triplet for $(T_{\min})^* = T_{\max}$ from Theorem 8.4.1, and let Λ be the mapping in (8.5.4). Then $\{L^2(\partial\Omega), \Lambda\}$ is a boundary pair for $T_{\min}$ corresponding to the Kreĭn type extension $S_{K,\eta}$ which is compatible with the boundary triplet $\{L^2(\partial\Omega), \Gamma_0, \Gamma_1\}$. Moreover, one has

$$(T_{\max} f, g)_{L^2(\Omega)} = t_{S_{K,\eta}}[f, g] + (\Gamma_1 f, \Lambda g)_{L^2(\partial\Omega)} \qquad (8.5.5)$$

for all $f \in \text{dom}\, T_{\max}$ and $g \in \text{dom}\, t_{S_{K,\eta}}$.

Proof. According to Lemma 5.6.5 (ii), it suffices to show that for some $a < \eta$ the mapping Λ in (8.5.4) is bounded from the Hilbert space

$$\mathfrak{H}_{t_{S_{\mathrm{K},\eta}}-a} = \big(\mathrm{dom}\, t_{S_{\mathrm{K},\eta}}, (\cdot,\cdot)_{t_{S_{\mathrm{K},\eta}}-a}\big)$$

to $L^2(\partial\Omega)$ and that Λ extends the mapping Γ_0 in (8.5.2). In the present situation it is clear that the compatibility condition $A_1 = S_{\mathrm{K},\eta}$ is satisfied.

In order to show that Λ is bounded fix some $a < \eta$, recall first from (5.1.7) that the Hilbert space norm on $\mathfrak{H}_{t_{S_{\mathrm{K},\eta}}-a}$ is given by

$$\|f\|^2_{t_{S_{\mathrm{K},\eta}}-a} = t_{S_{\mathrm{K},\eta}}[f] - a\|f\|^2_{L^2(\Omega)}, \quad f \in \mathrm{dom}\, t_{S_{\mathrm{K},\eta}} = \mathfrak{H}_{t_{S_{\mathrm{K},\eta}}-a}.$$

It follows from Theorem 8.3.9 that the restriction $\iota_-\widetilde{\tau}_{\mathrm{D}} : \mathfrak{N}_\eta(T_{\max}) \to L^2(\partial\Omega)$ is bounded and hence for $f = f_\eta + f_{\mathrm{F}} \in \mathrm{dom}\, t_{S_{\mathrm{K},\eta}}$, decomposed according to (8.5.3) in $f_\eta \in \mathfrak{N}_\eta(T_{\max})$ and $f_{\mathrm{F}} \in \mathrm{dom}\, t_{A_{\mathrm{D}}}$, one has the estimate

$$\|\Lambda f\|^2_{L^2(\partial\Omega)} = \|\iota_-\widetilde{\tau}_{\mathrm{D}} f_\eta\|^2_{L^2(\partial\Omega)} \leq C\|f_\eta\|^2_{L^2(\Omega)}. \tag{8.5.6}$$

Now the orthogonal sum decomposition

$$\mathrm{dom}\, t_{S_{\mathrm{K},\eta}} = \mathfrak{N}_a(T_{\max}) \oplus_{t_{S_{\mathrm{K},\eta}}-a} \mathrm{dom}\, t_{A_{\mathrm{D}}} \tag{8.5.7}$$

from Corollary 5.4.15 will be used. To this end, define

$$f_a := \big(I + (a-\eta)(A_{\mathrm{D}} - a)^{-1}\big)f_\eta$$

and note that $f = f_a + h_{\mathrm{F}}$ with $f_a \in \mathfrak{N}_a(T_{\max})$ and $h_{\mathrm{F}} = f_\eta - f_a + f_{\mathrm{F}} \in \mathrm{dom}\, t_{A_{\mathrm{D}}}$. Then one has

$$f_\eta = \big(I + (\eta-a)(A_{\mathrm{D}} - \eta)^{-1}\big)f_a$$

and Proposition 1.4.6 leads to the estimate

$$\|f_\eta\|_{L^2(\Omega)} \leq \frac{m(A_{\mathrm{D}}) - a}{m(A_{\mathrm{D}}) - \eta} \|f_a\|_{L^2(\Omega)}. \tag{8.5.8}$$

Furthermore, it follows from (5.1.9) and the orthogonal sum decomposition (8.5.7) that

$$(\eta - a)\|f_a\|^2_{L^2(\Omega)} \leq \|f_a\|^2_{t_{S_{\mathrm{K},\eta}}-a} \leq \|f_a\|^2_{t_{S_{\mathrm{K},\eta}}-a} + \|h_{\mathrm{F}}\|^2_{t_{S_{\mathrm{K},\eta}}-a} = \|f\|^2_{t_{S_{\mathrm{K},\eta}}-a}.$$

From this estimate, (8.5.6), and (8.5.8) one concludes that $\Lambda : \mathfrak{H}_{t_{S_{\mathrm{K},\eta}}-a} \to L^2(\partial\Omega)$ is bounded.

From the definition of Λ in (8.5.4) and the decompositions (8.5.1) and (8.5.3) it is clear that Λ is an extension of the mapping Γ_0 in (8.5.2). Moreover, by construction, the condition $A_1 = S_{\mathrm{K},\eta}$ is satisfied. Therefore, Lemma 5.6.5 (ii) shows that $\{L^2(\partial\Omega), \Lambda\}$ is a boundary pair for $T_{\min}$ corresponding to $S_{\mathrm{K},\eta}$ which is compatible with the boundary triplet $\{L^2(\partial\Omega), \Gamma_0, \Gamma_1\}$. The identity (8.5.5) follows from Corollary 5.6.7. $\qquad\square$

The next theorem is a variant of Theorem 5.6.13 in the present situation.

Theorem 8.5.3. *Let $\Omega \subset \mathbb{R}^n$ be a bounded C^2-domain, let A_{D} be the self-adjoint Dirichlet realization of $-\Delta + V$ with lower bound $m(A_{\mathrm{D}})$, and fix $\eta < m(A_{\mathrm{D}})$. Let $\{L^2(\partial\Omega), \Gamma_0, \Gamma_1\}$ be the boundary triplet for $(T_{\min})^* = T_{\max}$ from Theorem 8.4.1 and let $\{L^2(\partial\Omega), \Lambda\}$ be the compatible boundary pair in Lemma 8.5.2. Furthermore, let Θ be a semibounded self-adjoint relation in $L^2(\partial\Omega)$ and let A_Θ be the corresponding semibounded self-adjoint extension of $T_{\min}$ in Proposition 8.5.1. Then the closed semibounded form ω_Θ in $L^2(\partial\Omega)$ corresponding to Θ and the densely defined closed semibounded form $\mathfrak{t}_{A_\Theta}$ corresponding to A_Θ are related by*

$$\mathfrak{t}_{A_\Theta}[f, g] = \mathfrak{t}_{S_{\mathrm{K},\eta}}[f, g] + \omega_\Theta[\Lambda f, \Lambda g],$$
$$\operatorname{dom} \mathfrak{t}_{A_\Theta} = \{f \in \operatorname{dom} \mathfrak{t}_{S_{\mathrm{K},\eta}} : \Lambda f \in \operatorname{dom} \omega_\Theta\}. \tag{8.5.9}$$

For completeness, the form $\mathfrak{t}_{A_\Theta}$ in Theorem 8.5.3 will be made more explicit using Corollary 5.6.14. First note that, by the definition of the boundary map Λ in (8.5.4) and the decomposition (8.5.3), one can rewrite (8.5.9) as

$$\mathfrak{t}_{A_\Theta}[f, g] = \mathfrak{t}_{S_{\mathrm{K},\eta}}[f, g] + \omega_\Theta[\iota_- \widetilde{\tau}_{\mathrm{D}} f_\eta, \iota_- \widetilde{\tau}_{\mathrm{D}} g_\eta],$$
$$\operatorname{dom} \mathfrak{t}_{A_\Theta} = \{f = f_\eta + f_{\mathrm{F}} \in \operatorname{dom} \mathfrak{t}_{S_{\mathrm{K},\eta}} : \iota_- \widetilde{\tau}_{\mathrm{D}} f_\eta \in \operatorname{dom} \omega_\Theta\}. \tag{8.5.10}$$

If $m(\Theta)$ denotes the lower bound of the semibounded self-adjoint relation Θ and $\mu \leq m(\Theta)$ is fixed, then the closed semibounded form $\mathfrak{t}_{A_\Theta}$ in (8.5.9)–(8.5.10) corresponding to A_Θ is given by

$$\mathfrak{t}_{A_\Theta}[f, g] = \mathfrak{t}_{S_{\mathrm{K},\eta}}[f, g] + \left((\Theta_{\mathrm{op}} - \mu)^{\frac{1}{2}} \iota_- \widetilde{\tau}_{\mathrm{D}} f_\eta, (\Theta_{\mathrm{op}} - \mu)^{\frac{1}{2}} \iota_- \widetilde{\tau}_{\mathrm{D}} g_\eta\right)_{L^2(\partial\Omega)}$$
$$+ \mu \left(\iota_- \widetilde{\tau}_{\mathrm{D}} f_\eta, \iota_- \widetilde{\tau}_{\mathrm{D}} g_\eta\right)_{L^2(\partial\Omega)},$$
$$\operatorname{dom} \mathfrak{t}_{A_\Theta} = \{f = f_\eta + f_{\mathrm{F}} \in \operatorname{dom} \mathfrak{t}_{S_{\mathrm{K},\eta}} : \iota_- \widetilde{\tau}_{\mathrm{D}} f_\eta \in \operatorname{dom} (\Theta_{\mathrm{op}} - \mu)^{\frac{1}{2}}\};$$

as usual, here Θ_{op} denotes the semibounded self-adjoint operator part of Θ acting in $L^2(\partial\Omega)_{\mathrm{op}} = \overline{\operatorname{dom}} \Theta$. In the special case where $\Theta_{\mathrm{op}} \in \mathbf{B}(L^2(\partial\Omega)_{\mathrm{op}})$ one has

$$\mathfrak{t}_{A_\Theta}[f, g] = \mathfrak{t}_{S_{\mathrm{K},\eta}}[f, g] + \left(\Theta_{\mathrm{op}} \iota_- \widetilde{\tau}_{\mathrm{D}} f_\eta, \iota_- \widetilde{\tau}_{\mathrm{D}} g_\eta\right)_{L^2(\partial\Omega)},$$
$$\operatorname{dom} \mathfrak{t}_{A_\Theta} = \{f = f_\eta + f_{\mathrm{F}} \in \operatorname{dom} \mathfrak{t}_{S_{\mathrm{K},\eta}} : \iota_- \widetilde{\tau}_{\mathrm{D}} f_\eta \in \operatorname{dom} \Theta_{\mathrm{op}}\},$$

and if $\Theta \in \mathbf{B}(L^2(\partial\Omega))$, then

$$\mathfrak{t}_{A_\Theta}[f, g] = \mathfrak{t}_{S_{\mathrm{K},\eta}}[f, g] + \left(\Theta \iota_- \widetilde{\tau}_{\mathrm{D}} f_\eta, \iota_- \widetilde{\tau}_{\mathrm{D}} g_\eta\right)_{L^2(\partial\Omega)}, \qquad \operatorname{dom} \mathfrak{t}_{A_\Theta} = \operatorname{dom} \mathfrak{t}_{S_{\mathrm{K},\eta}}.$$

Recall also from Corollary 5.4.15 that the form $\mathfrak{t}_{S_{\mathrm{K},\eta}}$ can be expressed in terms of the form $\mathfrak{t}_{A_{\mathrm{D}}}$ and the resolvent of A_{D}.

Finally, the special case $V \geq 0$ will be briefly considered. In this situation the minimal operator $T_{\min}$ and the Dirichlet operator A_{D} are both uniformly positive and hence in the above construction of a boundary triplet and corresponding boundary pair one may choose $\eta = 0$. More precisely, Theorem 8.4.1 has the following form.

Corollary 8.5.4. *Let $\Omega \subset \mathbb{R}^n$ be a bounded C^2-domain, let A_D be the self-adjoint Dirichlet realization of $-\Delta + V$ in $L^2(\Omega)$ with $V \geq 0$, and decompose $f \in \mathrm{dom}\, T_{\max}$ according to (8.4.1) with $\eta = 0$ in the form $f = f_D + f_0$, where $f_D \in \mathrm{dom}\, A_D$ and $f_0 \in \ker T_{\max}$. Then $\{L^2(\partial\Omega), \Gamma_0, \Gamma_1\}$, where*

$$\Gamma_0 f = \iota_- \widetilde{\tau}_D f \quad and \quad \Gamma_1 f = -\iota_+ \tau_N f_D, \qquad f = f_D + f_0 \in \mathrm{dom}\, T_{\max},$$

is a boundary triplet for $(T_{\min})^* = T_{\max}$ such that

$$A_0 = A_D \qquad and \qquad A_1 = T_{\min} \widehat{+} \widehat{\mathfrak{N}}_0(T_{\max})$$

coincide with the Friedrichs extension and the Kreĭn–von Neumann extension of $T_{\min}$, respectively.

It is clear from Proposition 8.4.4 that for all $\lambda \in \rho(A_D)$ the γ-field and Weyl function corresponding to the boundary triplet $\{L^2(\partial\Omega), \Gamma_0, \Gamma_1\}$ in Corollary 8.5.4 have the form

$$\gamma(\lambda)\varphi = \bigl(I + \lambda(A_D - \lambda)^{-1}\bigr) f_0(\varphi), \quad \varphi \in L^2(\partial\Omega),$$

and

$$M(\lambda)\varphi = -\iota_+ \tau_N \lambda (A_D - \lambda)^{-1} f_0(\varphi), \quad \varphi \in L^2(\partial\Omega), \tag{8.5.11}$$

respectively, where $f_0(\varphi)$ is the unique element in $\mathfrak{N}_0(T_{\max})$ with the property that $\Gamma_0 f_0(\varphi) = \iota_- \widetilde{\tau}_D f_0(\varphi) = \varphi$.

The next proposition is a variant of Proposition 8.5.1 for nonnegative extensions.

Proposition 8.5.5. *Let $\Omega \subset \mathbb{R}^n$ be a bounded C^2-domain, assume that $V \geq 0$, and let $\{L^2(\partial\Omega), \Gamma_0, \Gamma_1\}$ be the boundary triplet for $(T_{\min})^* = T_{\max}$ from Corollary 8.5.4. Let*

$$A_\Theta = -\Delta + V,$$
$$\mathrm{dom}\, A_\Theta = \bigl\{ f \in \mathrm{dom}\, T_{\max} : \{\Gamma_0 f, \Gamma_1 f\} \in \Theta \bigr\},$$

be a self-adjoint extension of $T_{\min}$ in $L^2(\Omega)$ corresponding to a self-adjoint relation Θ in $L^2(\partial\Omega)$ as in (8.4.13). Then

$$A_\Theta \text{ is nonnegative} \quad \Leftrightarrow \quad \Theta \text{ is nonnegative.}$$

Proof. Note that the Weyl function M in (8.5.11) satisfies $M(0) = 0$ and that $T_{\min}$ is uniformly positive. Therefore, if A_Θ is a nonnegative self-adjoint extension of $T_{\min}$, then Proposition 5.5.6 with $x = 0$ shows that the self-adjoint relation Θ in $L^2(\partial\Omega)$ is nonnegative. Conversely, if Θ is a nonnegative self-adjoint relation in $L^2(\partial\Omega)$, then it follows from Corollary 5.5.15 and $A_1 = S_{K,0} \geq 0$ that A_Θ is a nonnegative self-adjoint extension of $T_{\min}$. $\qquad\Box$

In the nonnegative case the boundary mapping Λ in (8.5.4) is given by

$$\Lambda : \operatorname{dom} \mathsf{t}_{S_{K,0}} \to L^2(\partial\Omega), \quad f \mapsto \Lambda f = \iota_- \widetilde{\tau}_{\mathrm{D}} f_0, \tag{8.5.12}$$

where one has the direct sum decomposition

$$\operatorname{dom} \mathsf{t}_{S_{K,0}} = \mathfrak{N}_0(T_{\max}) + \operatorname{dom} \mathsf{t}_{A_{\mathrm{D}}} = \mathfrak{N}_0(T_{\max}) + H_0^1(\Omega),$$

and according to Lemma 8.5.2 $\{L^2(\partial\Omega), \Lambda\}$ is a boundary pair that is compatible with the boundary triplet $\{L^2(\partial\Omega), \Gamma_0, \Gamma_1\}$ in Corollary 8.5.4.

In the nonnegative case a description of the nonnegative extensions and their form domains is of special interest. In the present situation Corollary 5.6.18 reads as follows.

Corollary 8.5.6. *Let $\Omega \subset \mathbb{R}^n$ be a bounded C^2-domain, let A_{D} be the self-adjoint Dirichlet realization of $-\Delta + V$ with $V \geq 0$, let $\{L^2(\partial\Omega), \Gamma_0, \Gamma_1\}$ be the boundary triplet for $(T_{\min})^* = T_{\max}$ from Corollary 8.5.4, and let $\{L^2(\partial\Omega), \Lambda\}$ be the compatible boundary pair in (8.5.12). Then the formula*

$$\mathsf{t}_{A_\Theta}[f, g] = \mathsf{t}_{S_{K,0}}[f, g] + \left(\Theta_{\mathrm{op}}^{\frac{1}{2}} \iota_- \widetilde{\tau}_{\mathrm{D}} f_0, \Theta_{\mathrm{op}}^{\frac{1}{2}} \iota_- \widetilde{\tau}_{\mathrm{D}} g_0\right)_{L^2(\partial\Omega)},$$

$$\operatorname{dom} \mathsf{t}_{A_\Theta} = \left\{ f = f_0 + f_{\mathrm{F}} \in \operatorname{dom} \mathsf{t}_{S_{K,0}} : \iota_- \widetilde{\tau}_{\mathrm{D}} f_0 \in \operatorname{dom} \Theta_{\mathrm{op}}^{\frac{1}{2}} \right\},$$

establishes a one-to-one correspondence between all closed nonnegative forms t_{A_Θ} corresponding to nonnegative self-adjoint extension A_Θ of $T_{\min}$ in $L^2(\Omega)$ and all closed nonnegative forms ω_Θ corresponding to nonnegative self-adjoint relations Θ in $L^2(\partial\Omega)$.

8.6   Coupling of Schrödinger operators

The aim of this section is to interpret the *natural* self-adjoint Schrödinger operator

$$A = -\Delta + V, \qquad \operatorname{dom} A = H^2(\mathbb{R}^n), \tag{8.6.1}$$

in $L^2(\mathbb{R}^n)$ with a real potential $V \in L^\infty(\mathbb{R}^n)$ as a coupling of Schrödinger operators on a bounded C^2-domain and its complement, that is, A is identified as a self-adjoint extension of the orthogonal sum of the minimal Schrödinger operators on the subdomains and its resolvent is expressed in a Kreĭn type resolvent formula. The present treatment is a multidimensional variant of the discussion in Section 6.5 and is based on the abstract coupling construction in Section 4.6.

Let $\Omega_+ \subset \mathbb{R}^n$ be a bounded C^2-domain and let $\Omega_- := \mathbb{R}^n \setminus \overline{\Omega}_+$ be the corresponding exterior domain. Since $\mathcal{C} := \partial\Omega_- = \partial\Omega_+$ is C^2-smooth in the sense of Definition 8.2.1, the term C^2-domain will be used here for Ω_-, although Ω_- is unbounded. In the following the common boundary $\mathcal{C}$ is sometimes referred to as an interface, linking the two domains Ω_+ and Ω_-. Note that one has the identification

$$L^2(\mathbb{R}^n) = L^2(\Omega_+) \oplus L^2(\Omega_-). \tag{8.6.2}$$

Consider the Schrödinger operator $A = -\Delta + V$, $\operatorname{dom} A = H^2(\mathbb{R}^n)$, in (8.6.1) with $V \in L^\infty(\mathbb{R}^n)$ real. Since the Laplacian $-\Delta$ defined on $H^2(\mathbb{R}^n)$ is unitarily equivalent in $L^2(\mathbb{R}^n)$ via the Fourier transform to the maximal multiplication operator with the function $x \mapsto |x|^2$, it is clear that $-\Delta$, and hence A in (8.6.1), is self-adjoint in $L^2(\mathbb{R}^n)$. Moreover, for $f \in C_0^\infty(\mathbb{R}^n)$ integration by parts shows that

$$(Af, f)_{L^2(\mathbb{R}^n)} = (\nabla f, \nabla f)_{L^2(\mathbb{R}^n;\mathbb{C}^n)} + (Vf, f)_{L^2(\mathbb{R}^n)} \geq v_- \|f\|^2_{L^2(\mathbb{R}^n)},$$

where $v_- = \operatorname{essinf} V$. As $C_0^\infty(\mathbb{R}^n)$ is dense in $H^2(\mathbb{R}^n)$, this estimate extends to $H^2(\mathbb{R}^n)$. Therefore, A is semibounded from below and v_- is a lower bound.

The restriction of the real function $V \in L^\infty(\Omega)$ to $\Omega_\pm$ is denoted by $V_\pm$ and the same $\pm$-index notation will be used for the restriction $f_\pm \in L^2(\Omega_\pm)$ of an element $f \in L^2(\mathbb{R}^n)$. The minimal and maximal operator associated with $-\Delta + V_+$ in $L^2(\Omega_+)$ will be denoted by $T_{\min}^+$ and $T_{\max}^+$, respectively, and the self-adjoint Dirichlet realization in $L^2(\Omega_+)$ will be denoted by A_D^+; cf. Proposition 8.3.1, Proposition 8.3.2, and Theorem 8.3.4. For the minimal operator

$$T_{\min}^- = -\Delta + V_-, \qquad \operatorname{dom} T_{\min}^- = H_0^2(\Omega_-),$$

and the maximal operator

$$T_{\max}^- = -\Delta + V_-,$$
$$\operatorname{dom} T_{\max}^- = \left\{ f_- \in L^2(\Omega_-) : -\Delta f_- + V_- f_- \in L^2(\Omega_-) \right\},$$

on the unbounded C^2-domain one can show in the same way as in the proof of Proposition 8.3.1 that $(T_{\min}^-)^* = T_{\max}^-$ and $T_{\min}^- = (T_{\max}^-)^*$. Furthermore, since Ω_- has a compact C^2-smooth boundary, it follows by analogy to Theorem 8.3.4 that the self-adjoint Dirichlet realization A_D^- corresponding to the densely defined closed semibounded form

$$\mathfrak{t}_D^-[f_-, g_-] = (\nabla f_-, \nabla g_-)_{L^2(\Omega_-;\mathbb{C}^n)} + (V_- f_-, g_-)_{L^2(\Omega_-)}, \quad \operatorname{dom} \mathfrak{t}_D^- = H_0^1(\Omega_-),$$

via the first representation theorem (Theorem 5.1.18) is given by

$$A_D^- = -\Delta + V_-, \quad \operatorname{dom} A_D^- = \left\{ f_- \in H^2(\Omega_-) : \tau_D^- f_- = 0 \right\},$$

where τ_D^- denotes the Dirichlet trace operator on Ω_-; cf. (8.2.13). The operator A_D^- is semibounded from below and $v_- = \operatorname{essinf} V$ is a lower bound. In contrast to the Dirichlet operator A_D^+, the resolvent of A_D^- is not compact since Rellich's theorem is not valid on the unbounded domain Ω_-; cf. the proof of Proposition 8.3.2. Note also that the Dirichlet trace operator $\tau_D^- : H^2(\Omega_-) \to H^{3/2}(\mathbb{C})$ and Neumann trace operator $\tau_{\mathrm{N}}^- : H^2(\Omega_-) \to H^{1/2}(\mathcal{C})$ have the same mapping properties as on a bounded domain. Moreover, both trace operators admit continuous extensions to $\operatorname{dom} T_{\max}^-$ as in Theorem 8.3.9 and Theorem 8.3.10. With the identification (8.6.2) it is clear that the orthogonal sum

$$\widetilde{A}_{\mathrm{D}} = \begin{pmatrix} A_{\mathrm{D}}^+ & 0 \\ 0 & A_{\mathrm{D}}^- \end{pmatrix} \tag{8.6.3}$$

is a self-adjoint operator in $L^2(\mathbb{R}^n)$ with Dirichlet boundary conditions on $\mathcal{C}$. The goal of the following considerations is to identify the self-adjoint Schrödinger operator A in (8.6.1) as a self-adjoint extension of the orthogonal sum of the minimal operators $T_{\min}^{\pm}$ and to compare A with the orthogonal sum $\widetilde{A}_{\mathrm{D}}$ in (8.6.3) using a Kreĭn type resolvent formula.

From now on it is assumed that $\eta < \operatorname{essinf} V$ is fixed, so that, in particular, $\eta \in \rho(A_{\mathrm{D}}^+) \cap \rho(A_{\mathrm{D}}^-) \cap \mathbb{R}$. Consider the boundary triplet $\{L^2(\mathcal{C}), \Gamma_0^+, \Gamma_1^+\}$ for $T_{\max}^+$ in Theorem 8.4.1, that is,

$$\Gamma_0^+ f_+ = \iota_- \widetilde{\tau}_{\mathrm{D}}^+ f_+ \quad \text{and} \quad \Gamma_1^+ f_+ = -\iota_+ \tau_{\mathrm{N}}^+ f_{\mathrm{D},+}, \quad f_+ \in \operatorname{dom} T_{\max}^+,$$

where $f_+ = f_{\mathrm{D},+} + f_{\eta,+}$ with $f_{\mathrm{D},+} \in \operatorname{dom} A_{\mathrm{D}}^+$ and $f_{\eta,+} \in \mathfrak{N}_\eta(T_{\max}^+)$. In the same way as in the proof of Theorem 8.4.1 one verifies that $\{L^2(\mathcal{C}), \Gamma_0^-, \Gamma_1^-\}$, where

$$\Gamma_0^- f_- = \iota_- \widetilde{\tau}_{\mathrm{D}}^- f_- \quad \text{and} \quad \Gamma_1^- f_- = -\iota_+ \tau_{\mathrm{N}}^- f_{\mathrm{D},-}, \quad f_- \in \operatorname{dom} T_{\max}^-,$$

where $f_- = f_{\mathrm{D},-} + f_{\eta,-}$ with $f_{\mathrm{D},-} \in \operatorname{dom} A_{\mathrm{D}}^-$ and $f_{\eta,-} \in \mathfrak{N}_\eta(T_{\max}^-)$, is a boundary triplet for $T_{\max}^-$ such that $\operatorname{dom} A_{\mathrm{D}}^- = \ker \Gamma_0^-$. The γ-fields and Weyl functions in Proposition 8.4.4 corresponding to the boundary triplets $\{L^2(\mathcal{C}), \Gamma_0^\pm, \Gamma_1^\pm\}$ are denoted by $\gamma_\pm$ and $M_\pm$, respectively.

In analogy to Section 4.6, the orthogonal coupling of the boundary triplets $\{L^2(\mathcal{C}), \Gamma_0^+, \Gamma_1^+\}$ and $\{L^2(\mathcal{C}), \Gamma_0^-, \Gamma_1^-\}$ leads to the boundary triplet

$$\{L^2(\mathcal{C}) \oplus L^2(\mathcal{C}), \widetilde{\Gamma}_0, \widetilde{\Gamma}_1\} \tag{8.6.4}$$

for the orthogonal sum $T_{\max} := T_{\max}^+ \widehat{\oplus} T_{\max}^-$ of the maximal operator $T_{\max}^\pm$, where

$$\widetilde{\Gamma}_0 f = \begin{pmatrix} \Gamma_0^+ f_+ \\ \Gamma_0^- f_- \end{pmatrix} = \begin{pmatrix} \iota_- \widetilde{\tau}_{\mathrm{D}}^+ f_+ \\ \iota_- \widetilde{\tau}_{\mathrm{D}}^- f_- \end{pmatrix}, \quad f = \begin{pmatrix} f_+ \\ f_- \end{pmatrix}, \quad f_\pm \in \operatorname{dom} T_{\max}^\pm, \tag{8.6.5}$$

and

$$\widetilde{\Gamma}_1 f = \begin{pmatrix} \Gamma_1^+ f_+ \\ \Gamma_1^- f_- \end{pmatrix} = \begin{pmatrix} -\iota_+ \tau_{\mathrm{N}}^+ f_{\mathrm{D},+} \\ -\iota_+ \tau_{\mathrm{N}}^- f_{\mathrm{D},-} \end{pmatrix}, \quad f = \begin{pmatrix} f_+ \\ f_- \end{pmatrix}, \quad f_\pm \in \operatorname{dom} T_{\max}^\pm. \tag{8.6.6}$$

It is clear that

$$\operatorname{dom} A_{\mathrm{D}}^+ \times \operatorname{dom} A_{\mathrm{D}}^- = \ker \widetilde{\Gamma}_0,$$

and hence the self-adjoint operator in (8.6.3) coincides with the self-adjoint extension of $T_{\min} := T_{\min}^+ \,\widehat{\oplus}\, T_{\min}^-$ corresponding to the boundary condition $\ker \widetilde{\Gamma}_0$. Note also that the corresponding γ-field $\widetilde{\gamma}$ and Weyl function $\widetilde{M}$ have the form

$$\widetilde{\gamma}(\lambda) = \begin{pmatrix} \gamma_+(\lambda) & 0 \\ 0 & \gamma_-(\lambda) \end{pmatrix} \quad \text{and} \quad \widetilde{M}(\lambda) = \begin{pmatrix} M_+(\lambda) & 0 \\ 0 & M_-(\lambda) \end{pmatrix} \tag{8.6.7}$$

for $\lambda \in \rho(A_{\mathrm{D}}^+) \cap \rho(A_{\mathrm{D}}^-)$.

In Lemma 8.6.2 it will be shown that a certain relation $\widetilde{\Theta}$ is self-adjoint in $L^2(\mathcal{C}) \oplus L^2(\mathcal{C})$. This relation will turn out to be the boundary parameter that corresponds to the Schrödinger operator A in (8.6.1) via the boundary triplet (8.6.4). The following lemma on the sum of the Dirichlet-to-Neumann maps is preparatory.

Lemma 8.6.1. *Let $\eta < \operatorname{essinf} V$ and let $D_\pm(\lambda) : H^{3/2}(\mathcal{C}) \to H^{1/2}(\mathcal{C})$ be the Dirichlet-to-Neumann maps as in Definition 8.3.6 corresponding to $-\Delta + V_\pm$. Then for all $\lambda \in \mathbb{C} \setminus [\eta, \infty)$ the operator*

$$D_+(\lambda) + D_-(\lambda) : H^{3/2}(\mathcal{C}) \to H^{1/2}(\mathcal{C}) \tag{8.6.8}$$

is bijective.

Proof. First it will be shown that the operator in (8.6.8) is injective. Assume that $(D_+(\lambda) + D_-(\lambda))\varphi = 0$ for some $\varphi \in H^{3/2}(\mathcal{C})$ and some $\lambda \in \mathbb{C} \setminus [\eta, \infty)$. Then there exist $f_{\lambda,\pm} \in H^2(\Omega_\pm)$ such that

$$(-\Delta + V_\pm)f_{\lambda,\pm} = \lambda f_{\lambda,\pm}, \qquad \tau_{\mathrm{D}}^+ f_{\lambda,+} = \tau_{\mathrm{D}}^- f_{\lambda,-} = \varphi, \tag{8.6.9}$$

and

$$0 = \bigl(D_+(\lambda) + D_-(\lambda)\bigr)\varphi = \bigl(D_+(\lambda) + D_-(\lambda)\bigr)\tau_{\mathrm{D}}^\pm f_{\lambda,\pm} = \tau_{\mathrm{N}}^+ f_{\lambda,+} + \tau_{\mathrm{N}}^- f_{\lambda,-}.$$

As $\tau_{\mathrm{D}}^+ f_{\lambda,+} = \tau_{\mathrm{D}}^- f_{\lambda,-}$ and $\tau_{\mathrm{N}}^+ f_{\lambda,+} = -\tau_{\mathrm{N}}^- f_{\lambda,-}$ this implies that

$$f_\lambda = \begin{pmatrix} f_{\lambda,+} \\ f_{\lambda,-} \end{pmatrix} \in H^2(\mathbb{R}^n). \tag{8.6.10}$$

In fact, for each $h = (h_+, h_-)^\top \in \operatorname{dom} A = H^2(\mathbb{R}^n)$ one also has $\tau_{\mathrm{D}}^+ h_+ = \tau_{\mathrm{D}}^- h_-$ and $\tau_{\mathrm{N}}^+ h_+ = -\tau_{\mathrm{N}}^- h_-$ (note that the different signs are due to the fact that the Neumann trace on each domain is taken with respect to the outward normal vector) and hence

$$
\begin{aligned}
(Ah, & f_\lambda)_{L^2(\mathbb{R}^n)} - (h, T_{\max} f_\lambda)_{L^2(\mathbb{R}^n)} \\
&= (T_{\max}^+ h_+, f_{\lambda,+})_{L^2(\Omega_+)} - (h_+, T_{\max}^+ f_{\lambda,+})_{L^2(\Omega_+)} \\
&\qquad + (T_{\max}^- h_-, f_{\lambda,-})_{L^2(\Omega_-)} - (h_-, T_{\max}^- f_{\lambda,-})_{L^2(\Omega_-)} \\
&= (\tau_{\mathrm{D}}^+ h_+, \tau_{\mathrm{N}}^+ f_{\lambda,+})_{L^2(\mathcal{C})} - (\tau_{\mathrm{N}}^+ h_+, \tau_{\mathrm{D}}^+ f_{\lambda,+})_{L^2(\mathcal{C})} \\
&\qquad + (\tau_{\mathrm{D}}^- h_-, \tau_{\mathrm{N}}^- f_{\lambda,-})_{L^2(\mathcal{C})} - (\tau_{\mathrm{N}}^- h_-, \tau_{\mathrm{D}}^- f_{\lambda,-})_{L^2(\mathcal{C})} \\
&= 0.
\end{aligned}
$$

As the operator A is self-adjoint this shows, in particular, $f_\lambda \in \mathrm{dom}\, A$ and hence (8.6.10) holds. Furthermore, from (8.6.9) it follows that

$$Af_\lambda = (-\Delta + V)f_\lambda = \lambda f_\lambda.$$

Since $\sigma(A) \subset [v_-, \infty) \subset [\eta, \infty)$ and $\lambda \in \mathbb{C} \setminus [\eta, \infty)$, this implies that $f_\lambda = 0$ and hence $\varphi = \tau_D^\pm f_{\lambda,\pm} = 0$. Thus, it has been shown that the operator in (8.6.8) is injective for all $\lambda \in \mathbb{C} \setminus [\eta, \infty)$.

Next it will be shown that the operator in (8.6.8) is surjective. For this consider the space

$$H_{\mathcal{C}}(\mathbb{R}^n) := \left\{ f = \begin{pmatrix} f_+ \\ f_- \end{pmatrix} : f_\pm \in H^2(\Omega_\pm),\ \tau_D^+ f_+ = \tau_D^- f_- \right\}$$

and observe that as a consequence of (8.2.12) the mapping

$$\tau_N^{\mathcal{C}} : H_{\mathcal{C}}(\mathbb{R}^n) \to H^{1/2}(\mathcal{C}), \qquad f \mapsto \tau_N^{\mathcal{C}} f := \tau_N^+ f_+ + \tau_N^- f_-,$$

is surjective. For $\lambda \in \mathbb{C} \setminus [\eta, \infty)$ it will be shown now that the direct sum decomposition

$$H_{\mathcal{C}}(\mathbb{R}^n) = \mathrm{dom}\, A + \left\{ f_\lambda = \begin{pmatrix} f_{\lambda,+} \\ f_{\lambda,-} \end{pmatrix} : \begin{array}{l} f_{\lambda,\pm} \in H^2(\Omega_\pm),\ \tau_D^+ f_{\lambda,+} = \tau_D^- f_{\lambda,-}, \\ (-\Delta + V_\pm) f_{\lambda,\pm} = \lambda f_{\lambda,\pm} \end{array} \right\}$$

holds. In fact, the inclusion ($\supset$) is clear since $\mathrm{dom}\, A = H^2(\mathbb{R}^n)$ and the second summand on the right-hand side is obviously contained in $H_{\mathcal{C}}(\mathbb{R}^n)$. The inclusion ($\subset$) follows from Theorem 1.7.1 applied to $T = -\Delta + V$, $\mathrm{dom}\, T = H_{\mathcal{C}}(\mathbb{R}^n)$, after observing that the space

$$\left\{ f_\lambda = \begin{pmatrix} f_{\lambda,+} \\ f_{\lambda,-} \end{pmatrix} : \begin{array}{l} f_{\lambda,\pm} \in H^2(\Omega_\pm),\ \tau_D^+ f_{\lambda,+} = \tau_D^- f_{\lambda,-}, \\ (-\Delta + V_\pm) f_{\lambda,\pm} = \lambda f_{\lambda,\pm} \end{array} \right\} \tag{8.6.11}$$

coincides with $\mathfrak{N}_\lambda(T) = \ker(T - \lambda)$ and $\lambda \in \rho(A)$. Note also that $\lambda \in \rho(A)$ implies that the sum is direct.

Next observe that for $f \in \mathrm{dom}\, A$ one has $\tau_N^{\mathcal{C}} f = 0$ and hence also the restriction of $\tau_N^{\mathcal{C}}$ to the space (8.6.11) maps onto $H^{1/2}(\mathcal{C})$. Therefore, for $\psi \in H^{1/2}(\mathcal{C})$ there exists $f_\lambda = (f_{\lambda,+}, f_{\lambda,-})^\top$ such that $f_{\lambda,\pm} \in H^2(\Omega_\pm)$, $(-\Delta + V_\pm) f_{\lambda,\pm} = \lambda f_{\lambda,\pm}$,

$$\tau_D^+ f_{\lambda,+} = \tau_D^- f_{\lambda,-} =: \varphi \in H^{3/2}(\mathcal{C}) \quad \text{and} \quad \tau_N^{\mathcal{C}} f_\lambda = \tau_N^+ f_{\lambda,+} + \tau_N^- f_{\lambda,-} = \psi.$$

It follows that

$$\big(D_+(\lambda) + D_-(\lambda)\big)\varphi = D_+(\lambda)\tau_D^+ f_{\lambda,+} + D_-(\lambda)\tau_D^- f_{\lambda,-} = \tau_N^+ f_{\lambda,+} + \tau_N^- f_{\lambda,-} = \psi,$$

and hence the operator in (8.6.8) is surjective.

Consequently, it has been shown that (8.6.8) is a bijective operator for all $\lambda \in \mathbb{C} \setminus [\eta, \infty)$. $\qquad \square$

Lemma 8.6.1 will be used to prove the following lemma on the self-adjointness of a particular relation $\widetilde{\Theta}$ in $L^2(\mathcal{C}) \oplus L^2(\mathcal{C})$.

Lemma 8.6.2. *Let $\eta <$ essinf V and let $D_\pm(\eta) : H^{3/2}(\mathcal{C}) \to H^{1/2}(\mathcal{C})$ be the Dirichlet-to-Neumann maps as in Definition 8.3.6 corresponding to $-\Delta + V_\pm$. Then the relation*

$$\widetilde{\Theta} = \left\{ \left\{ \begin{pmatrix} \xi \\ \xi \end{pmatrix}, \begin{pmatrix} \varphi \\ \psi \end{pmatrix} \right\} : \xi \in H^2(\mathcal{C}), \, \varphi + \psi = \iota_+\big(D_+(\eta) + D_-(\eta)\big)\iota_-^{-1}\xi \right\}$$

is self-adjoint in $L^2(\mathcal{C}) \oplus L^2(\mathcal{C})$.

Proof. Recall first from Example 8.4.7 that $\iota_+ D_+(\eta)\iota_-^{-1}$ and $\iota_+ D_-(\eta)\iota_-^{-1}$ are both unbounded bijective self-adjoint operators in $L^2(\mathcal{C})$ with domain $H^2(\mathcal{C})$. Since $\eta <$ essinf V, one also sees from (8.4.18) that these operators are nonnegative. It follows, in particular, that $\iota_+\big(D_+(\eta) + D_-(\eta)\big)\iota_-^{-1}$ is a symmetric operator in $L^2(\mathcal{C})$. Since

$$D_+(\eta) + D_-(\eta) : H^{3/2}(\mathcal{C}) \to H^{1/2}(\mathcal{C})$$

is bijective by Lemma 8.6.1 and the restricted operators $\iota_-^{-1} : H^2(\mathcal{C}) \to H^{3/2}(\mathcal{C})$ and $\iota_+ : H^{1/2}(\mathcal{C}) \to L^2(\mathcal{C})$ are also bijective, one concludes that

$$\iota_+\big(D_+(\eta) + D_-(\eta)\big)\iota_-^{-1} \tag{8.6.12}$$

is a uniformly positive self-adjoint operator in $L^2(\mathcal{C})$ defined on $H^2(\mathcal{C})$.

To show that $\widetilde{\Theta} \subset \widetilde{\Theta}^*$, consider two arbitrary elements

$$\left\{ \begin{pmatrix} \xi \\ \xi \end{pmatrix}, \begin{pmatrix} \varphi \\ \psi \end{pmatrix} \right\}, \left\{ \begin{pmatrix} \xi' \\ \xi' \end{pmatrix}, \begin{pmatrix} \varphi' \\ \psi' \end{pmatrix} \right\} \in \widetilde{\Theta},$$

that is, $\xi, \xi' \in H^2(\mathcal{C})$,

$$\varphi + \psi = \iota_+\big(D_+(\eta) + D_-(\eta)\big)\iota_-^{-1}\xi \quad \text{and} \quad \varphi' + \psi' = \iota_+\big(D_+(\eta) + D_-(\eta)\big)\iota_-^{-1}\xi'.$$

Then one computes

$$\left(\begin{pmatrix} \xi \\ \xi \end{pmatrix}, \begin{pmatrix} \varphi' \\ \psi' \end{pmatrix} \right)_{(L^2(\mathcal{C}))^2} - \left(\begin{pmatrix} \varphi \\ \psi \end{pmatrix}, \begin{pmatrix} \xi' \\ \xi' \end{pmatrix} \right)_{(L^2(\mathcal{C}))^2}$$

$$= (\xi, \varphi' + \psi')_{L^2(\mathcal{C})} - (\varphi + \psi, \xi')_{L^2(\mathcal{C})}$$

$$= \big(\xi, \iota_+\big(D_+(\eta) + D_-(\eta)\big)\iota_-^{-1}\xi'\big)_{L^2(\mathcal{C})} - \big(\iota_+\big(D_+(\eta) + D_-(\eta)\big)\iota_-^{-1}\xi, \xi'\big)_{L^2(\mathcal{C})}$$

$$= 0,$$

where in the last step it was used that (8.6.12) is a symmetric operator in $L^2(\mathcal{C})$. Hence, the relation $\widetilde{\Theta}$ is symmetric in $L^2(\mathcal{C})$. For the opposite inclusion $\widetilde{\Theta}^* \subset \widetilde{\Theta}$ consider an element

$$\left\{ \begin{pmatrix} \alpha \\ \beta \end{pmatrix}, \begin{pmatrix} \gamma \\ \delta \end{pmatrix} \right\} \in \widetilde{\Theta}^*, \tag{8.6.13}$$

that is,

$$\left(\begin{pmatrix}\alpha\\\beta\end{pmatrix},\begin{pmatrix}\varphi\\\psi\end{pmatrix}\right)_{(L^2(\mathcal{C}))^2}=\left(\begin{pmatrix}\gamma\\\delta\end{pmatrix},\begin{pmatrix}\xi\\\xi\end{pmatrix}\right)_{(L^2(\mathcal{C}))^2}\tag{8.6.14}$$

holds for all

$$\left\{\begin{pmatrix}\xi\\\xi\end{pmatrix},\begin{pmatrix}\varphi\\\psi\end{pmatrix}\right\}\in\widetilde{\Theta}.$$

The special choice $\xi=0$ yields $\varphi+\psi=0$ by the definition of $\widetilde{\Theta}$ and hence $(\alpha-\beta,\varphi)_{L^2(\mathcal{C})}=0$ for all $\varphi\in L^2(\mathcal{C})$. This shows $\alpha=\beta$ and therefore (8.6.14) becomes

$$\left(\alpha,\iota_+\big(D_+(\eta)+D_-(\eta)\big)\iota_-^{-1}\xi\right)_{L^2(\mathcal{C})}=(\alpha,\varphi+\psi)_{L^2(\mathcal{C})}=(\gamma+\delta,\xi)_{L^2(\mathcal{C})}$$

for all $\xi\in H^2(\mathcal{C})$. Since $\iota_+(D_+(\eta)+D_-(\eta))\iota_-^{-1}$ is a self-adjoint operator in $L^2(\mathcal{C})$ defined on $H^2(\mathcal{C})$, it follows that $\alpha\in H^2(\mathcal{C})$ and

$$\iota_+\big(D_+(\eta)+D_-(\eta)\big)\iota_-^{-1}\alpha=\gamma+\delta.$$

This implies that the element in (8.6.13) belongs to $\widetilde{\Theta}$. Thus, $\widetilde{\Theta}$ is a self-adjoint relation in $L^2(\mathcal{C})\oplus L^2(\mathcal{C})$. $\qquad\square$

The following theorem is the main result in this section. It turns out that the self-adjoint operator corresponding to $\widetilde{\Theta}$ in Lemma 8.6.2 coincides with the Schrödinger operator A.

Theorem 8.6.3. *Let* $\{L^2(\mathcal{C})\oplus L^2(\mathcal{C}),\widetilde{\Gamma}_0,\widetilde{\Gamma}_1\}$ *be the boundary triplet for* $T_{\max}^+\,\widehat{\oplus}\,T_{\max}^-$ *from* (8.6.4) *with* γ-*field* $\widetilde{\gamma}$, *let* $\widetilde{\Theta}$ *be the self-adjoint relation in Lemma 8.6.2, and let* $D_\pm(\lambda)$ *be the Dirichlet-to-Neumann maps corresponding to* $-\Delta+V_\pm$. *Then the self-adjoint operator* $\widetilde{A}_{\widetilde{\Theta}}$ *corresponding to the parameter* $\widetilde{\Theta}$ *coincides with the Schrödinger operator* A *in* (8.6.1) *and for all* $\lambda\in\mathbb{C}\setminus[\eta,\infty)$ *one has the resolvent formula*

$$(A-\lambda)^{-1}=(\widetilde{A}_{\mathrm{D}}-\lambda)^{-1}+\widetilde{\gamma}(\lambda)\widetilde{\Lambda}(\lambda)\widetilde{\gamma}(\bar{\lambda})^*,$$

where $\widetilde{\Lambda}(\lambda)\in\mathbf{B}(L^2(\mathcal{C})\oplus L^2(\mathcal{C}))$ *has the form*

$$\widetilde{\Lambda}(\lambda)=\begin{pmatrix}\iota_-(D_+(\lambda)+D_-(\lambda))^{-1}\iota_+^{-1}&\iota_-(D_+(\lambda)+D_-(\lambda))^{-1}\iota_+^{-1}\\\iota_-(D_+(\lambda)+D_-(\lambda))^{-1}\iota_+^{-1}&\iota_-(D_+(\lambda)+D_-(\lambda))^{-1}\iota_+^{-1}\end{pmatrix}.$$

Proof. First it will be shown that the self-adjoint extension $\widetilde{A}_{\widetilde{\Theta}}$ and the self-adjoint Schrödinger operator A in (8.6.1) coincide. Since both operators are self-adjoint, it suffices to verify the inclusion $A\subset\widetilde{A}_{\widetilde{\Theta}}$. For this, consider $f\in\operatorname{dom}A=H^2(\mathbb{R}^n)$ and note that $f=(f_+,f_-)^\top$ satisfies $\tau_{\mathrm{D}}^+f_+=\tau_{\mathrm{D}}^-f_-$ and $\tau_{\mathrm{N}}^+f_+=-\tau_{\mathrm{N}}^-f_-$. It will be shown that $\{\widetilde{\Gamma}_0f,\widetilde{\Gamma}_1f\}\in\widetilde{\Theta}$. By the definition of the boundary mappings $\widetilde{\Gamma}_0$ and $\widetilde{\Gamma}_1$ in (8.6.5)–(8.6.6), one has

$$\widetilde{\Gamma}_0f=\begin{pmatrix}\iota_-\widetilde{\tau}_{\mathrm{D}}^+f_+\\\iota_-\widetilde{\tau}_{\mathrm{D}}^-f_-\end{pmatrix}\quad\text{and}\quad\widetilde{\Gamma}_1f=\begin{pmatrix}-\iota_+\tau_{\mathrm{N}}^+f_{\mathrm{D},+}\\-\iota_+\tau_{\mathrm{N}}^-f_{\mathrm{D},-}\end{pmatrix}=:\begin{pmatrix}\varphi\\\psi\end{pmatrix}$$

and, as $f \in H^2(\mathbb{R}^n)$, it follows that

$$\xi := \iota_- \tau_{\mathrm{D}}^+ f_+ = \iota_- \widetilde{\tau}_{\mathrm{D}}^+ f_+ = \iota_- \widetilde{\tau}_{\mathrm{D}}^- f_- = \iota_- \tau_{\mathrm{D}}^- f_- \in H^2(\mathbb{C}).$$

Since $f_\pm = f_{\mathrm{D},\pm} + f_{\eta,\pm}$ with $f_{\mathrm{D},\pm} \in \operatorname{dom} A_{\mathrm{D}}^\pm$ and $f_{\eta,\pm} \in \mathfrak{N}_\eta(T_{\max}^\pm)$, one has $\tau_{\mathrm{D}}^\pm f_\pm = \tau_{\mathrm{D}}^\pm f_{\eta,\pm}$ and one concludes that

$$
\begin{aligned}
\iota_+ \big(D_+(\eta) + D_-(\eta) \big) \iota_-^{-1} \xi
&= \iota_+ \big(D_+(\eta) \tau_{\mathrm{D}}^+ f_{\eta,+} + D_-(\eta) \tau_{\mathrm{D}}^- f_{\eta,-} \big) \\
&= \iota_+ \big(\tau_{\mathrm{N}}^+ f_{\eta,+} + \tau_{\mathrm{N}}^- f_{\eta,-} \big) \\
&= \iota_+ \big(\tau_{\mathrm{N}}^+ f_+ + \tau_{\mathrm{N}}^- f_- - \tau_{\mathrm{N}}^+ f_{\mathrm{D},+} - \tau_{\mathrm{N}}^- f_{\mathrm{D},-} \big) \\
&= -\iota_+ \tau_{\mathrm{N}}^+ f_{\mathrm{D},+} - \iota_+ \tau_{\mathrm{N}}^- f_{\mathrm{D},-} \\
&= \varphi + \psi,
\end{aligned}
$$

where the property $\tau_{\mathrm{N}}^+ f_+ = -\tau_{\mathrm{N}}^- f_-$ for $f \in \operatorname{dom} A$ was used. These considerations imply $\{\widetilde{\Gamma}_0 f, \widetilde{\Gamma}_1 f\} \in \widetilde{\Theta}$ and thus $f \in \operatorname{dom} \widetilde{A}_{\widetilde{\Theta}}$. Therefore, $\operatorname{dom} A = H^2(\mathbb{R}^n)$ is contained in $\operatorname{dom} \widetilde{A}_{\widetilde{\Theta}}$, and since both operators are self-adjoint, it follows that they coincide, that is, $A = \widetilde{A}_{\widetilde{\Theta}}$.

As a consequence of Theorem 2.6.1 one has for $\lambda \in \rho(A) \cap \rho(\widetilde{A}_{\mathrm{D}})$ that

$$(A - \lambda)^{-1} = (\widetilde{A}_{\mathrm{D}} - \lambda)^{-1} + \widetilde{\gamma}(\lambda) \big(\widetilde{\Theta} - \widetilde{M}(\lambda) \big)^{-1} \widetilde{\gamma}(\bar{\lambda})^*,$$

where $\widetilde{\gamma}$ and $\widetilde{M}$ are the γ-field and Weyl function, respectively, of the boundary triplet $\{L^2(\mathbb{C}) \oplus L^2(\mathbb{C}), \widetilde{\Gamma}_0, \widetilde{\Gamma}_1\}$ in (8.6.7). Here it is also clear from Theorem 2.6.1 that

$$\big(\widetilde{\Theta} - \widetilde{M}(\lambda) \big)^{-1} \in \mathbf{B}(L^2(\mathbb{C}) \oplus L^2(\mathbb{C})), \qquad \lambda \in \rho(A) \cap \rho(\widetilde{A}_{\mathrm{D}}).$$

From now consider only $\lambda \in \mathbb{C} \setminus [\eta, \infty)$. It follows from Lemma 8.6.2 and (8.6.7) that

$$
\big(\widetilde{\Theta} - \widetilde{M}(\lambda) \big)^{-1}
= \left\{ \left\{ \begin{pmatrix} \varphi - M_+(\lambda)\xi \\ \psi - M_-(\lambda)\xi \end{pmatrix}, \begin{pmatrix} \xi \\ \xi \end{pmatrix} \right\} : \begin{array}{l} \xi \in H^2(\mathbb{C}), \\ \varphi + \psi = \iota_+ \big(D_+(\eta) + D_-(\eta) \big) \iota_-^{-1} \xi \end{array} \right\},
$$

and setting $\vartheta_1 = \varphi - M_+(\lambda)\xi$ and $\vartheta_2 = \psi - M_-(\lambda)\xi$ one obtains

$$
\begin{aligned}
\vartheta_1 + \vartheta_2 &= \varphi + \psi - M_+(\lambda)\xi - M_-(\lambda)\xi \\
&= \iota_+ \big(D_+(\eta) + D_-(\eta) \big) \iota_-^{-1} \xi - M_+(\lambda)\xi - M_-(\lambda)\xi.
\end{aligned}
$$

Since $M_\pm(\lambda)\xi = \iota_+ \big(D_\pm(\eta) - D_\pm(\lambda) \big) \iota_-^{-1} \xi$ for $\xi \in H^2(\mathbb{C})$ by Lemma 8.4.5, it follows that

$$\vartheta_1 + \vartheta_2 = \iota_+ \big(D_+(\lambda) + D_-(\lambda) \big) \iota_-^{-1} \xi.$$

Lemma 8.6.1 implies that $\iota_+ \big(D_+(\lambda) + D_-(\lambda) \big) \iota_-^{-1}$ is a bijective operator in $L^2(\mathbb{C})$ for $\lambda \in \mathbb{C} \setminus [\eta, \infty)$ and hence

$$\iota_- \big(D_+(\lambda) + D_-(\lambda) \big)^{-1} \iota_+^{-1} \vartheta_1 + \iota_- \big(D_+(\lambda) + D_-(\lambda) \big)^{-1} \iota_+^{-1} \vartheta_2 = \xi.$$

Therefore, one has

$$(\widetilde{\Theta} - \widetilde{M}(\lambda))^{-1} = \begin{pmatrix} \iota_-(D_+(\lambda) + D_-(\lambda))^{-1}\iota_+^{-1} & \iota_-(D_+(\lambda) + D_-(\lambda))^{-1}\iota_+^{-1} \\ \iota_-(D_+(\lambda) + D_-(\lambda))^{-1}\iota_+^{-1} & \iota_-(D_+(\lambda) + D_-(\lambda))^{-1}\iota_+^{-1} \end{pmatrix}.$$

This completes the proof of Theorem 8.6.3. □

Finally, the boundary triplet in (8.6.4) is modified in the same way as in Proposition 4.6.4 to interpret the Schrödinger operator A as the self-adjoint extension corresponding to the boundary mapping $\widehat{\Gamma}_0$. More precisely, the boundary triplets $\{L^2(\mathcal{C}), \Gamma_0^+, \Gamma_1^+\}$ and $\{L^2(\mathcal{C}), \Gamma_0^-, \Gamma_1^-\}$ lead to the boundary triplet

$$\{L^2(\mathcal{C}) \oplus L^2(\mathcal{C}), \widehat{\Gamma}_0, \widehat{\Gamma}_1\} \tag{8.6.15}$$

for $T_{\max} = T_{\max}^+ \,\widehat{\oplus}\, T_{\max}^-$, where

$$\widehat{\Gamma}_0 f = \begin{pmatrix} -\Gamma_1^+ f_+ - \Gamma_1^- f_- \\ \Gamma_0^+ f_+ - \Gamma_0^- f_- \end{pmatrix} = \begin{pmatrix} \iota_+(\tau_N^+ f_{D,+} + \tau_N^- f_{D,-}) \\ \iota_-(\widetilde{\tau}_D^+ f_+ - \widetilde{\tau}_D^- f_-) \end{pmatrix}$$

and

$$\widehat{\Gamma}_1 f = \begin{pmatrix} \Gamma_0^+ f_+ \\ -\Gamma_1^- f_- \end{pmatrix} = \begin{pmatrix} \iota_- \widetilde{\tau}_D^+ f_+ \\ \iota_+ \tau_N^- f_{D,-} \end{pmatrix}$$

for $f = (f_+, f_-)^\top$ with $f_\pm \in \operatorname{dom} T_{\max}^\pm$. It follows from Proposition 4.6.4 that the Schrödinger operator $A = -\Delta + V$ in (8.6.1) coincides with the self-adjoint extension defined on $\ker \widehat{\Gamma}_0$ and that the Weyl function corresponding to the boundary triplet in (8.6.15) is given by

$$\widehat{M}(\lambda) = -\begin{pmatrix} M_+(\lambda) & -I \\ -I & -M_-(\lambda)^{-1} \end{pmatrix}^{-1}, \qquad \lambda \in \mathbb{C} \setminus \mathbb{R},$$

where $M_\pm(\lambda) = \iota_+(\widetilde{D}_\pm(\eta) - \widetilde{D}_\pm(\lambda))\iota_-^{-1}$ is the Weyl function corresponding to the boundary triplet $\{L^2(\mathcal{C}), \Gamma_0^\pm, \Gamma_1^\pm\}$; cf. Proposition 8.4.4 and Lemma 8.4.5. In particular, the results in Section 3.5 and Section 3.6 can be used to describe the isolated and embedded eigenvalues, continuous, and absolutely continuous spectrum of A with the help of the limit properties of the Dirichlet-to-Neumann maps $\widetilde{D}_\pm$. For this, however, one has to ensure that the underlying minimal operator $T_{\min} = T_{\min}^+ \,\widehat{\oplus}\, T_{\min}^-$ is simple, which follows from Proposition 8.3.13 and [120, Proposition 2.2].

8.7 Bounded Lipschitz domains

In this last section Schrödinger operators $-\Delta + V$ with a real function $V \in L^\infty(\Omega)$ on bounded Lipschitz domains are briefly discussed. This situation is more general than the setting of bounded C^2-domains treated in the previous sections. The main objective here is to highlight the differences to the C^2-case and to indicate which methods have to be adapted in order to obtain results of similar nature as above.

The notions of a Lipschitz hypograph and a bounded Lipschitz domain are defined in the same way as C^2-hypographs and bounded C^2-domains in Section 8.2. More precisely, for a Lipschitz continuous function $\phi : \mathbb{R}^{n-1} \to \mathbb{R}$ the domain

$$\Omega_\phi := \left\{ (x', x_n)^\top \in \mathbb{R}^n : x_n < \phi(x') \right\}$$

is called a *Lipschitz hypograph* with boundary $\partial\Omega$. The surface integral and surface measure on $\partial\Omega_\phi$ are defined in the same way as in (8.2.4), and this leads to the L^2-space $L^2(\partial\Omega_\phi)$ on $\partial\Omega_\phi$. For $s \in [0,1]$ define the Sobolev space of order s on $\partial\Omega_\phi$ by

$$H^s(\partial\Omega_\phi) := \left\{ h \in L^2(\partial\Omega_\phi) : x' \mapsto h(x', \phi(x')) \in H^s(\mathbb{R}^{n-1}) \right\}$$

and equip $H^s(\partial\Omega_\phi)$ with the corresponding scalar product (8.2.6).

Definition 8.7.1. A bounded nonempty open subset $\Omega \subset \mathbb{R}^n$ is called a *Lipschitz domain* if there exist open sets $U_1, \ldots, U_l \subset \mathbb{R}^n$ and (possibly up to rotations of coordinates) Lipschitz hypographs $\Omega_1, \ldots, \Omega_l \subset \mathbb{R}^n$, such that

$$\partial\Omega \subset \bigcup_{j=1}^{l} U_j \quad \text{and} \quad \Omega \cap U_j = \Omega_j \cap U_j, \quad j = 1, \ldots, l.$$

For a bounded Lipschitz domain $\Omega \subset \mathbb{R}^n$ the boundary $\partial\Omega \subset \mathbb{R}^n$ is compact. Using a partition of unity subordinate to the open cover $\{U_j\}$ of $\partial\Omega$ one defines the surface integral, surface measure, and the L^2-space $L^2(\partial\Omega)$ in the same way as in Section 8.2. The Sobolev space $H^s(\partial\Omega)$ for $s \in [0,1]$ is then defined by

$$H^s(\partial\Omega) := \left\{ h \in L^2(\partial\Omega) : \eta_j h \in H^s(\partial\Omega_j), j = 1, \ldots, l \right\}$$

and equipped with the corresponding Hilbert space scalar product (8.2.7). It follows that $H^s(\partial\Omega)$, $s \in [0,1]$, is densely and continuously embedded in $L^2(\partial\Omega)$, and the embedding $H^t(\partial\Omega) \hookrightarrow H^s(\partial\Omega)$ is compact for $s < t \leq 1$. As in Section 8.2, the spaces $H^s(\partial\Omega)$, $s \in [0,1]$, can be defined in an equivalent way via interpolation. The dual space of the antilinear continuous functionals on $H^s(\partial\Omega)$ is denoted by $H^{-s}(\partial\Omega)$, $s \in [0,1]$.

For a bounded Lipschitz domain Ω define the spaces

$$H^s_\Delta(\Omega) := \left\{ f \in H^s(\Omega) : \Delta f \in L^2(\Omega) \right\}, \qquad s \geq 0,$$

and equip them with the Hilbert space scalar product

$$(f,g)_{H^s_\Delta(\Omega)} := (f,g)_{H^s(\Omega)} + (\Delta f, \Delta g)_{L^2(\Omega)}, \qquad f,g \in H^s_\Delta(\Omega). \tag{8.7.1}$$

It is clear that $H^s(\Omega) = H^s_\Delta(\Omega)$ for $s \geq 2$ and that $H^0_\Delta(\Omega) = \operatorname{dom} T_{\max}$ for $s = 0$, with (8.7.1) as the graph norm; cf. (8.3.3). The unit normal vector field pointing outwards on $\partial\Omega$ will again be denoted by ν. It is known that the Dirichlet trace mapping $C^\infty(\overline{\Omega}) \ni f \mapsto f|_{\partial\Omega}$ extends by continuity to a continuous surjective mapping

$$\tau_{\mathrm{D}} : H^s_\Delta(\Omega) \to H^{s-1/2}(\partial\Omega), \qquad \frac{1}{2} \le s \le \frac{3}{2},$$

and that the Neumann trace mapping $C^\infty(\overline{\Omega}) \ni f \mapsto \nu \cdot \nabla f|_{\partial\Omega}$ extends by continuity to a continuous surjective mapping

$$\tau_{\mathrm{N}} : H^s_\Delta(\Omega) \to H^{s-3/2}(\partial\Omega), \qquad \frac{1}{2} \le s \le \frac{3}{2};$$

cf. [92, 326]. For the present purposes it is particularly useful to note that the mappings

$$\tau_{\mathrm{D}} : H^{3/2}_\Delta(\Omega) \to H^1(\partial\Omega) \quad \text{and} \quad \tau_{\mathrm{N}} : H^{3/2}_\Delta(\Omega) \to L^2(\partial\Omega) \tag{8.7.2}$$

are both continuous and surjective. Furthermore, the first and second Green identities remain true in the natural form, that is,

$$(-\Delta f, g)_{L^2(\Omega)} = (\nabla f, \nabla g)_{L^2(\Omega;\mathbb{C}^n)} - \big\langle \tau_{\mathrm{N}} f, \tau_{\mathrm{D}} g \big\rangle_{H^{-1/2}(\partial\Omega) \times H^{1/2}(\partial\Omega)}$$

and

$$(-\Delta f, g)_{L^2(\Omega)} - (f, -\Delta g)_{L^2(\Omega)}$$
$$= \big\langle \tau_{\mathrm{D}} f, \tau_{\mathrm{N}} g \big\rangle_{H^{1/2}(\partial\Omega) \times H^{-1/2}(\partial\Omega)} - \big\langle \tau_{\mathrm{N}} f, \tau_{\mathrm{D}} g \big\rangle_{H^{-1/2}(\partial\Omega) \times H^{1/2}(\partial\Omega)}$$

hold for all $f, g \in H^1_\Delta(\Omega)$.

The minimal operator $T_{\min}$ and maximal operator $T_{\max}$ associated with $-\Delta + V$ on a bounded Lipschitz domain are defined in exactly the same way as in the beginning of Section 8.3. The assertions $T^*_{\min} = T_{\max}$ and $T_{\min} = T^*_{\max}$ in Proposition 8.3.1 remain valid in the present situation. Furthermore, the Dirichlet realization A_{D} and Neumann realization A_{N} of $-\Delta + V$ are defined as in Section 8.3, and their properties are the same as in Proposition 8.3.2 and Proposition 8.3.3. The first remarkable and substantial difference for Schrödinger operators on a bounded Lipschitz domain appears in connection with the regularity of the domains of A_{D} and A_{N} when comparing with Theorem 8.3.4. In the present case one has the following regularity result from [431, 432], see also [92, 323].

Theorem 8.7.2. *Assume that $\Omega \subset \mathbb{R}^n$ is a bounded Lipschitz domain. Then one has*

$$A_{\mathrm{D}} f = -\Delta f + V f, \quad \mathrm{dom}\, A_{\mathrm{D}} = \{f \in H^{3/2}_\Delta(\Omega) : \tau_{\mathrm{D}} f = 0\},$$

and

$$A_{\mathrm{N}} f = -\Delta f + V f, \quad \mathrm{dom}\, A_{\mathrm{N}} = \{f \in H^{3/2}_\Delta(\Omega) : \tau_{\mathrm{N}} f = 0\}.$$

The same reasoning as in Section 8.3 one obtains the following useful decomposition of the space $H^{3/2}_\Delta(\Omega)$.

Corollary 8.7.3. *Assume that $\Omega \subset \mathbb{R}^n$ is a bounded Lipschitz domain. Then for $\lambda \in \rho(A_D)$ one has the direct sum decomposition*

$$H_{\Delta}^{3/2}(\Omega) = \operatorname{dom} A_D + \{ f_\lambda \in H_{\Delta}^{3/2}(\Omega) : (-\Delta + V)f_\lambda = \lambda f_\lambda \}$$
$$= \ker \tau_D + \{ f_\lambda \in H_{\Delta}^{3/2}(\Omega) : (-\Delta + V)f_\lambda = \lambda f_\lambda \},$$

and for $\lambda \in \rho(A_N)$ one has the direct sum decomposition

$$H_{\Delta}^{3/2}(\Omega) = \operatorname{dom} A_N + \{ f_\lambda \in H_{\Delta}^{3/2}(\Omega) : (-\Delta + V)f_\lambda = \lambda f_\lambda \}$$
$$= \ker \tau_N + \{ f_\lambda \in H_{\Delta}^{3/2}(\Omega) : (-\Delta + V)f_\lambda = \lambda f_\lambda \}.$$

For a bounded Lipschitz domain and $\lambda \in \rho(A_D)$ the Dirichlet-to-Neumann map is defined as

$$D(\lambda) : H^1(\partial\Omega) \to L^2(\partial\Omega), \qquad \tau_D f_\lambda \mapsto \tau_N f_\lambda, \tag{8.7.3}$$

where $f_\lambda \in H_{\Delta}^{3/2}(\Omega)$ is such that $(-\Delta + V)f_\lambda = \lambda f_\lambda$. This definition is the natural analog of Definition 8.3.6, taking into account the decomposiion in (8.7.3). As before, it follows that for $\lambda \in \rho(A_D) \cap \rho(A_N)$ the Dirichlet-to-Neumann map (8.7.3) is a bijective operator.

For completeness the following a priori estimates are stated.

Corollary 8.7.4. *Assume that $\Omega \subset \mathbb{R}^n$ is a bounded Lipschitz domain. Then there exist constants $C_D > 0$ and $C_N > 0$ such that*

$$\|f\|_{H_{\Delta}^{3/2}(\Omega)} \leq C_D \big(\|f\|_{L^2(\Omega)} + \|A_D f\|_{L^2(\Omega)} \big), \quad f \in \operatorname{dom} A_D,$$

and

$$\|g\|_{H_{\Delta}^{3/2}(\Omega)} \leq C_N \big(\|g\|_{L^2(\Omega)} + \|A_N g\|_{L^2(\Omega)} \big) \quad g \in \operatorname{dom} A_N.$$

Next a variant of Theorem 8.3.9 and Theorem 8.3.10 on the extensions of the Dirichlet and Neumann trace operators to $\operatorname{dom} T_{\max} = H_{\Delta}^0(\Omega)$ for bounded Lipschitz domains is formulated. For this consider the spaces

$$\mathscr{G}_0 := \{ \tau_D f : f \in \operatorname{dom} A_N \} \quad \text{and} \quad \mathscr{G}_1 := \{ \tau_N g : g \in \operatorname{dom} A_D \}, \tag{8.7.4}$$

and note that for the special case of a bounded C^2-domain the spaces $\mathscr{G}_0$ and $\mathscr{G}_1$ coincide with the spaces $H^{3/2}(\partial\Omega)$ and $H^{1/2}(\partial\Omega)$, respectively. The spaces $\mathscr{G}_0$ and $\mathscr{G}_1$ are dense in $L^2(\partial\Omega)$ and, equipped with the scalar products

$$(\varphi, \psi)_{\mathscr{G}_0} := (\Sigma^{-1/2}\varphi, \Sigma^{-1/2}\psi)_{L^2(\partial\Omega)}, \qquad \Sigma = \operatorname{Im}(D(i)^{-1}),$$
$$(\varphi, \psi)_{\mathscr{G}_1} := (\Lambda^{-1/2}\varphi, \Lambda^{-1/2}\psi)_{L^2(\partial\Omega)}, \qquad \Lambda = -\overline{\operatorname{Im} D(i)}, \tag{8.7.5}$$

they are Hilbert spaces, as was shown in [92, 115]; here both $\Sigma^{-1/2}$ and $\Lambda^{-1/2}$ are unbounded nonnegative self-adjoint operators in $L^2(\partial\Omega)$. The corresponding dual spaces of antilinear continuous functionals are denoted by $\mathscr{G}_0'$ and $\mathscr{G}_1'$, respectively, and one obtains Gelfand triples $\{\mathscr{G}_i, L^2(\partial\Omega), \mathscr{G}_i'\}$, $i = 0, 1$, which serve as the counterparts of $\{H^s(\partial\Omega), L^2(\partial\Omega), H^s(\partial\Omega)\}$, $s = 1/2, 3/2$. Now one can prove the variant of Theorem 8.3.9 and Theorem 8.3.10 alluded to above.

Theorem 8.7.5. *Assume that $\Omega \subset \mathbb{R}^n$ is a bounded Lipschitz domain. Then the Dirichlet and Neumann trace operators in* (8.7.2) *admit unique extensions to continuous surjective operators*

$$\widetilde{\tau}_D : \operatorname{dom} T_{\max} \to \mathscr{G}_1' \quad and \quad \widetilde{\tau}_N : \operatorname{dom} T_{\max} \to \mathscr{G}_0',$$

where $\operatorname{dom} T_{\max}$ *is equipped with the graph norm. Furthermore,*

$$\ker \widetilde{\tau}_D = \ker \tau_D = \operatorname{dom} A_D \quad and \quad \ker \widetilde{\tau}_N = \ker \tau_N = \operatorname{dom} A_N.$$

By analogy to Corollary 8.3.11, the second Green identity extends to elements $f \in \operatorname{dom} T_{\max}$ and $g \in \operatorname{dom} A_D$ in the form

$$(T_{\max}f, g)_{L^2(\Omega)} - (f, T_{\max}g)_{L^2(\Omega)} = \langle \widetilde{\tau}_D f, \tau_N g \rangle_{\mathscr{G}_1' \times \mathscr{G}_1},$$

and for $f \in \operatorname{dom} T_{\max}$ and $g \in \operatorname{dom} A_N$ the second Green identity reads

$$(T_{\max}f, g)_{L^2(\Omega)} - (f, T_{\max}g)_{L^2(\Omega)} = -\langle \widetilde{\tau}_N f, \tau_D g \rangle_{\mathscr{G}_0' \times \mathscr{G}_0}.$$

It will also be used that for $\lambda \in \rho(A_D)$ the Dirichlet-to-Neumann map in (8.7.3) admits an extension to a bounded operator

$$\widetilde{D}(\lambda) : \mathscr{G}_1' \to \mathscr{G}_0', \qquad \widetilde{\tau}_D f_\lambda \mapsto \widetilde{\tau}_N f_\lambda, \tag{8.7.6}$$

where $f_\lambda \in \mathfrak{N}_\lambda(T_{\max})$.

With the preparations above one can now follow the strategy in Section 8.4 and construct a boundary triplet for the maximal operator $T_{\max}$ under the assumption that $\Omega \subset \mathbb{R}^n$ is a bounded Lipschitz domain. Consider the Gelfand triple $\{\mathscr{G}_1, L^2(\partial\Omega), \mathscr{G}_1'\}$ and the corresponding isometric isomorphisms $\iota_+ : \mathscr{G}_1 \to L^2(\partial\Omega)$ and $\iota_- : \mathscr{G}_1' \to L^2(\partial\Omega)$ such that

$$\langle \varphi, \psi \rangle_{\mathscr{G}_1' \times \mathscr{G}_1} = (\iota_-\varphi, \iota_+\psi)_{L^2(\partial\Omega)}, \qquad \varphi \in \mathscr{G}_1', \ \psi \in \mathscr{G}_1;$$

cf. Lemma 8.1.2. When comparing (8.1.6) and (8.7.5) it is clear that $\iota_+ = \Lambda^{-1/2}$ and ι_- is the extension of $\Lambda^{1/2}$ onto $\mathscr{G}_1'$. Recall also the definition and the properties of the Dirichlet operator A_D in Theorem 8.7.2 and the direct sum decomposition (8.4.1).

Theorem 8.7.6. *Let $\Omega \subset \mathbb{R}^n$ be a bounded Lipschitz domain and let A_D be the self-adjoint Dirichlet realization of $-\Delta + V$ in $L^2(\Omega)$ in Theorem 8.7.2. Fix a number $\eta \in \rho(A_D) \cap \mathbb{R}$ and decompose $f \in \operatorname{dom} T_{\max}$ according to* (8.4.1) *in the form*

$f = f_D + f_\eta$, where $f_D \in \operatorname{dom} A_D$ and $f_\eta \in \mathfrak{N}_\eta(T_{\max})$. Then $\{L^2(\partial\Omega), \Gamma_0, \Gamma_1\}$, where

$$\Gamma_0 f = \iota_- \widetilde{\tau}_D f \quad and \quad \Gamma_1 f = -\iota_+ \tau_N f_D, \qquad f = f_D + f_\eta \in \operatorname{dom} T_{\max},$$

is a boundary triplet for $(T_{\min})^* = T_{\max}$ such that

$$A_0 = A_D \qquad and \qquad A_1 = T_{\min} \,\widehat{+}\, \widehat{\mathfrak{N}}_\eta(T_{\max}).$$

The γ-field and Weyl function corresponding to the boundary triplet in Theorem 8.7.6 are formally the same as in Proposition 8.4.4. In fact, if $f_\eta(\varphi)$ denotes the unique element in $\mathfrak{N}_\eta(T_{\max})$ such that $\Gamma_0 f_\eta(\varphi) = \varphi$, then for all $\lambda \in \rho(A_D)$ the γ-field is given by

$$\gamma(\lambda)\varphi = \big(I + (\lambda - \eta)(A_D - \lambda)^{-1}\big) f_\eta(\varphi), \quad \varphi \in L^2(\partial\Omega),$$

where $f_\lambda(\varphi) := \gamma(\lambda)\varphi$ is the unique element in $\mathfrak{N}_\lambda(T_{\max})$ such that $\Gamma_0 f_\lambda(\varphi) = \varphi$. As in Proposition 8.4.4 one also has

$$\gamma(\lambda)^* = -\iota_+ \tau_N (A_D - \bar{\lambda})^{-1}, \quad \lambda \in \rho(A_D).$$

Moreover, the Weyl function M is given by

$$M(\lambda)\varphi = (\eta - \lambda)\iota_+ \tau_N (A_D - \lambda)^{-1} f_\eta(\varphi), \quad \varphi \in L^2(\partial\Omega).$$

As in the case of bounded C^2-domains, the Weyl function can be expressed via the Dirichlet-to-Neumann map; here the extended mapping $\widetilde{D}(\lambda)$ in (8.7.6) is used. In the same way as in Lemma 8.4.5 one verifies the relation

$$M(\lambda) = \iota_+ \big(\widetilde{D}(\eta) - \widetilde{D}(\lambda)\big) \iota_-^{-1}.$$

With the boundary triplet $\{L^2(\partial\Omega), \Gamma_0, \Gamma_1\}$ in Theorem 8.7.6 and the corresponding γ-field and Weyl function the self-adjoint realizations of $-\Delta + V$ on a bounded Lipschitz domain $\Omega \subset \mathbb{R}^n$ can be parametrized and the spectral properties can be described in a similar form as in Section 8.4. The discussion of the semibounded extensions and of the corresponding sesquilinear forms with the help of a compatible boundary pair is parallel to the considerations in Section 8.5 and is not provided here. Finally, the coupling technique of Schrödinger operators from Section 8.6 also extends under appropriate modifications to the general situation of Lipschitz domains.

Appendix A

Operator-Valued Nevanlinna Functions and their Integral Representations

Operator-valued Nevanlinna functions and their integral representations are presented in this appendix. First the case of scalar Nevanlinna functions is considered. Then follows a short introduction to operator-valued integrals; by interpreting these integrals as improper integrals the methods are kept as simple as possible. The general operator-valued Nevanlinna functions are treated based on the previous notions. Special operator-valued Nevanlinna functions such as Kac functions, Stieltjes functions, and inverse Stieltjes functions are discussed in detail.

A.1 Borel transforms and their Stieltjes inversion

This preparatory section contains a brief discussion of the Stieltjes inversion formula for the Borel transform. The form of the transform and the conditions have been chosen so that the results are easy to apply. In particular, with the inversion formula one can prove a weak form of the Stone inversion formula, a useful denseness property, and a general form of the Stieltjes inversion formula for Nevanlinna functions.

Let $\tau : \mathbb{R} \to \mathbb{R}$ be a nondecreasing function and let $g : \mathbb{R} \to \mathbb{C}$ be a measurable function such that

$$\int_{\mathbb{R}} \frac{|g(t)|}{|t| + 1} \, d\tau(t) < \infty. \tag{A.1.1}$$

The *Borel transform* $G : \mathbb{C} \setminus \mathbb{R} \to \mathbb{C}$ of the combination $g \, d\tau$ is defined by

$$G(\lambda) = \int_{\mathbb{R}} \frac{1}{t - \lambda} \, g(t) \, d\tau(t), \quad \lambda \in \mathbb{C} \setminus \mathbb{R}. \tag{A.1.2}$$

Observe that the Borel transform G of $g \, d\tau$ in (A.1.2) is well defined and holomorphic on $\mathbb{C} \setminus \mathbb{R}$. The following result is the *Stieltjes inversion formula* for the Borel transform in (A.1.1).

Proposition A.1.1. *Let $\tau : \mathbb{R} \to \mathbb{R}$ be a nondecreasing function, let $g : \mathbb{R} \to \mathbb{C}$ be a measurable function which satisfies (A.1.1), and let G be the Borel transform of $g\,d\tau$. Then*

$$\lim_{\varepsilon \downarrow 0} \frac{1}{2\pi i} \int_a^b \big(G(s + i\varepsilon) - G(s - i\varepsilon)\big)\, ds$$

$$= \frac{1}{2} \int_{\{a\}} g(t)\, d\tau(t) + \int_{a+}^{b-} g(t)\, d\tau(t) + \frac{1}{2} \int_{\{b\}} g(t)\, d\tau(t)$$

holds for each compact interval $[a,b] \subset \mathbb{R}$. Furthermore, if $g \in L_{d\tau}^1(\mathbb{R})$, then for $0 < \varepsilon < 1$:

$$\frac{1}{2\pi} \left| \int_a^b \big(G(s + i\varepsilon) - G(s - i\varepsilon)\big)\, ds \right| \le \int_{\mathbb{R}} |g(t)|\, d\tau(t). \tag{A.1.3}$$

Proof. For $\varepsilon > 0$ and $s \in \mathbb{R}$ one has

$$\frac{G(s + i\varepsilon) - G(s - i\varepsilon)}{2\pi i} = \frac{1}{\pi} \int_{\mathbb{R}} \frac{\varepsilon}{(s-t)^2 + \varepsilon^2} g(t)\, d\tau(t).$$

Integration of the left-hand side over the interval $[a,b]$ and Fubini's theorem lead to

$$\frac{1}{2\pi i} \int_a^b \big(G(s + i\varepsilon) - G(s - i\varepsilon)\big)\, ds$$

$$= \frac{1}{\pi} \int_{\mathbb{R}} \left(\int_a^b \frac{\varepsilon}{(s-t)^2 + \varepsilon^2}\, ds \right) g(t)\, d\tau(t) \tag{A.1.4}$$

$$= \frac{1}{\pi} \int_{\mathbb{R}} \left(\arctan\left(\frac{b-t}{\varepsilon}\right) - \arctan\left(\frac{a-t}{\varepsilon}\right) \right) g(t)\, d\tau(t).$$

In order to justify the use of Fubini's theorem in (A.1.4) note first that the functions

$$\arctan\left(\frac{b-t}{\varepsilon}\right) - \arctan\left(\frac{a-t}{\varepsilon}\right), \qquad 0 < \varepsilon < 1, \tag{A.1.5}$$

are nonnegative and bounded on $\mathbb{R}$. Furthermore, observe that for $0 < \varepsilon < 1$ and $x > 0$ one has $\arctan x < \arctan x/\varepsilon < \pi/2$ and hence the functions in (A.1.5) have the upper bound

$$k(t) = \begin{cases} \frac{\pi}{2} - \arctan(a - t), & t \le a, \\ \pi, & a < t < b, \\ \arctan(b - t) + \frac{\pi}{2}, & t \ge b. \end{cases}$$

Since

$$t\left(\frac{\pi}{2} - \arctan(a - t)\right) \to -1, \quad t \to -\infty,$$
$$t\left(\arctan(b - t) + \frac{\pi}{2}\right) \to 1, \quad t \to \infty,$$

one has $kg \in L^1_{d\tau}(\mathbb{R})$ by (A.1.1), and it follows that the integral on the right-hand side in (A.1.4) is finite. Thus, the interchange of integration in (A.1.4) is justified.

Now the dominated convergence theorem will be applied to (A.1.4). For $\varepsilon \downarrow 0$ one has

$$\arctan\left(\frac{b - t}{\varepsilon}\right) - \arctan\left(\frac{a - t}{\varepsilon}\right) \to \begin{cases} 0, & t < a, \\ \pi/2, & t = a, \\ \pi, & a < t < b, \\ \pi/2, & t = b, \\ 0, & t > b, \end{cases}$$

and as an integrable majorant one can use $k|g|$. The right-hand side of (A.1.4) then shows that

$$\lim_{\varepsilon \downarrow 0} \frac{1}{\pi} \int_{\mathbb{R}} \left(\arctan\left(\frac{b - t}{\varepsilon}\right) - \arctan\left(\frac{a - t}{\varepsilon}\right)\right) g(t)\, d\tau(t)$$
$$= \frac{1}{2} \int_{\{a\}} g(t)\, d\tau(t) + \int_{a+}^{b-} g(t)\, d\tau(t) + \frac{1}{2} \int_{\{b\}} g(t)\, d\tau(t),$$

which leads to the assertion.

Finally, assume that $g \in L^1_{d\tau}(\mathbb{R})$ and $0 < \varepsilon < 1$. Then the estimate (A.1.3) follows due to (A.1.4) and (A.1.5). $\qquad\square$

Observe that when the function g in (A.1.1) is real, then the function G in (A.1.2) satisfies the symmetry property $\overline{G(\lambda)} = G(\bar{\lambda})$, in which case one has that

$$\frac{1}{2\pi i} \int_a^b \left(G(s + i\varepsilon) - G(s - i\varepsilon)\right) ds = \frac{1}{\pi} \int_a^b \operatorname{Im} G(s + i\varepsilon)\, ds.$$

The special case $g(t) = 1$ in Proposition A.1.1 is of particular interest; see for instance Chapter 3.

Corollary A.1.2. *Let $\tau : \mathbb{R} \to \mathbb{R}$ be a nondecreasing function which satisfies the integrability condition*

$$\int_{\mathbb{R}} \frac{1}{|t| + 1}\, d\tau(t) < \infty \qquad (A.1.6)$$

and let G be given by

$$G(\lambda) = \int_{\mathbb{R}} \frac{1}{t - \lambda}\, d\tau(t), \quad \lambda \in \mathbb{C} \setminus \mathbb{R}.$$

Then the inversion formula

$$\lim_{\varepsilon \downarrow 0} \frac{1}{2\pi i} \int_a^b \big(G(s+i\varepsilon) - G(s-i\varepsilon)\big)\, ds = \frac{\tau(b+) + \tau(b-)}{2} - \frac{\tau(a+) + \tau(a-)}{2} \quad (A.1.7)$$

holds for every compact interval $[a,b] \subset \mathbb{R}$. In particular, if $\tau : \mathbb{R} \to \mathbb{R}$ is a nondecreasing function which is bounded, then (A.1.6) is satisfied and (A.1.7) holds.

The Stieltjes inversion result in Proposition A.1.1 and Corollary A.1.2 has a number of interesting consequences. A first observation concerns functions of bounded variation. Recall that any function of bounded variation on $\mathbb{R}$ is a linear combination of four bounded nondecreasing functions. Hence, the following corollary is straightforward.

Corollary A.1.3. *Let $\tau : \mathbb{R} \to \mathbb{C}$ be a function of bounded variation. Then the function*

$$H(\lambda) = \int_{\mathbb{R}} \frac{1}{t - \lambda}\, d\tau(t), \quad \lambda \in \mathbb{C} \setminus \mathbb{R},$$

is well defined and holomorphic, and for each compact interval $[a,b] \subset \mathbb{R}$ one has

$$\lim_{\varepsilon \downarrow 0} \frac{1}{2\pi i} \int_a^b \big(H(s + i\varepsilon) - H(s - i\varepsilon)\big)\, ds = \frac{\tau(b+) + \tau(b-)}{2} - \frac{\tau(a+) + \tau(a-)}{2}.$$

Proposition A.1.1 and Corollary A.1.2 can also be used to compute the spectral projection of a self-adjoint relation via Stone's formula; see Chapter 1.

Example A.1.4. Let H be a self-adjoint relation in a Hilbert space $\mathfrak{H}$ and let $E(\cdot)$ be the corresponding spectral measure. For $f \in \mathfrak{H}$ consider the function

$$G(\lambda) = \big((H - \lambda)^{-1} f, f\big), \quad \lambda \in \mathbb{C} \setminus \mathbb{R}.$$

It is clear that $\tau(t) = (E(-\infty, t) f, f)$, $t \in \mathbb{R}$, is a bounded nondecreasing function and that

$$G(\lambda) = \int_{\mathbb{R}} \frac{1}{t - \lambda}\, d\tau(t).$$

By Proposition A.1.1 with $g(t) = 1$, $t \in \mathbb{R}$, one has for $[a,b] \subset \mathbb{R}$

$$\lim_{\varepsilon \downarrow 0} \frac{1}{2\pi i} \int_a^b \big(\big(H - (s + i\varepsilon)\big)^{-1} - \big(H - (s - i\varepsilon)\big)^{-1}\big) f, f)\, ds$$

$$= \frac{1}{2}(E(\{a\}) f, f) + (E((a,b)) f, f) + \frac{1}{2}(E(\{b\}) f, f),$$

which is Stone's formula in the weak sense; cf. (1.5.4) and (1.5.7) in Section 1.5.

Another consequence is the Stieltjes inversion formula for Nevanlinna functions; see Lemma A.2.7 and Corollary A.2.8. Moreover, there is the following denseness statement for a space of the form $L^2_{d\sigma}(\mathbb{R})$ which is used in Section 4.3.

Corollary A.1.5. *Assume that the function $\sigma : \mathbb{R} \to \mathbb{R}$ is nondecreasing and that it satisfies*

$$\int_{\mathbb{R}} \frac{1}{t^2 + 1} \, d\sigma(t) < \infty. \tag{A.1.8}$$

Let g be an element in $L^2_{d\sigma}(\mathbb{R})$ such that

$$\int_{\mathbb{R}} \frac{1}{t - \lambda} g(t) \, d\sigma(t) = 0, \qquad \lambda \in \mathbb{C} \setminus \mathbb{R}. \tag{A.1.9}$$

Then $g = 0$ in $L^2_{d\sigma}(\mathbb{R})$.

Proof. The conditions (A.1.8) and $g \in L^2_{d\sigma}(\mathbb{R})$ show, using the Cauchy–Schwarz inequality, that

$$\int_{\mathbb{R}} \frac{|g(t)|}{\sqrt{t^2 + 1}} \, d\sigma(t) \leq \left(\int_{\mathbb{R}} |g(t)|^2 \, d\sigma(t) \right) \left(\int_{\mathbb{R}} \frac{1}{t^2 + 1} \, d\sigma(t) \right) < \infty,$$

which implies that the condition (A.1.1) is satisfied. It follows from (A.1.9) and Proposition A.1.1 that for each compact interval $[a, b] \subset \mathbb{R}$ one has

$$\frac{1}{2} \int_{\{a\}} g(t) \, d\sigma(t) + \int_{a+}^{b-} g(t) \, d\sigma(t) + \frac{1}{2} \int_{\{b\}} g(t) \, d\sigma(t) = 0. \tag{A.1.10}$$

Next it will be shown that the contribution of the endpoints a and b in (A.1.10) is trivial. To see this, suppose that a is a point mass of $d\sigma$ and choose $\eta_k > 0$, $k = 1, 2, \dots$, such that $\eta_k \to 0$, $k \to \infty$, and $a \pm \eta_k$ are not point masses of $d\sigma$, which is possible since the point masses of $d\sigma$ form a countable subset of $\mathbb{R}$. Now Proposition A.1.1 for the compact interval $[a - \eta_k, a + \eta_k] \subset \mathbb{R}$, (A.1.9), and dominated convergence lead to

$$0 = \int_{a-\eta_k}^{a+\eta_k} g(t) \, d\sigma(t) = \lim_{k \to \infty} \int_{a-\eta_k}^{a+\eta_k} g(t) \, d\sigma(t) = \int_{\{a\}} g(t) \, d\sigma(t).$$

The same argument shows that the contribution of the endpoint b in (A.1.10) is trivial, and hence (A.1.10) reduces to

$$\int_a^b g(t) \, d\sigma(t) = 0 \quad \text{for all } a < b.$$

In other words, g is orthogonal to all characteristic functions in $L^2_{d\sigma}(\mathbb{R})$ and hence $g = 0$ in $L^2_{d\sigma}(\mathbb{R})$. $\qquad\square$

Remark A.1.6. Sometimes it is useful to have a matrix-valued version of the Borel transform in (A.1.2) and of the Stieltjes inversion formula in Proposition A.1.1; see also Remark A.2.10. More precisely, if g is a measurable $n \times n$ matrix function on $\mathbb{R}$ such that the entries of g satisfy the integrability condition (A.1.1), then Proposition A.1.1 remains valid with the integrals interpreted in the matrix sense.

A.2   Scalar Nevanlinna functions

This section contains a brief treatment of the integral representation of scalar Nevanlinna functions and its consequences.

Lemma A.2.1. *Let $f : \mathbb{D} \to \mathbb{C}$ be a holomorphic function and let $r \in (0,1)$. Then the representation*

$$f(z) = i\operatorname{Im} f(0) + \frac{1}{2\pi} \int_0^{2\pi} \frac{re^{it} + z}{re^{it} - z} \operatorname{Re} f(re^{it})\, dt$$

holds for all $z \in \mathbb{D}$ with $|z| < r$.

Proof. Let $r \in (0,1)$ and $\mathbb{T}_r = \{z \in \mathbb{C} : |z| = r\}$. Then for $z \in \mathbb{D}$, $|z| < r$, one obtains by Cauchy's integral formula

$$f(z) = \frac{1}{2\pi i} \int_{\mathbb{T}_r} \frac{f(w)}{w - z}\, dw = \frac{1}{2\pi} \int_0^{2\pi} \frac{re^{it}}{re^{it} - z} f(re^{it})\, dt, \qquad (\text{A.2.1})$$

and, since $|r^2/\bar{z}| > r$, one obtains in a similar way

$$f(0) = \frac{1}{2\pi i} \int_{\mathbb{T}_r} \frac{f(w)}{w(1 - w(\bar{z}/r^2))}\, dw = \frac{1}{2\pi} \int_0^{2\pi} \frac{re^{-it}}{re^{-it} - \bar{z}} f(re^{it})\, dt,$$

and, by taking complex conjugates,

$$\overline{f(0)} = \frac{1}{2\pi} \int_0^{2\pi} \frac{re^{it}}{re^{it} - z} \overline{f(re^{it})}\, dt.$$

Furthermore, it is clear from (A.2.1) that

$$\operatorname{Re} f(0) = \frac{1}{2\pi} \int_0^{2\pi} \operatorname{Re} f(re^{it})\, dt.$$

Therefore,

$$\frac{1}{2\pi} \int_0^{2\pi} \frac{re^{it} + z}{re^{it} - z} \operatorname{Re} f(re^{it})\, dt$$

$$= \frac{1}{2\pi} \int_0^{2\pi} \left(\frac{2re^{it}}{re^{it} - z} - 1 \right) \frac{f(re^{it}) + \overline{f(re^{it})}}{2}\, dt$$

$$= \frac{1}{2\pi} \int_0^{2\pi} \frac{re^{it}}{re^{it} - z} f(re^{it})\, dt + \frac{1}{2\pi} \int_0^{2\pi} \frac{re^{it}}{re^{it} - z} \overline{f(re^{it})}\, dt - \operatorname{Re} f(0)$$

$$= f(z) + \overline{f(0)} - \operatorname{Re} f(0) = f(z) - i\operatorname{Im} f(0),$$

which implies the assertion of the lemma. $\qquad \square$

If $f : \mathbb{D} \to \mathbb{C}$ is a holomorphic function such that $\operatorname{Re} f(re^{it}) \geq 0$, then the expression

$$\operatorname{Re} f(re^{it}) \, dt/2\pi$$

in Lemma A.2.1 leads to a measure. This observation is important in the proof of the next lemma.

Lemma A.2.2. *Let* $f : \mathbb{D} \to \mathbb{C}$ *be a function. Then the following statements are equivalent:*

(i) f *has an integral representation of the form*

$$f(z) = ic + \int_0^{2\pi} \frac{e^{it} + z}{e^{it} - z} \, d\tau(t), \qquad z \in \mathbb{D},$$

with $c \in \mathbb{R}$ *and a bounded nondecreasing function* $\tau : [0, 2\pi] \to \mathbb{R}$.

(ii) f *is holomorphic on* $\mathbb{D}$ *and* $\operatorname{Re} f(z) \geq 0$ *for all* $z \in \mathbb{D}$.

Proof. (i) $\Rightarrow$ (ii) It is clear that f is holomorphic on $\mathbb{D}$. For $z \in \mathbb{D}$ a straightforward calculation shows that

$$\operatorname{Re} f(z) = \int_0^{2\pi} \frac{1 - |z|^2}{|e^{it} - z|^2} \, d\tau(t) \geq 0.$$

(ii) $\Rightarrow$ (i) Since $\operatorname{Re} f(z) \geq 0$, $z \in \mathbb{D}$, it is obvious that for any $r \in (0, 1)$ the function

$$\tau_r : [0, 2\pi] \to \mathbb{R}, \qquad t \mapsto \frac{1}{2\pi} \int_0^t \operatorname{Re} f(re^{is}) \, ds,$$

is nondecreasing. Furthermore, Cauchy's integral formula shows that

$$\tau_r(2\pi) = \frac{1}{2\pi} \int_0^{2\pi} \operatorname{Re} f(re^{is}) \, ds = \operatorname{Re}\left(\frac{1}{2\pi i} \int_{\mathbb{T}_r} \frac{f(w)}{w} \, dw \right) = \operatorname{Re} f(0),$$

and so

$$0 = \tau_r(0) \leq \tau_r(t) \leq \tau_r(2\pi) = \operatorname{Re} f(0) < \infty \tag{A.2.2}$$

for $t \in (0, 2\pi)$ and $r \in (0, 1)$. Therefore, the Borel measure induced by τ_r on $[0, 2\pi]$ is finite and since $\operatorname{Re} f$ is continuous, it follows that τ_r, $r \in (0, 1)$, is a regular Borel measure on $[0, 2\pi]$. Observe that, by Lemma A.2.1,

$$\begin{aligned}
f(z) &= i\operatorname{Im} f(0) + \frac{1}{2\pi} \int_0^{2\pi} \frac{re^{it} + z}{re^{it} - z} \operatorname{Re} f(re^{it}) \, dt \\
&= i\operatorname{Im} f(0) + \int_0^{2\pi} \frac{re^{it} + z}{re^{it} - z} \, d\tau_r(t)
\end{aligned} \tag{A.2.3}$$

for all $z \in \mathbb{D}$, $|z| < r$. Next it will be verified that the above formula remains valid when r tends to 1.

By the Helly selection principle (cf. [763, Theorem 16.2]) and (A.2.2), there exists a nonnegative nondecreasing sequence (r_k), $k = 1, 2, \dots$, tending to 1 and a function τ such that $\tau_{r_k}(t) \to \tau(t)$, $0 \le t \le 2\pi$. Moreover, according to the Helly–Bray theorem (cf. [763, Theorem 16.4]), one has

$$\lim_{k \to \infty} \int_0^{2\pi} h(t)\, d\tau_{r_k}(t) = \int_0^{2\pi} h(t)\, d\tau(t)$$

for all continuous functions $h : [0, 2\pi] \to \mathbb{C}$. Observe that, by (A.2.2), in particular

$$\tau(2\pi) - \tau(0) = \operatorname{Re} f(0).$$

Therefore, using the Helly–Bray theorem and the fact that $t \mapsto \frac{r_k e^{it} + z}{r_k e^{it} - z}$ converges uniformly to $t \mapsto \frac{e^{it} + z}{e^{it} - z}$, one finds that

$$
\begin{aligned}
&\lim_{k \to \infty} \int_0^{2\pi} \frac{r_k e^{it} + z}{r_k e^{it} - z}\, d\tau_{r_k}(t) \\
&= \int_0^{2\pi} \frac{e^{it} + z}{e^{it} - z}\, d\tau(t) + \lim_{k \to \infty} \int_0^{2\pi} \left(\frac{r_k e^{it} + z}{r_k e^{it} - z} - \frac{e^{it} + z}{e^{it} - z} \right) d\tau_{r_k}(t) \\
&= \int_0^{2\pi} \frac{e^{it} + z}{e^{it} - z}\, d\tau(t).
\end{aligned}
$$

Hence, (A.2.3) yields

$$f(z) = ic + \int_0^{2\pi} \frac{e^{it} + z}{e^{it} - z}\, d\tau(t) \quad \text{and} \quad c = \operatorname{Im} f(0),$$

as needed. □

Here is the definition of a scalar Nevanlinna function. The operator-valued version will be considered in Definition A.4.1.

Definition A.2.3. A function $F : \mathbb{C} \setminus \mathbb{R} \to \mathbb{C}$ is called a *Nevanlinna function* if

(i) F is holomorphic on $\mathbb{C} \setminus \mathbb{R}$;
(ii) $\overline{F(\lambda)} = F(\bar{\lambda})$, $\lambda \in \mathbb{C} \setminus \mathbb{R}$;
(iii) $\operatorname{Im} F(\lambda) / \operatorname{Im} \lambda \ge 0$, $\lambda \in \mathbb{C} \setminus \mathbb{R}$.

The next result provides an integral representation for Nevanlinna functions.

Theorem A.2.4. *Let $F : \mathbb{C} \setminus \mathbb{R} \to \mathbb{C}$. Then the following statements are equivalent:*

(i) *F has an integral representation of the form*

$$F(\lambda) = \alpha + \beta\lambda + \int_{\mathbb{R}} \frac{1 + t\lambda}{t - \lambda}\, d\theta(t), \quad \lambda \in \mathbb{C} \setminus \mathbb{R}, \tag{A.2.4}$$

with $\alpha \in \mathbb{R}$, $\beta \ge 0$, and a bounded nondecreasing function $\theta : \mathbb{R} \to \mathbb{R}$.

(ii) *F is a Nevanlinna function.*

Proof. (i) $\Rightarrow$ (ii) It is clear from the representation (A.2.4) that F is holomorphic on $\mathbb{C} \setminus \mathbb{R}$ and that $\overline{F(\lambda)} = F(\bar{\lambda})$. Moreover it follows that

$$\frac{\operatorname{Im} F(\lambda)}{\operatorname{Im} \lambda} = \beta + \int_{\mathbb{R}} \frac{t^2 + 1}{|t - \lambda|^2} \, d\theta(t), \quad \lambda \in \mathbb{C} \setminus \mathbb{R}.$$

Hence, F is a Nevanlinna function.

(ii) $\Rightarrow$ (i) Assume that F is a Nevanlinna function and consider the following transformations

$$\lambda = i \frac{1 + z}{1 - z}, \quad z \in \mathbb{D}, \qquad F(\lambda) = i f(z).$$

Note that $z \mapsto \lambda$ is a bijective mapping from $\mathbb{D}$ onto $\mathbb{C}^+$ and that the function f is holomorphic on $\mathbb{D}$. Furthermore, observe that

$$\operatorname{Im} F(\lambda) \geq 0 \quad \Rightarrow \quad \operatorname{Re} f(z) \geq 0.$$

Hence, according to Lemma A.2.2, there exist $c \in \mathbb{R}$ and a bounded nondecreasing function $\tau : [0, 2\pi] \to \mathbb{R}$, such that the function f has the representation

$$
\begin{aligned}
f(z) &= ic + \int_0^{2\pi} \frac{e^{is} + z}{e^{is} - z} \, d\tau(s) \\
&= ic + \int_{0+}^{2\pi-} \frac{e^{is} + z}{e^{is} - z} \, d\tau(s) \\
&\quad + \frac{1 + z}{1 - z} \big(\tau(2\pi) - \tau(2\pi-) + \tau(0+) - \tau(0) \big) \\
&= ic + \beta \frac{1 + z}{1 - z} + \int_{0+}^{2\pi-} \frac{e^{is} + z}{e^{is} - z} \, d\tau(s),
\end{aligned}
$$

where $\beta = \tau(2\pi) - \tau(2\pi-) + \tau(0+) - \tau(0) \geq 0$. Thus, the function F has the integral representation

$$F(\lambda) = -c + \beta\lambda + i \int_{0+}^{2\pi-} \frac{e^{is} + z}{e^{is} - z} \, d\tau(s).$$

Since $z = (\lambda - i)/(\lambda + i)$, one sees that

$$F(\lambda) = -c + \beta\lambda + \int_{0+}^{2\pi-} \frac{\lambda \cot s/2 - 1}{\cot s/2 + \lambda} \, d\tau(s).$$

With the substitutions $\alpha = -c$, $-\cot s/2 = t$, and the function θ defined by $\tau(s) = \theta(t)$, one finds that

$$F(\lambda) = \alpha + \beta\lambda + \int_{\mathbb{R}} \frac{1 + t\lambda}{t - \lambda} \, d\theta(t).$$

Note that the function $\theta : \mathbb{R} \to \mathbb{R}$ is bounded since the function $\tau : [0, 2\pi] \to \mathbb{R}$ is bounded. $\qquad\square$

There is an equivalent formulation of this theorem involving the possibly unbounded measure $d\sigma(t) = (t^2 + 1)d\theta(t)$ defined by

$$\sigma(t) = \int_0^t (s^2 + 1)d\theta(s),$$

which is equivalent to

$$\theta(b) - \theta(a) = \int_a^b \frac{d\sigma(t)}{t^2 + 1}$$

for every compact interval $[a, b] \subset \mathbb{R}$. Hence, one obtains the following variant of Theorem A.2.4.

Theorem A.2.5. *Let $F : \mathbb{C} \setminus \mathbb{R} \to \mathbb{C}$. Then the following statements are equivalent:*

(i) *F has an integral representation of the form*

$$F(\lambda) = \alpha + \beta\lambda + \int_{\mathbb{R}} \left(\frac{1}{t - \lambda} - \frac{t}{t^2 + 1} \right) d\sigma(t), \quad \lambda \in \mathbb{C} \setminus \mathbb{R}, \qquad (A.2.5)$$

with $\alpha \in \mathbb{R}$, $\beta \geq 0$, and a nondecreasing function $\sigma : \mathbb{R} \to \mathbb{R}$ such that

$$\int_{\mathbb{R}} \frac{d\sigma(t)}{t^2 + 1} < \infty.$$

(ii) *F is a Nevanlinna function.*

It follows from the integral representation (A.2.5) that the imaginary part of the function F satisfies

$$\frac{\operatorname{Im} F(\lambda)}{\operatorname{Im} \lambda} = \beta + \int_{\mathbb{R}} \frac{1}{|t - \lambda|^2} d\sigma(t), \quad \lambda \in \mathbb{C} \setminus \mathbb{R}. \qquad (A.2.6)$$

With (A.2.5) and (A.2.6) it is possible to recover the ingredients in the integral formula (A.2.5) directly in terms of the function F. These results are used in Chapter 3.

Lemma A.2.6. *Let F be a Nevanlinna function as in Theorem A.2.5. Then*

$$\alpha = \operatorname{Re} F(i) \quad \text{and} \quad \beta = \lim_{y \to \infty} \frac{F(iy)}{iy} = \lim_{y \to \infty} \frac{\operatorname{Im} F(iy)}{y}. \qquad (A.2.7)$$

Moreover, for all $x \in \mathbb{R}$

$$\lim_{y \downarrow 0} y \operatorname{Im} F(x + iy) = \sigma(x+) - \sigma(x-) \qquad (A.2.8)$$

and

$$\lim_{y \downarrow 0} y \operatorname{Re} F(x + iy) = 0. \qquad (A.2.9)$$

Proof. The statement concerning α in (A.2.7) is clear. It follows from (A.2.5) that

$$\frac{1}{iy}F(iy) = \frac{1}{iy}\alpha + \beta + \int_{\mathbb{R}} \frac{1}{iy}\frac{1+iyt}{(t-iy)}\frac{d\sigma(t)}{t^2+1}. \tag{A.2.10}$$

An application of the dominated convergence theorem shows the first identity for β in (A.2.7). The second identity in (A.2.7) follows from (A.2.6). The integral representation (A.2.6) also shows that

$$y\mathrm{Im}\, F(x+iy) = \beta y^2 + \int_{\mathbb{R}} \frac{y^2}{(t-x)^2+y^2}\, d\sigma(t),$$

and the identity (A.2.8) follows from the dominated convergence theorem. One also sees from (A.2.5) that

$$y\, \mathrm{Re}\, F(x+iy) = y\left(\alpha + \beta x\right) + \int_{\mathbb{R}} \frac{y[(t-x)(1+xt)-y^2 t]}{((t-x)^2+y^2)(t^2+1)}\, d\sigma(t). \tag{A.2.11}$$

By writing $xt = (t-x)x + x^2$ it follows that the numerator of the integrand is equal to

$$y\big[(t-x)\big(1+(t-x)x+x^2\big)-y^2 t\big]$$
$$= y\big[(t-x)\big(1+(t-x)x+x^2-y^2\big)-y^2 x\big]$$
$$= y(t-x)[1+x^2-y^2] + (t-x)^2 xy - y^3 x.$$

Now assume that $|y| \le 1$. Then one sees that for a fixed $x \in \mathbb{R}$ the integrand in (A.2.11) is dominated by

$$\frac{x^2+4|x|+2}{2(t^2+1)}.$$

Thus, the identity in (A.2.9) follows from the dominated convergence theorem. $\square$

    In addition to Lemma A.2.6, the following *Stieltjes inversion formula* helps to recover the essential parts of the function σ.

Lemma A.2.7. *Let $F : \mathbb{C}\setminus\mathbb{R} \to \mathbb{C}$ be a Nevanlinna function with the integral representation (A.2.5). Let $\mathfrak{U}$ be an open neighborhood in $\mathbb{C}$ of $[a,b] \subset \mathbb{R}$ and let $g : \mathfrak{U} \to \mathbb{C}$ be holomorphic. Then*

$$\lim_{\varepsilon \downarrow 0} \frac{1}{2\pi i} \int_a^b \big[(gF)(s+i\varepsilon) - (gF)(s-i\varepsilon)\big]\, ds$$
$$= \frac{1}{2}\int_{\{a\}} g(t)\, d\sigma(t) + \int_{a+}^{b-} g(t)\, d\sigma(t) + \frac{1}{2}\int_{\{b\}} g(t)\, d\sigma(t). \tag{A.2.12}$$

For any rectangle $R = [A,B] \times [-i\varepsilon_0, i\varepsilon_0] \subset \mathfrak{U}$ with $A < a < b < B$ there exists $M \ge 0$ such that, for $0 < \varepsilon \le \varepsilon_0$,

$$\left| \int_a^b \big[(gF)(s+i\varepsilon) - (gF)(s-i\varepsilon)\big]\, ds \right| \le M \sup\big\{|g(\lambda)|, |g'(\lambda)| : \lambda \in R\big\}, \tag{A.2.13}$$

where g' stands for the derivative of g.

Proof. Consider the Nevanlinna function F given by

$$F(\lambda) = \alpha + \beta\lambda + \int_{\mathbb{R}} \left(\frac{1}{t-\lambda} - \frac{t}{t^2+1} \right) d\sigma(t), \quad \lambda \in \mathbb{C} \setminus \mathbb{R},$$

where $\alpha \in \mathbb{R}$, $\beta \geq 0$, and σ is a nondecreasing function satisfying the integrability condition

$$\int_{\mathbb{R}} \frac{1}{t^2+1} d\sigma(t);$$

cf. Theorem A.2.5. Choose an interval $(A, B) \subset \mathbb{R}$ such that

$$[a, b] \subset (A, B) \subset [A, B] \subset \mathcal{U}$$

and choose $\varepsilon_0 > 0$ such that $R = [A, B] \times [-i\varepsilon_0, i\varepsilon_0] \subset \mathcal{U}$. Observe that the choice of A and B leads to the decomposition

$$g(\lambda)F(\lambda) = G(\lambda) + H(\lambda) + g(\lambda)K(\lambda), \quad \lambda \in (\mathbb{C} \setminus \mathbb{R}) \cap \mathcal{U}, \tag{A.2.14}$$

where the functions G and H are given by

$$G(\lambda) = \int_A^B \frac{1}{t-\lambda} g(t) \, d\sigma(t), \quad H(\lambda) = \int_A^B \frac{g(\lambda) - g(t)}{t-\lambda} \, d\sigma(t),$$

while the factor K is given by

$$K(\lambda) = \left[\alpha + \beta\lambda - \int_A^B \frac{t}{t^2+1} \, d\sigma(t) \right.$$
$$\left. + \left(\int_{-\infty}^A + \int_B^\infty \right) \left(\frac{1}{t-\lambda} - \frac{t}{t^2+1} \right) d\sigma(t) \right].$$

The contributions of the functions G, H, and K will be considered separately.

Denote by $\tilde{g}$ the extension of g on $[A, B]$ by zero to all of $\mathbb{R}$. Then G can be written as

$$G(\lambda) = \int_{\mathbb{R}} \frac{1}{t-\lambda} \tilde{g}(t) \, d\sigma(t), \quad \text{where} \quad \int_{\mathbb{R}} \frac{|\tilde{g}(t)|}{|t|+1} \, d\sigma(t) < \infty.$$

Hence, one concludes from Proposition A.1.1 that

$$\lim_{\varepsilon \downarrow 0} \frac{1}{2\pi i} \int_a^b \left(G(s+i\varepsilon) - G(s-i\varepsilon) \right) ds$$
$$= \frac{1}{2} \int_{\{a\}} \tilde{g}(t) \, d\sigma(t) + \int_{a+}^{b-} \tilde{g}(t) \, d\sigma(t) + \frac{1}{2} \int_{\{b\}} \tilde{g}(t) \, d\sigma(t) \tag{A.2.15}$$
$$= \frac{1}{2} \int_{\{a\}} g(t) \, d\sigma(t) + \int_{a+}^{b-} g(t) \, d\sigma(t) + \frac{1}{2} \int_{\{b\}} g(t) \, d\sigma(t).$$

Moreover, since $\widetilde{g} \in L^1_{d\sigma}(\mathbb{R})$, it follows from the estimate (A.1.3) in Proposition A.1.1 that

$$\left| \int_a^b \left(G(s + i\varepsilon) - G(s - i\varepsilon) \right) ds \right| \leq M' \sup \left\{ |g(\lambda)| : \lambda \in [A, B] \right\} \tag{A.2.16}$$

$$\leq M' \sup \left\{ |g(\lambda)| : \lambda \in R \right\}.$$

The function H is defined for $\lambda \in (\mathbb{C} \setminus \mathbb{R}) \cap \mathcal{U}$ and can be extended to $\mathcal{U}$ by setting

$$H(\lambda) = \int_A^B h(t, \lambda) \, d\sigma(t), \quad \text{where} \quad h(t, \lambda) = \begin{cases} \frac{g(\lambda) - g(t)}{t - \lambda}, & t \neq \lambda, \\ -g'(t), & t = \lambda. \end{cases} \tag{A.2.17}$$

Clearly, the function H in (A.2.17) is bounded on the rectangle R by

$$L \sup \left\{ |g'(\lambda)| : \lambda \in R \right\},$$

where L is a constant. Note that for all $s \in (a, b)$

$$H(s + i\varepsilon) - H(s - i\varepsilon) \to 0, \quad \varepsilon \to 0,$$

and hence dominated convergence yields

$$\lim_{\varepsilon \downarrow 0} \frac{1}{2\pi i} \int_a^b \left(H(s + i\varepsilon) - H(s - i\varepsilon) \right) ds = 0. \tag{A.2.18}$$

Note that it also follows from the above that there exists $M'' \geq 0$ such that for all $0 < \varepsilon \leq \varepsilon_0$

$$\left| \int_a^b \left(H(s + i\varepsilon) - H(s - i\varepsilon) \right) ds \right| \leq M'' \sup \{ |g'(\lambda)| : \lambda \in R \}. \tag{A.2.19}$$

The function K has a holomorphic extension to the set $(\mathbb{C} \setminus \mathbb{R}) \cup (A, B)$ and this extension is uniformly continuous on the rectangle $[a, b] \times [-i\varepsilon, i\varepsilon_0]$. It is clear that

$$(gK)(s + i\varepsilon) - (gK)(s - i\varepsilon) \to 0, \quad \varepsilon \to 0,$$

holds for all $s \in (a, b)$. Since $|g|$ is also bounded on $[a, b] \times [-i\varepsilon, i\varepsilon_0]$, dominated convergence shows that

$$\lim_{\varepsilon \downarrow 0} \frac{1}{2\pi i} \int_a^b \left((gK)(s + i\varepsilon) - (gK)(s - i\varepsilon) \right) ds = 0. \tag{A.2.20}$$

Furthermore, one sees that there exists $M''' \geq 0$ such that, for all $0 < \varepsilon \leq \varepsilon_0$,

$$\left| \int_a^b \left((gK)(s + i\varepsilon) - (gK)(s - i\varepsilon) \right) ds \right|$$

$$\leq M''' \sup \left\{ |g(\lambda)| : \lambda \in [a, b] \times [-i\varepsilon_0, i\varepsilon_0] \right\} \tag{A.2.21}$$

$$\leq M''' \sup \left\{ |g(\lambda)| : \lambda \in R \right\}.$$

Now the assertion (A.2.12) follows from (A.2.14), (A.2.15), (A.2.18), and (A.2.20); in a similar way the assertion (A.2.13) follows from (A.2.14), (A.2.16), (A.2.19), and (A.2.21). □

For the special case $g(t) = 1$ Lemma A.2.7 has the following form.

Corollary A.2.8. *Let* $F : \mathbb{C} \setminus \mathbb{R} \to \mathbb{C}$ *be a Nevanlinna function with the integral representation* (A.2.5). *Then*

$$\lim_{\varepsilon \downarrow 0} \frac{1}{\pi} \int_a^b \operatorname{Im} F(s + i\varepsilon) \, ds = \frac{\sigma(b+) + \sigma(b-)}{2} - \frac{\sigma(a+) + \sigma(a-)}{2}.$$

It may happen that a Nevanlinna function has an analytic continuation to a subinterval of $\mathbb{R}$.

Proposition A.2.9. *Let* F *be a Nevanlinna function as in Theorem A.2.5 and let* $(c, d) \subset \mathbb{R}$ *be an open interval. Then the following statements are equivalent:*

(i) F *is holomorphic on* $(\mathbb{C} \setminus \mathbb{R}) \cup (c, d)$;

(ii) σ *is constant on* (c, d).

In this case

$$F(x) = \alpha + x\beta + \int_{\mathbb{R}\setminus(c,d)} \left(\frac{1}{t - x} - \frac{t}{t^2 + 1} \right) d\sigma(t), \quad x \in (c, d), \qquad \text{(A.2.22)}$$

and F *is a real nondecreasing function on* (c, d).

Proof. (i) ⇒ (ii) By assumption, $F(x)$ is real for every $x \in (c, d)$. Now apply Corollary A.2.8 to any compact subinterval of (c, d). This implies that σ is constant on every compact subinterval of (c, d).

(ii) ⇒ (i) This is a direct consequence of Theorem A.2.5, since

$$F(\lambda) = \alpha + \lambda\beta + \int_{\mathbb{R}\setminus(c,d)} \left(\frac{1}{t - \lambda} - \frac{t}{t^2 + 1} \right) d\sigma(t), \quad \lambda \in \mathbb{C} \setminus \mathbb{R}.$$

If either (i) or (ii) holds, then (A.2.22) follows by a limit process. In particular, F is real on (c, d). From (A.2.22) one also concludes that for all $x \in (c, d)$ the function F is differentiable and

$$F'(x) = \beta + \int_{\mathbb{R}\setminus(c,d)} \frac{1}{(t - x)^2} \, d\sigma(t), \quad x \in (c, d),$$

is nonnegative. Hence, F is nondecreasing on (c, d). This completes the proof. □

Assume that $\operatorname{Im} F(\mu) = 0$ for some $\mu \in \mathbb{C} \setminus \mathbb{R}$. Then it follows from (A.2.6) that in the integral representation (A.2.5) $\beta = 0$ and $\sigma(t) = 0$, $t \in \mathbb{R}$. In other words, $\operatorname{Im} F(\lambda) = 0$ for all $\lambda \in \mathbb{C} \setminus \mathbb{R}$ and $F(\lambda) = \alpha$ for all $\lambda \in \mathbb{C} \setminus \mathbb{R}$. If F

admits an analytic continuation to $(c, d) \subset \mathbb{R}$ one concludes in the same way that $F(x) = \alpha$ for $x \in (c, d)$. If $F'(x_0) = 0$ for some $x_0 \in (c, d)$, then $F'(x) = 0$ for all $x \in (c, d)$ and $F(\lambda) = \alpha$ for all $\lambda \in \mathbb{C} \setminus \mathbb{R} \cup (c, d)$. Finally, observe that if $\operatorname{Im} F(\mu) \neq 0$ for some $\mu \in \mathbb{C} \setminus \mathbb{R}$, then

$$\operatorname{Im} \left(-\frac{1}{F(\lambda)} \right) = \frac{\operatorname{Im} F(\lambda)}{|F(\lambda)|^2}, \quad \lambda \in \mathbb{C} \setminus \mathbb{R},$$

and hence $-1/F$ is also a Nevanlinna function.

Remark A.2.10. This remark is a continuation of Remark A.1.6. If the function g in Lemma A.2.7 is a holomorphic $n \times n$ matrix function on $\mathcal{U}$, then the results (A.2.12) and (A.2.13) remain valid with the integrals interpreted in the matrix sense; cf. Remark A.1.6. Furthermore, the function g may be defined on $K \times \mathcal{U}$ with some compact space K such that $x \mapsto g_x(\lambda)$ is continuous for all $\lambda \in \mathcal{U}$ and $\lambda \mapsto g_x(\lambda)$ is holomorphic for all $x \in K$. In this case (A.2.12) remains valid for all $x \in K$, while the upper bound in (A.2.13) must be replaced by

$$M \sup \left\{ |g_x(\lambda)|, |g_x'(\lambda)| : x \in K, \, \lambda \in R \right\}.$$

A.3   Operator-valued integrals

This section is concerned with operator-valued integrals which will be used in the integral representation of operator-valued Nevanlinna functions. For this purpose the notion of an improper Riemann–Stieltjes integral of bounded continuous functions is carried over to the case of operator-valued distribution functions.

In order to treat the Riemann–Stieltjes integral in the operator-valued case one needs the following preparatory observations. A function $\Theta : \mathbb{R} \to \mathbf{B}(\mathcal{G})$ whose values are self-adjoint operators is said to be *nondecreasing* if $t_1 \leq t_2$ implies

$$(\Theta(t_1)\varphi, \varphi) \leq (\Theta(t_2)\varphi, \varphi), \quad \varphi \in \mathcal{G}.$$

In general, such a function has limits as $t \to \pm\infty$, that are self-adjoint relations; cf. Chapter 5. In the following Θ will be called a *self-adjoint nondecreasing operator function*. Furthermore, Θ will be called *uniformly bounded* if there exists M such that for all $t \in \mathbb{R}$

$$|(\Theta(t)\varphi, \varphi)| \leq M\|\varphi\|^2, \quad \varphi \in \mathcal{G}.$$

Lemma A.3.1. *Let $\Theta : \mathbb{R} \to \mathbf{B}(\mathcal{G})$ be a self-adjoint nondecreasing operator function. Then the one-sided limits*

$$\Theta(t\pm), \quad t \in \mathbb{R},$$

exist in the strong sense and are bounded self-adjoint operators. If, in addition, $\Theta : \mathbb{R} \to \mathbf{B}(\mathcal{G})$ is uniformly bounded, then $\Theta(\pm\infty)$ exist in the strong sense and are bounded self-adjoint operators.

Proof. Let $t \in \mathbb{R}$ and choose an increasing sequence $t_n \to t-$. It is no restriction to assume that $\Theta(t_n) \geq 0$ for all $n \in \mathbb{N}$. Then for every $\varphi \in \mathcal{G}$ the sequence $(\Theta(t_n)\varphi, \varphi)$ is nondecreasing and bounded by $(\Theta(t)\varphi, \varphi) \leq \|\Theta(t)\| \|\varphi\|^2$. Hence, $(\Theta(t_n)\varphi, \varphi)$ converges to some $\nu_{\varphi,\varphi} \in \mathbb{R}$. Via polarization one finds that $(\Theta(t_n)\varphi, \psi)$ converges for all $\varphi, \psi \in \mathcal{G}$ and therefore

$$\mathcal{G} \times \mathcal{G} \ni \varphi \times \psi \mapsto \lim_{n \to \infty} (\Theta(t_n)\varphi, \psi)$$

is a symmetric sesquilinear form which is continuous, because

$$\left| \lim_{n \to \infty} (\Theta(t_n)\varphi, \psi) \right| \leq \lim_{n \to \infty} (\Theta(t_n)\varphi, \varphi)^{1/2} (\Theta(t_n)\psi, \psi)^{1/2}$$
$$\leq \|\Theta(t)\| \|\varphi\| \|\psi\|,$$

where the Cauchy–Schwarz inequality was used for the nonnegative sesquilinear form $(\Theta(t_n)\cdot, \cdot)$. Thus, there exists a self-adjoint operator $\Omega \in \mathbf{B}(\mathcal{G})$ such that

$$\lim_{n \to \infty} (\Theta(t_n)\varphi, \psi) = (\Omega\varphi, \psi)$$

for all $\varphi, \psi \in \mathcal{G}$. Then $\|\Omega - \Theta(t_n)\| \leq \|\Omega\| + \|\Theta(t)\|$, while $\Omega - \Theta(t_n) \geq 0$. Recall the Cauchy–Schwarz inequality $\|Af\|^2 \leq \|A\| (Af, f)$, $f \in \mathcal{G}$, for nonnegative operators $A \in \mathbf{B}(\mathcal{G})$. Thus, one obtains

$$\|(\Omega - \Theta(t_n))\varphi\|^2 \leq (\|\Omega\| + \|\Theta(t)\|)((\Omega - \Theta(t_n))\varphi, \varphi) \to 0$$

for $n \to \infty$ and $\varphi \in \mathcal{G}$, i.e., $\Theta(t-)$ exists in the strong sense. Similar arguments show that $\Theta(t+)$, $t \in \mathbb{R}$, exists in the strong sense. If, in addition, Θ is uniformly bounded one verifies in the same way that the limits $\Theta(\pm\infty)$ exist in the strong sense and are bounded self-adjoint operators. $\qquad\square$

Corollary A.3.2. *Let $\Theta : \mathbb{R} \to \mathbf{B}(\mathcal{G})$ be a self-adjoint nondecreasing operator function which is uniformly bounded. Then for every compact interval $[a, b]$ one has*

$$0 \leq ((\Theta(b) - \Theta(a))\varphi, \varphi) \leq ((\Theta(+\infty) - \Theta(-\infty))\varphi, \varphi), \quad \varphi \in \mathcal{G},$$

and consequently

$$\|(\Theta(b) - \Theta(a))\| \leq \|\Theta(+\infty) - \Theta(-\infty)\|.$$

After these preliminaries the operator-valued integrals will be introduced. Let $[a, b]$ be a compact interval and let $\Theta : [a, b] \to \mathbf{B}(\mathcal{G})$ be a self-adjoint nondecreasing operator function. Let $f : [a, b] \to \mathbb{C}$ be a continuous function. For a finite partition $a = t_0 < t_1 < \cdots < t_n = b$ of the interval $[a, b]$, define the Riemann–Stieltjes sum

$$S_n := \sum_{i=1}^{n} f(t_i)(\Theta(t_i) - \Theta(t_{i-1})). \tag{A.3.1}$$

The bounded operators S_n converge in $\mathbf{B}(\mathcal{G})$ if $\max |t_i - t_{i-1}|$ tends to zero. The limit will be called the *operator Riemann–Stieltjes integral* of f with respect to Θ and will be denoted by

$$\int_a^b f(t) \, d\Theta(t) \in \mathbf{B}(\mathcal{G}). \tag{A.3.2}$$

Lemma A.3.3. *Let $[a, b]$ be a compact interval and let $\Theta : [a, b] \to \mathbf{B}(\mathcal{G})$ be a self-adjoint nondecreasing operator function. Let $f : [a, b] \to \mathbb{C}$ be a continuous function. Then for all $\varphi, \psi \in \mathcal{G}$*

$$\left(\left(\int_a^b f(t) \, d\Theta(t) \right) \varphi, \psi \right) = \int_a^b f(t) \, d(\Theta(t)\varphi, \psi). \tag{A.3.3}$$

Moreover, for all $\varphi \in \mathcal{G}$,

$$\left\| \left(\int_a^b f(t) \, d\Theta(t) \right) \varphi \right\|^2 \leq \|\Theta(b) - \Theta(a)\| \int_a^b |f(t)|^2 \, d(\Theta(t)\varphi, \varphi). \tag{A.3.4}$$

In particular, for all $\varphi \in \mathcal{G}$,

$$\left\| \left(\int_a^b f(t) \, d\Theta(t) \right) \varphi \right\| \leq \left(\sup_{t \in [a,b]} |f(t)| \right) \|\Theta(b) - \Theta(a)\| \|\varphi\|. \tag{A.3.5}$$

Proof. It is clear from the definition involving the Riemann–Stieltjes sums in (A.3.1) that the identity (A.3.3) holds.

To see that (A.3.4) holds, observe first that

$$\|T_1\varphi_1 + \cdots + T_n\varphi_n\|^2 \leq \|T_1 T_1^* + \cdots + T_n T_n^*\| \left(\|\varphi_1\|^2 + \cdots + \|\varphi_n\|^2 \right), \tag{A.3.6}$$

where $T_1, \ldots, T_n \in \mathbf{B}(\mathcal{G})$ and $\varphi_1, \ldots, \varphi_n \in \mathcal{G}$. One verifies (A.3.6) by interpreting the row $(T_1 \ldots T_n)$ as a bounded operator A from $\mathcal{G} \times \cdots \times \mathcal{G}$ to $\mathcal{G}$ and recalling that $\|A\|^2 = \|AA^*\|$. Now rewrite the following Riemann–Stieltjes sum as indicated:

$$\sum_{i=1}^n f(t_i) \left(\Theta(t_i) - \Theta(t_{i-1}) \right) \varphi$$

$$= \sum_{i=1}^n \left(\Theta(t_i) - \Theta(t_{i-1}) \right)^{\frac{1}{2}} f(t_i) \left(\Theta(t_i) - \Theta(t_{i-1}) \right)^{\frac{1}{2}} \varphi.$$

The right-hand side may be written as $T_1\varphi_1 + \cdots + T_n\varphi_n$, where

$$T_i = \left(\Theta(t_i) - \Theta(t_{i-1}) \right)^{\frac{1}{2}} \in \mathbf{B}(\mathcal{G}), \quad \varphi_i = f(t_i) \left(\Theta(t_i) - \Theta(t_{i-1}) \right)^{\frac{1}{2}} \varphi \in \mathcal{G}$$

for $i = 1, \ldots, n$. Furthermore, one has

$$T_1 T_1^* + \cdots + T_n T_n^* = \sum_{i=1}^n \left(\Theta(t_i) - \Theta(t_{i-1}) \right) = \Theta(b) - \Theta(a).$$

Hence, the general estimate (A.3.6) above gives

$$\left\| \sum_{i=1}^{n} f(t_i) \left(\Theta(t_i) - \Theta(t_{i-1}) \right) \varphi \right\|^2 = \| T_1 \varphi_1 + \cdots + T_n \varphi_n \|^2$$

$$\leq \| \Theta(b) - \Theta(a) \| \left(\| \varphi_1 \|^2 + \cdots + \| \varphi_n \|^2 \right).$$

Since

$$\| \varphi_1 \|^2 + \cdots + \| \varphi_n \|^2 = \sum_{i=1}^{n} |f(t_i)|^2 \left\| \left(\Theta(t_i) - \Theta(t_{i-1}) \right)^{\frac{1}{2}} \varphi \right\|^2$$

$$= \sum_{i=1}^{n} |f(t_i)|^2 \left(\left(\Theta(t_i) - \Theta(t_{i-1}) \right) \varphi, \varphi \right)$$

one concludes (A.3.4) with a limit argument. Finally, (A.3.5) is an immediate consequence of (A.3.4) and

$$\int_{a}^{b} d(\Theta(t)\varphi, \varphi) = \left((\Theta(b) - \Theta(a))\varphi, \varphi \right) \leq \| \Theta(b) - \Theta(a) \| \| \varphi \|^2.$$

This completes the proof.   □

The integral in (A.3.2) enjoys the usual linearity properties. The nonnegativity property

$$f(t) \geq 0, \ t \in [a, b] \quad \Rightarrow \quad \int_{a}^{b} f(t) \, d\Theta(t) \geq 0$$

is a direct consequence of (A.3.3) of the previous lemma. Moreover, if c is a point of the open interval (a, b), then

$$\int_{a}^{b} f(t) \, d\Theta(t) = \int_{a}^{c} f(t) \, d\Theta(t) + \int_{c}^{b} f(t) \, d\Theta(t). \tag{A.3.7}$$

It follows from (A.3.7) that the integral defined in (A.3.2) and the properties in Lemma A.3.3 remain valid for functions $f : [a, b] \to \mathbb{C}$ that are piecewise continuous.

Now the integral in (A.3.2) will be extended to an improper Riemann–Stieltjes integral on $\mathbb{R}$ under the assumption that the self-adjoint nondecreasing operator function $\Theta : \mathbb{R} \to \mathbf{B}(\mathcal{G})$ is uniformly bounded; cf. Lemma A.3.1. Note that the restriction to bounded continuous functions guarantees the existence of the improper integral. However, it is clear that the results remain valid for bounded functions f that are continuous up to finitely many points.

Proposition A.3.4. *Let* $\Theta : \mathbb{R} \to \mathbf{B}(\mathcal{G})$ *be a self-adjoint nondecreasing operator function which is uniformly bounded and let* $f : \mathbb{R} \to \mathbb{C}$ *be a bounded continuous function. Then there exists a unique linear operator*

$$\int_{\mathbb{R}} f(t)\, d\Theta(t) \in \mathbf{B}(\mathcal{G}) \tag{A.3.8}$$

such that

$$\left(\int_{\mathbb{R}} f(t)\, d\Theta(t) \right)\varphi = \lim_{a \to -\infty} \lim_{b \to \infty} \left(\int_a^b f(t)\, d\Theta(t) \right)\varphi \tag{A.3.9}$$

for all $\varphi \in \mathcal{G}$, *and*

$$\left(\left(\int_{\mathbb{R}} f(t)\, d\Theta(t) \right)\varphi, \psi \right) = \int_{\mathbb{R}} f(t)\, d(\Theta(t)\varphi, \psi) \tag{A.3.10}$$

for all $\varphi, \psi \in \mathcal{G}$. *Moreover,*

$$\left\| \left(\int_{\mathbb{R}} f(t)\, d\Theta(t) \right)\varphi \right\|^2 \leq \|\Theta(+\infty) - \Theta(-\infty)\| \int_{\mathbb{R}} |f(t)|^2\, d(\Theta(t)\varphi, \varphi) \tag{A.3.11}$$

for all $\varphi \in \mathcal{G}$. *In particular,*

$$\left\| \left(\int_{\mathbb{R}} f(t)\, d\Theta(t) \right)\varphi \right\| \leq \sup_{t \in \mathbb{R}} |f(t)| \|\Theta(+\infty) - \Theta(-\infty)\| \|\varphi\| \tag{A.3.12}$$

for all $\varphi \in \mathcal{G}$.

Proof. By assumption, there exists some $M > 0$ such that for all $\varphi \in \mathcal{G}$

$$|(\Theta(t)\varphi, \varphi)| \leq M\|\varphi\|^2, \qquad t \in \mathbb{R}; \tag{A.3.13}$$

cf. Lemma A.3.1. First the existence of the limit in (A.3.9) will be verified. With the estimate (A.3.13) the inequality (A.3.4) may be written as

$$\left\| \left(\int_a^b f(t)\, d\Theta(t) \right)\varphi \right\|^2 \leq 2M \int_a^b |f(t)|^2\, d(\Theta(t)\varphi, \varphi). \tag{A.3.14}$$

Now consider two compact intervals $[a, b] \subset [a', b']$ and observe from (A.3.7) that

$$\int_{a'}^{b'} f(t)\, d\Theta(t) - \int_a^b f(t)\, d\Theta(t) = \int_{a'}^a f(t)\, d\Theta(t) + \int_b^{b'} f(t)\, d\Theta(t).$$

Hence, for every $\varphi \in \mathcal{G}$ one has

$$\left\| \int_{a'}^{b'} f(t)\, d\Theta(t)\varphi - \int_a^b f(t)\, d\Theta(t)\varphi \right\|^2$$

$$\leq 2 \left\| \int_{a'}^a f(t)\, d\Theta(t)\varphi \right\|^2 + 2 \left\| \int_b^{b'} f(t)\, d\Theta(t)\varphi \right\|^2 \tag{A.3.15}$$

$$\leq 4M \int_{a'}^a |f(t)|^2\, d(\Theta(t)\varphi, \varphi) + 4M \int_b^{b'} |f(t)|^2\, d(\Theta(t)\varphi, \varphi),$$

where the estimate (A.3.14) has been used. The right-hand side of (A.3.15) gives a Cauchy sequence for $a, a' \to -\infty$ and $b, b' \to +\infty$, since for all $\varphi \in \mathcal{G}$

$$
\begin{aligned}
\int_{\mathbb{R}} |f(t)|^2 \, d(\Theta(t)\varphi, \varphi) &\leq \|f\|_\infty^2 \int_{\mathbb{R}} d(\Theta(t)\varphi, \varphi) \\
&= \|f\|_\infty^2 \left((\Theta(+\infty)\varphi, \varphi) - (\Theta(-\infty)\varphi, \varphi) \right) \\
&\leq 2M \|f\|_\infty^2 \|\varphi\|^2 < \infty.
\end{aligned}
$$

Therefore, the strong limit on the right-hand side of (A.3.9) exists. This limit is denoted by the left-hand side of (A.3.9).

To verify (A.3.10), observe that the left-hand side of (A.3.10) is given by

$$
\lim_{a \to -\infty} \lim_{b \to \infty} \left(\left(\int_a^b f(t) \, d\Theta(t) \right) \varphi, \psi \right) = \lim_{a \to -\infty} \lim_{b \to \infty} \int_a^b f(t) \, d(\Theta(t)\varphi, \psi),
$$

where (A.3.3) was used. The statement now follows from the dominated convergence theorem.

Finally, as to (A.3.11) and (A.3.8), recall from (A.3.4) that for every $\varphi \in \mathcal{G}$ and for every compact interval $[a, b]$ one has the estimate

$$
\begin{aligned}
\left\| \left(\int_a^b f(t) \, d\Theta(t) \right) \varphi \right\|^2 &\leq \|\Theta(b) - \Theta(a)\| \int_a^b |f(t)|^2 \, d(\Theta(t)\varphi, \varphi) \\
&\leq \|\Theta(+\infty) - \Theta(-\infty)\| \int_{\mathbb{R}} |f(t)|^2 \, d(\Theta(t)\varphi, \varphi),
\end{aligned}
\tag{A.3.16}
$$

where in the last inequality Corollary A.3.2 and the dominated convergence theorem have been used. Clearly, (A.3.11) follows from (A.3.16). This also leads to (A.3.12), which implies (A.3.8). $\qquad \square$

The linearity and nonnegativity properties are preserved for the improper Riemann–Stieltjes integral. The adjoint of $\int_{\mathbb{R}} f(t) \, d\Theta(t)$ is given by

$$
\left(\int_{\mathbb{R}} f(t) \, d\Theta(t) \right)^* = \int_{\mathbb{R}} \overline{f(t)} \, d\Theta(t).
\tag{A.3.17}
$$

Another immediate consequence is the following limit result.

Corollary A.3.5. *Let $\Theta : \mathbb{R} \to \mathbf{B}(\mathcal{G})$ be a self-adjoint nondecreasing operator function which is uniformly bounded. Let $f_n : \mathbb{R} \to \mathbb{C}$ be a sequence of continuous functions which are uniformly bounded. Let $f : \mathbb{R} \to \mathbb{C}$ be a bounded continuous function such that $\lim_{n \to \infty} f_n(t) = f(t)$ for all $t \in \mathbb{R}$. Then for all $\varphi \in \mathcal{G}$*

$$
\left(\int_{\mathbb{R}} f_n(t) \, d\Theta(t) \right) \varphi \to \left(\int_{\mathbb{R}} f(t) \, d\Theta(t) \right) \varphi.
$$

Proof. Use the linearity property and Proposition A.3.4 to conclude that for all $\varphi \in \mathcal{G}$

$$\left\| \left(\int_{\mathbb{R}} f(t) \, d\Theta(t) \right) \varphi - \left(\int_{\mathbb{R}} f_n(t) \, d\Theta(t) \right) \varphi \right\|^2$$

$$\leq \| \Theta(+\infty) - \Theta(-\infty) \| \int_{\mathbb{R}} |f(t) - f_n(t)|^2 \, d(\Theta(t)\varphi, \varphi).$$

Now apply the dominated convergence theorem. $\qquad\square$

The next goal is to extend the operator-valued Riemann–Stieltjes integral to the case where the self-adjoint nondecreasing $\mathbf{B}(\mathcal{G})$-valued function appearing in the integral is not uniformly bounded. In principle this more general situation will be reduced to the case discussed above. In the following consider a (not necessarily uniformly bounded) self-adjoint nondecreasing operator function $\Sigma : \mathbb{R} \to \mathbf{B}(\mathcal{G})$. Let $\omega : \mathbb{R} \to \mathbb{R}$ be a continuous positive function with a positive lower bound, and define the operator function $\Theta : \mathbb{R} \to \mathbf{B}(\mathcal{G})$ by

$$\Theta(t) = \int_0^t \frac{d\Sigma(s)}{\omega(s)}, \quad t \in \mathbb{R}. \tag{A.3.18}$$

The function Θ can be used to extend the definition of the integral in Proposition A.3.4. First a preliminary lemma is needed.

Lemma A.3.6. *Let $\Sigma : \mathbb{R} \to \mathbf{B}(\mathcal{G})$ be a self-adjoint nondecreasing operator function and let $\omega : \mathbb{R} \to \mathbb{R}$ be a continuous positive function with a positive lower bound. Then Θ in (A.3.18) defines a self-adjoint nondecreasing operator function from $\mathbb{R}$ to $\mathbf{B}(\mathcal{G})$. Moreover, for every bounded continuous function $f : \mathbb{R} \to \mathbb{C}$ and for every compact interval $[a, b]$ one has*

$$\int_a^b f(t) \frac{d\Sigma(t)}{\omega(t)} = \int_a^b f(t) \, d\Theta(t) \in \mathbf{B}(\mathcal{G}). \tag{A.3.19}$$

Proof. It follows from Lemma A.3.3 that $\Theta(t)$, $t \in \mathbb{R}$, in (A.3.18) is well defined and that $\Theta(t) \in \mathbf{B}(\mathcal{G})$. Moreover,

$$(\Theta(t)\varphi, \varphi) = \int_0^t \frac{d(\Sigma(s)\varphi, \varphi)}{\omega(s)}, \quad t \in \mathbb{R}, \tag{A.3.20}$$

for all $\varphi \in \mathcal{G}$; cf. Lemma A.3.3. Thus, $\Theta(t)$ is self-adjoint and it follows that

$$(\Theta(t_2)\varphi, \varphi) - (\Theta(t_1)\varphi, \varphi) = \int_{t_1}^{t_2} \frac{d(\Sigma(s)\varphi, \varphi)}{\omega(s)}, \quad t_1 \leq t_2,$$

so that Θ is a nondecreasing operator function. Since the functions f/ω and f are continuous on $[a, b]$, Lemma A.3.3 shows that both integrals

$$\int_a^b f(t) \frac{d\Sigma(t)}{\omega(t)} \quad \text{and} \quad \int_a^b f(t) \, d\Theta(t)$$

belong to $\mathbf{B}(\mathcal{G})$. Observe that, for all $\varphi \in \mathcal{G}$,

$$\left(\left(\int_a^b f(t)\frac{d\Sigma(t)}{\omega(t)}\right)\varphi, \varphi\right) = \int_a^b f(t)\frac{d(\Sigma(t)\varphi, \varphi)}{\omega(t)}$$

$$= \int_a^b f(t)\, d(\Theta(t)\varphi, \varphi)$$

$$= \left(\left(\int_a^b f(t)d\Theta(t)\right)\varphi, \varphi\right).$$

Here the first and the third equality follow from Lemma A.3.3, while the second equality uses the Radon–Nikodým derivative in (A.3.20). By polarization,

$$\left(\left(\int_a^b f(t)\frac{d\Sigma(t)}{\omega(t)}\right)\varphi, \psi\right) = \left(\left(\int_a^b f(t)d\Theta(t)\right)\varphi, \psi\right)$$

for all $\varphi, \psi \in \mathcal{G}$, and hence

$$\left(\int_a^b f(t)\frac{d\Sigma(t)}{\omega(t)}\right)\varphi = \left(\int_a^b f(t)d\Theta(t)\right)\varphi \qquad (A.3.21)$$

holds for all $\varphi \in \mathcal{G}$. This implies (A.3.19).   □

As a consequence of Lemma A.3.6 one may recover the function Σ from the function Θ, since for every compact interval $[a, b] \subset \mathbb{R}$ one has

$$\int_a^b d\Sigma(t) = \int_a^b \omega(t)\, d\Theta(t).$$

Now the definition of the integral in Proposition A.3.4 is extended to "unbounded" measures by means of a "Radon–Nikodým" derivative.

Proposition A.3.7. *Let $\Sigma : \mathbb{R} \to \mathbf{B}(\mathcal{G})$ be a self-adjoint nondecreasing operator function, let $\omega : \mathbb{R} \to \mathbb{R}$ be a continuous positive function with a positive lower bound, and let $\Theta : \mathbb{R} \to \mathbf{B}(\mathcal{G})$ be defined by (A.3.18). Then the following statements are equivalent:*

(i) *For all $\varphi \in \mathcal{G}$ one has*

$$\int_{\mathbb{R}} \frac{d(\Sigma(s)\varphi, \varphi)}{\omega(s)} < \infty. \qquad (A.3.22)$$

(ii) *In the sense of strong limits one has*

$$\int_{\mathbb{R}} \frac{d\Sigma(t)}{\omega(t)} \in \mathbf{B}(\mathcal{G}).$$

(iii) *The nondecreasing operator function $\Theta : \mathbb{R} \to \mathbf{B}(\mathcal{G})$ in (A.3.18) is uniformly bounded.*

Assume that either of the above conditions is satisfied. Let $f : \mathbb{R} \to \mathbb{C}$ be a bounded continuous function. Then

$$\int_{\mathbb{R}} f(t) \frac{d\Sigma(t)}{\omega(t)} = \int_{\mathbb{R}} f(t)\, d\Theta(t), \qquad (A.3.23)$$

where each side belongs to $\mathbf{B}(\mathcal{G})$ and is meant as the strong limit of the correponding operators in the identity (A.3.19), respectively.

Proof. (i) $\Rightarrow$ (iii) Assume the condition (A.3.22). Then it follows from (A.3.18) and the monotone convergence theorem that

$$\|((\Theta(b) - \Theta(a))^{\frac{1}{2}}\varphi\|^2 = \int_a^b \frac{d(\Sigma(s)\varphi, \varphi)}{\omega(s)} \le \int_{\mathbb{R}} \frac{d(\Sigma(s)\varphi, \varphi)}{\omega(s)}$$

for every compact interval $[a, b]$. The assumption (A.3.22) and the uniform boundedness principle show the existence of a constant M such that

$$\|(\Theta(b) - \Theta(a))^{\frac{1}{2}}\| \le M$$

for every compact interval $[a, b]$. In particular, this leads to the inequality

$$|(\Theta(b) - \Theta(a))\varphi, \varphi)| \le M^2 \|\varphi\|^2,$$

which implies that the nondecreasing operator function Θ is uniformly bounded. This gives (iii).

(iii) $\Rightarrow$ (ii) Assume that $\Theta : \mathbb{R} \to \mathbf{B}(\mathcal{G})$ in (A.3.18) is uniformly bounded. It follows from (A.3.19) that for every compact interval $[a, b] \subset \mathbb{R}$

$$\int_a^b \frac{d\Sigma(t)}{\omega(t)} = \int_a^b d\Theta(t) = \Theta(b) - \Theta(a),$$

where each side belongs to $\mathbf{B}(\mathcal{G})$. The result now follows by taking strong limits for $[a, b] \to \mathbb{R}$.

(ii) $\Rightarrow$ (i) The assumption means that for all $\varphi \in \mathcal{G}$

$$\left(\int_{\mathbb{R}} \frac{d\Sigma(t)}{\omega(t)}\right)\varphi = \lim_{a \to -\infty} \lim_{b \to \infty} \left(\int_a^b \frac{d\Sigma(t)}{\omega(t)}\right)\varphi,$$

with convergence in $\mathcal{G}$. In particular, this gives that

$$\left(\left(\int_{\mathbb{R}} \frac{d\Sigma(t)}{\omega(t)}\right)\varphi, \varphi\right) = \lim_{a \to -\infty} \lim_{b \to \infty} \left(\left(\int_a^b \frac{d\Sigma(t)}{\omega(t)}\right)\varphi, \varphi\right)$$

$$= \lim_{a \to -\infty} \lim_{b \to \infty} \int_a^b \frac{d(\Sigma(t)\varphi, \varphi)}{\omega(t)}$$

$$= \int_{\mathbb{R}} \frac{d(\Sigma(t)\varphi, \varphi)}{\omega(t)},$$

where the second step is justified by Lemma A.3.3 and the last step by the monotone convergence theorem. This leads to (A.3.22).

By Proposition A.3.4, the integral on the right-hand side of (A.3.21) converges strongly to

$$\int_{\mathbb{R}} f(t) d\Theta(t) \in \mathbf{B}(\mathcal{G}).$$

Together with the identity (A.3.19) this leads to the assertion in (ii).   □

For the reader's convenience the following facts are mentioned in terms of $\Sigma : \mathbb{R} \to \mathbf{B}(\mathcal{G})$ and $\omega : \mathbb{R} \to \mathbb{R}$ from Proposition A.3.7. They can be easily verified via the identity (A.3.23). Let $f : \mathbb{R} \to \mathbb{C}$ be a bounded continuous function. Then

$$\left(\int_{\mathbb{R}} f(t) \frac{d\Sigma(t)}{\omega(t)} \right)^{*} = \int_{\mathbb{R}} \overline{f(t)} \, \frac{d\Sigma(t)}{\omega(t)} \in \mathbf{B}(\mathcal{G}); \qquad (A.3.24)$$

cf. (A.3.17), and for all $\varphi \in \mathcal{G}$

$$\left(\int_{\mathbb{R}} f(t) \frac{d\Sigma(t)}{\omega(t)} \varphi, \varphi \right) = \int_{\mathbb{R}} f(t) \frac{d(\Sigma(t)\varphi, \varphi)}{\omega(t)}; \qquad (A.3.25)$$

cf. (A.3.10). Furthermore, if $f_n : \mathbb{R} \to \mathbb{C}$ is a sequence of continuous functions which is uniformly bounded and $f : \mathbb{R} \to \mathbb{C}$ is a bounded continuous function such that $\lim_{n\to\infty} f_n(t) = f(t)$ for all $t \in \mathbb{R}$, then for all $\varphi \in \mathcal{G}$

$$\left(\int_{\mathbb{R}} f_n(t) \frac{d\Sigma(t)}{\omega(t)} \right) \varphi \to \left(\int_{\mathbb{R}} f(t) \frac{d\Sigma(t)}{\omega(t)} \right) \varphi; \qquad (A.3.26)$$

cf. Corollary A.3.5. It is also clear that Proposition A.3.7 and the above properties of the integral remain true for bounded functions $f : \mathbb{R} \to \mathbb{C}$ with finitely many discontinuities.

The following notation will be used later: Consider an interval $I \subset \mathbb{R}$, let χ_I be the corresponding characteristic function, and let $f : \mathbb{R} \to \mathbb{C}$ be a bounded continuous function. Then $\chi_I f$ is a piecewise continuous function and one defines

$$\int_{I} f(t) \frac{d\Sigma(t)}{\omega(t)} = \int_{\mathbb{R}} \chi_I(t) f(t) \frac{d\Sigma(t)}{\omega(t)}. \qquad (A.3.27)$$

In particular, for $c \in \mathbb{R}$ integrals of the form

$$\int_{[c,\infty)} f(t) \frac{d\Sigma(t)}{\omega(t)} = \int_{\mathbb{R}} \chi_{[c,\infty)}(t) f(t) \frac{d\Sigma(t)}{\omega(t)} \qquad (A.3.28)$$

will appear in the context of Nevanlinna functions that admit an analytic continuation to $(-\infty, c)$; see Section A.6.

A.4   Operator-valued Nevanlinna functions

The notion of scalar Nevanlinna function in Definition A.2.3 is easily carried over to the operator-valued case. The present section is concerned with developing the corresponding operator-valued integral representations. The main tool is provided by Proposition A.3.7.

Definition A.4.1. Let $\mathcal{G}$ be a Hilbert space and let $F : \mathbb{C} \setminus \mathbb{R} \to \mathbf{B}(\mathcal{G})$ be an operator function. Then F is called a $\mathbf{B}(\mathcal{G})$-valued *Nevanlinna function* if

(i) F is holomorphic on $\mathbb{C} \setminus \mathbb{R}$;

(ii) $F(\lambda) = F(\bar{\lambda})^*$, $\lambda \in \mathbb{C} \setminus \mathbb{R}$;

(iii) $\operatorname{Im} F(\lambda)/\operatorname{Im} \lambda \geq 0$, $\lambda \in \mathbb{C} \setminus \mathbb{R}$.

Operator-valued Nevanlinna functions admit integral representations as in the scalar case; cf. Theorem A.2.5.

Theorem A.4.2. *Let $\mathcal{G}$ be a Hilbert space and let $F : \mathbb{C} \setminus \mathbb{R} \to \mathbf{B}(\mathcal{G})$ be an operator function. Then the following statements are equivalent:*

(i) *F has an integral representation of the form*

$$F(\lambda) = \alpha + \lambda\beta + \int_{\mathbb{R}} \left(\frac{1}{t - \lambda} - \frac{t}{t^2 + 1} \right) d\Sigma(t), \quad \lambda \in \mathbb{C} \setminus \mathbb{R}, \qquad (A.4.1)$$

with self-adjoint operators $\alpha, \beta \in \mathbf{B}(\mathcal{G})$, $\beta \geq 0$, and a nondecreasing self-adjoint operator function $\Sigma : \mathbb{R} \to \mathbf{B}(\mathcal{G})$ such that

$$\int_{\mathbb{R}} \frac{d\Sigma(t)}{t^2 + 1} \in \mathbf{B}(\mathcal{G}), \qquad (A.4.2)$$

where the integrals in (A.4.1) and (A.4.2) converge in the strong topology.

(ii) *F is a Nevanlinna function.*

Note that for $\lambda \in \mathbb{C} \setminus \mathbb{R}$ the identity (A.4.1) can be rewritten as

$$F(\lambda) = \alpha + \lambda\beta + \int_{\mathbb{R}} f_\lambda(t) \, d\Theta(t), \quad \lambda \in \mathbb{C} \setminus \mathbb{R}, \qquad (A.4.3)$$

where the bounded continuous function $f_\lambda : \mathbb{R} \to \mathbb{C}$ and the "bounded measure" Θ are given by

$$f_\lambda(t) = \frac{1 + \lambda t}{t - \lambda} \quad \text{and} \quad \Theta(t) = \int_0^t \frac{d\Sigma(s)}{s^2 + 1}, \quad t \in \mathbb{R}, \qquad (A.4.4)$$

respectively; cf. (A.3.18) and (A.3.23) with $\omega(t) = t^2 + 1$.

Proof. (i) $\Rightarrow$ (ii) Due to the condition (A.4.2) it follows that $F(\lambda)$ in (A.4.1) is well defined and represents an element in $\mathbf{B}(\mathcal{G})$; cf. (A.4.3), (A.4.4), and Proposition A.3.7. Moreover, it follows from (A.4.3) and (A.3.25) that for $\varphi \in \mathcal{G}$

$$(F(\lambda)\varphi, \varphi) = (\alpha\varphi, \varphi) + \lambda(\beta\varphi, \varphi) + \int_{\mathbb{R}} f_\lambda(t)\, d(\Theta(t)\varphi, \varphi), \quad \lambda \in \mathbb{C} \setminus \mathbb{R}.$$

Therefore, one sees that the function F is holomorphic on $\mathbb{C} \setminus \mathbb{R}$. It is clear that $F(\lambda)^* = F(\bar{\lambda})$; cf. (A.3.24). Since $(\alpha\varphi, \varphi) \in \mathbb{R}$ and $(\beta\varphi, \varphi) \in \mathbb{R}$, one also sees that

$$\operatorname{Im}(F(\lambda)\varphi, \varphi) = (\operatorname{Im}\lambda)\left[(\beta\varphi, \varphi) + \int_{\mathbb{R}} \frac{t^2+1}{|t-\lambda|^2} d(\Theta(t)\varphi, \varphi)\right], \quad \lambda \in \mathbb{C} \setminus \mathbb{R}.$$

Therefore, $\operatorname{Im}(F(\lambda)\varphi, \varphi)/\operatorname{Im}\lambda \geq 0$ for all $\lambda \in \mathbb{C} \setminus \mathbb{R}$. This implies that F is a Nevanlinna function.

(ii) $\Rightarrow$ (i) Let $F : \mathbb{C} \setminus \mathbb{R} \to \mathbf{B}(\mathcal{G})$ be an operator-valued Nevanlinna function. Then $M(\lambda) = F(\lambda) - \operatorname{Re} F(i)$ is also a Nevanlinna function and $\operatorname{Re} M(i) = 0$. For $\varphi \in \mathcal{G}$ the function $(M(\lambda)\varphi, \varphi)$ is a scalar Nevanlinna function with the integral representation

$$(M(\lambda)\varphi, \varphi) = \lambda\beta_{\varphi,\varphi} + \int_{\mathbb{R}} \frac{1+\lambda t}{t-\lambda}\, d\theta_{\varphi,\varphi}(t), \quad \lambda \in \mathbb{C} \setminus \mathbb{R},$$

where $\beta_{\varphi,\varphi} \geq 0$ and $\theta_{\varphi,\varphi} : \mathbb{R} \to \mathbb{R}$ is a nondecreasing function; cf. Theorem A.2.4 and Lemma A.2.6. Then it follows that

$$\operatorname{Im}(M(i)\varphi, \varphi) = \beta_{\varphi,\varphi} + \int_{\mathbb{R}} d\theta_{\varphi,\varphi}(t),$$

where each term on the right-hand side is nonnegative, so that

$$0 \leq \beta_{\varphi,\varphi} + \int_{\mathbb{R}} d\theta_{\varphi,\varphi}(t) = \operatorname{Im}(M(i)\varphi, \varphi) \leq \|M(i)\|\|\varphi\|^2, \quad \varphi \in \mathcal{G}.$$

Without loss of generality it will be assumed that

$$\theta_{\varphi,\varphi}(-\infty) = 0 \quad \text{and} \quad \theta_{\varphi,\varphi}(t) = \frac{\theta_{\varphi,\varphi}(t+) - \theta_{\varphi,\varphi}(t-)}{2}, \quad t \in \mathbb{R}, \tag{A.4.5}$$

(see also the proof of Theorem A.2.4) and thus

$$0 \leq \beta_{\varphi,\varphi} \leq \|M(i)\|\|\varphi\|^2 \quad \text{and} \quad 0 \leq \theta_{\varphi,\varphi}(t) \leq \|M(i)\|\|\varphi\|^2 \tag{A.4.6}$$

for all $t \in \mathbb{R}$. For $\varphi, \psi \in \mathcal{G}$ one defines by polarization

$$\beta_{\varphi,\psi} = \frac{1}{4}\left(\beta_{\varphi+\psi,\varphi+\psi} - \beta_{\varphi-\psi,\varphi-\psi} + i\beta_{\varphi+i\psi,\varphi+i\psi} - i\beta_{\varphi-i\psi,\varphi-i\psi}\right),$$

and similarly

$$\theta_{\varphi,\psi} = \frac{1}{4}\big(\theta_{\varphi+\psi,\varphi+\psi} - \theta_{\varphi-\psi,\varphi-\psi} + i\theta_{\varphi+i\psi,\varphi+i\psi} - i\theta_{\varphi-i\psi,\varphi-i\psi}\big). \tag{A.4.7}$$

Then it follows that

$$(M(\lambda)\varphi,\psi) = \lambda\beta_{\varphi,\psi} + \int_{\mathbb{R}} \frac{1+\lambda t}{t-\lambda}\, d\theta_{\varphi,\psi}(t), \quad \lambda \in \mathbb{C}\setminus\mathbb{R}, \tag{A.4.8}$$

where $\beta_{\varphi,\psi}$ is determined by

$$\beta_{\varphi,\psi} = \lim_{y\to\infty} \frac{(M(iy)\varphi,\psi)}{iy}$$

(see (A.2.10) in the proof of Lemma A.2.6) and $\theta_{\varphi,\psi} : \mathbb{R} \to \mathbb{C}$ is a function of bounded variation.

Now the representation (A.4.8) will be used to verify that the bounded forms

$$\{\varphi,\psi\} \mapsto \beta_{\varphi,\psi}, \quad \{\varphi,\psi\} \mapsto \theta_{\varphi,\psi}(t), \tag{A.4.9}$$

are sesquilinear. For instance, with $\varphi, \varphi', \psi \in \mathcal{G}$ it follows that

$$\lambda\beta_{\varphi+\varphi',\psi} + \int_{\mathbb{R}} \frac{1+\lambda t}{t-\lambda}\, d\theta_{\varphi+\varphi',\psi}(t)$$
$$= \big(M(\lambda)(\varphi+\varphi'),\psi\big)$$
$$= (M(\lambda)\varphi,\psi) + (M(\lambda)\varphi',\psi)$$
$$= \lambda\beta_{\varphi,\psi} + \int_{\mathbb{R}} \frac{1+\lambda t}{t-\lambda}\, d\theta_{\varphi,\psi}(t) + \lambda\beta_{\varphi',\psi} + \int_{\mathbb{R}} \frac{1+\lambda t}{t-\lambda}\, d\theta_{\varphi',\psi}(t)$$
$$= \lambda(\beta_{\varphi,\psi} + \beta_{\varphi',\psi}) + \int_{\mathbb{R}} \frac{1+\lambda t}{t-\lambda}\, d(\theta_{\varphi,\psi}(t) + \theta_{\varphi',\psi}(t)).$$

After dividing this equality by $\lambda = iy$ and letting $y \to \infty$, one concludes that

$$\beta_{\varphi+\varphi',\psi} = \beta_{\varphi,\psi} + \beta_{\varphi',\psi}. \tag{A.4.10}$$

Setting $\widetilde{\theta} = \theta_{\varphi+\varphi',\psi} - \theta_{\varphi,\psi} - \theta_{\varphi',\psi}$, one obtains

$$0 = \int_{\mathbb{R}} \frac{1+\lambda t}{t-\lambda}\, d\widetilde{\theta}(t) = \int_{\mathbb{R}} \left(\lambda + \frac{1+\lambda^2}{t-\lambda}\right) d\widetilde{\theta}(t), \quad \lambda \in \mathbb{C}\setminus\mathbb{R},$$

and hence

$$\int_{\mathbb{R}} \frac{1}{t-\lambda}\, d\widetilde{\theta}(t) = -\int_{\mathbb{R}} \frac{\lambda}{1+\lambda^2}\, d\widetilde{\theta}(t), \quad \lambda \in \mathbb{C}\setminus\mathbb{R}.$$

Now it follows from Corollary A.1.3 that for every compact subinterval $[a,b] \subset \mathbb{R}$ one has

$$0 = \frac{\widetilde{\theta}(b+) + \widetilde{\theta}(b-)}{2} - \frac{\widetilde{\theta}(a+) + \widetilde{\theta}(a-)}{2} = \widetilde{\theta}(b) - \widetilde{\theta}(a),$$

where (A.4.5) and (A.4.7) were used in the second equality. With the help of (A.4.5) and (A.4.7) via polarization one also concludes that $\widetilde{\theta}(a) \to 0$ for $a \to -\infty$ and therefore $\widetilde{\theta}(b) = 0$ for all $b \in \mathbb{R}$. This implies

$$\theta_{\varphi+\varphi',\psi}(t) = \theta_{\varphi,\psi}(t) + \theta_{\varphi',\psi}(t), \qquad t \in \mathbb{R}. \tag{A.4.11}$$

It follows from (A.4.10), (A.4.11), and similar considerations that the forms in (A.4.9) are both sesquilinear. Furthermore, these forms are nonnegative and from the Cauchy–Schwarz inequality and (A.4.6) it follows that they are bounded. Hence, there exists a uniquely determined nonnegative operator $\beta \in \mathbf{B}(\mathcal{G})$ such that

$$\beta_{\varphi,\psi} = (\beta\varphi, \psi), \qquad \varphi, \psi \in \mathcal{G},$$

and for each fixed $t \in \mathbb{R}$ there exists a uniquely determined bounded operator $\Theta(t) \in \mathbf{B}(\mathcal{G})$ such that

$$\theta_{\varphi,\psi}(t) = (\Theta(t)\varphi, \psi) \qquad \varphi, \psi \in \mathcal{G}.$$

From

$$(\Theta(t)\varphi, \varphi) = \theta_{\varphi,\varphi}(t) = \overline{\theta_{\varphi,\varphi}(t)} = (\varphi, \Theta(t)\varphi), \qquad \varphi \in \mathcal{G},$$

one concludes that $(\Theta(t)\varphi, \psi) = (\varphi, \Theta(t)\psi)$, so that $\Theta(t)$ is a self-adjoint operator in $\mathbf{B}(\mathcal{G})$. Furthermore, for $t \leq t'$ one sees that

$$\theta_\varphi(t) \leq \theta_\varphi(t') \quad \Rightarrow \quad \Theta(t) \leq \Theta(t')$$

and therefore $\Theta : \mathbb{R} \to \mathbf{B}(\mathcal{G})$, $t \mapsto \Theta(t)$, is a nondecreasing self-adjoint operator function. Thus, one obtains for all $\varphi, \psi \in \mathcal{G}$

$$\left(\left(\lambda\beta + \int_{\mathbb{R}} \frac{1+\lambda t}{t-\lambda} \, d\Theta(t) \right) \varphi, \psi \right) = \lambda\beta_{\varphi,\psi} + \int_{\mathbb{R}} \frac{1+\lambda t}{t-\lambda} \, d\theta_{\varphi,\psi}(t)$$
$$= ((F(\lambda) - \operatorname{Re} F(i))\varphi, \psi),$$

which gives (A.4.3) with $\alpha = \operatorname{Re} F(i)$. $\qquad\square$

The imaginary part $\operatorname{Im} F(\lambda) \in \mathbf{B}(\mathcal{G})$ of the Nevanlinna function N in Theorem A.4.2 admits the representation

$$\frac{\operatorname{Im} F(\lambda)}{\operatorname{Im} \lambda} = \beta + \int_{\mathbb{R}} \frac{1}{|t-\lambda|^2} \, d\Sigma(t), \qquad \lambda \in \mathbb{C} \setminus \mathbb{R}, \tag{A.4.12}$$

where the integral exists in the strong sense. In particular, one has for all $\varphi \in \mathcal{G}$:

$$\frac{(\operatorname{Im} F(\lambda)\varphi, \varphi)}{\operatorname{Im} \lambda} = (\beta\varphi, \varphi) + \int_{\mathbb{R}} \frac{1}{|t-\lambda|^2} \, d(\Sigma(t)\varphi, \varphi), \qquad \lambda \in \mathbb{C} \setminus \mathbb{R};$$

cf. (A.3.25). Hence, $\operatorname{Im} F(\lambda)/\operatorname{Im} \lambda$ is a nonnegative operator in $\mathbf{B}(\mathcal{G})$.

The next lemma is the counterpart of Lemma A.2.6 for operator-valued Nevanlinna functions.

Lemma A.4.3. *Let F be a $\mathbf{B}(\mathcal{G})$-valued Nevanlinna function with integral representation (A.4.1). The operators α, β, and the nondecreasing self-adjoint operator function $\Sigma : \mathbb{R} \to \mathbf{B}(\mathcal{G})$ are related to the function F by the following identities:*

$$\alpha = \operatorname{Re} F(i), \tag{A.4.13}$$

$$\beta\varphi = \lim_{y\to\infty} \frac{F(iy)}{iy}\varphi = \lim_{y\to\infty} \frac{\operatorname{Im} F(iy)}{y}\varphi, \quad \varphi \in \mathcal{G}. \tag{A.4.14}$$

Moreover, for all $x \in \mathbb{R}$

$$\lim_{y\downarrow 0} y\big(\operatorname{Im} F(x+iy)\varphi, \varphi\big) = (\Sigma(x+)\varphi, \varphi) - (\Sigma(x-)\varphi, \varphi), \quad \varphi \in \mathcal{G}. \tag{A.4.15}$$

Proof. It is clear from (A.4.1) that (A.4.13) holds. The identities for β in (A.4.14) follow with the help of (A.3.26) in the same was as in the proof of Lemma A.2.6. Finally, note that the statement (A.4.15) is a direct consequence of Lemma A.2.6 and Lemma A.3.1. $\qquad\square$

In the present context the Stieltjes inversion formula has the following form; cf. Lemma A.2.7.

Lemma A.4.4. *Let $F : \mathbb{C} \setminus \mathbb{R} \to \mathbb{C}$ be a $\mathbf{B}(\mathcal{G})$-valued Nevanlinna function with the integral representation (A.4.1) and let $[a, b] \subset \mathbb{R}$. Assume that $\mathcal{U}$ is an open neighborhood of $[a, b]$ in $\mathbb{C}$ and that $g : \mathcal{U} \to \mathbb{C}$ is holomorphic. Then*

$$\lim_{\varepsilon\downarrow 0} \frac{1}{2\pi i} \int_a^b \big(((gF)(s+i\varepsilon) - (gF)(s-i\varepsilon))\varphi, \varphi\big)\, ds$$

$$= \frac{1}{2}\int_{\{a\}} g(t)\, d(\Sigma(t)\varphi, \varphi) + \int_{a+}^{b-} g(t)\, d(\Sigma(t)\varphi, \varphi) + \frac{1}{2}\int_{\{b\}} g(t)\, d(\Sigma(t)\varphi, \varphi) \tag{A.4.16}$$

holds for all $\varphi \in \mathcal{G}$. If the function g is entire, then (A.4.16) is valid for any compact interval $[a, b]$. In particular,

$$\lim_{\varepsilon\downarrow 0} \frac{1}{\pi} \int_a^b (\operatorname{Im} F(s+i\varepsilon)\varphi, \varphi)\, ds$$

$$= \frac{(\Sigma(b+)\varphi, \varphi) + (\Sigma(b-)\varphi, \varphi)}{2} - \frac{(\Sigma(a+)\varphi, \varphi) + (\Sigma(a-)\varphi, \varphi)}{2}$$

for all $\varphi \in \mathcal{G}$.

For operator-valued Nevanlinna functions Proposition A.2.9 has the following form.

Proposition A.4.5. *Let F be a Nevanlinna function as in Theorem A.4.2 and let $(c, d) \subset \mathbb{R}$ be an open interval. Then the following statements are equivalent:*

(i) *F is holomorphic on $(\mathbb{C} \setminus \mathbb{R}) \cup (c, d)$.*

(ii) *For every $\varphi \in \mathcal{G}$ the function $t \mapsto \Sigma(t)\varphi$ is constant on (c, d).*

In this case

$$F(x) = \alpha + x\beta + \int_{\mathbb{R}\setminus(c,d)} \left(\frac{1}{t-x} - \frac{t}{t^2+1} \right) d\Sigma(t), \quad x \in (c,d), \qquad (A.4.17)$$

is self-adjoint; the integral in (A.4.17) converges in the strong topology. Moreover, F is nondecreasing on (c, d):

$$(F(x_1)\varphi, \varphi) \le (F(x_2)\varphi, \varphi), \quad c < x_1 < x_2 < d.$$

Proof. The implication (i) $\Rightarrow$ (ii) follows from the Stieltjes inversion formula in Lemma A.4.4, which implies that $t \mapsto (\Sigma(t)\varphi, \varphi)$ is constant for $t \in (c,d)$ and $\varphi \in \mathcal{G}$. For $c < t_1 < t_2 < d$ one concludes from the inequality

$$\left| ((\Sigma(t_2) - \Sigma(t_1))\varphi, \psi) \right| \le ((\Sigma(t_2) - \Sigma(t_1))\varphi, \varphi)((\Sigma(t_2) - \Sigma(t_1))\psi, \psi) = 0$$

that the function $t \mapsto \Sigma(t)\varphi$ is constant on (c,d) for all $\varphi \in \mathcal{G}$. For the implication (ii) $\Rightarrow$ (i) consider the integral representation (A.4.1) in Theorem A.4.2, where the integral converges in the strong sense. Since $t \mapsto \Sigma(t)\varphi$ is constant on (c,d), this representation takes the form

$$\begin{aligned}
F(\lambda) &= \alpha + \lambda\beta + \int_{\mathbb{R}} \chi_{\mathbb{R}\setminus(c,d)}(t) \left(\frac{1}{t-\lambda} - \frac{t}{t^2+1} \right) d\Sigma(t) \\
&= \alpha + \lambda\beta + \int_{\mathbb{R}\setminus(c,d)} \left(\frac{1}{t-\lambda} - \frac{t}{t^2+1} \right) d\Sigma(t)
\end{aligned} \qquad (A.4.18)$$

for all $\lambda \in \mathbb{C} \setminus \mathbb{R}$; cf. (A.3.27). It follows that $\lambda \mapsto (F(\lambda)\varphi, \varphi)$ is holomorphic on $(\mathbb{C} \setminus \mathbb{R}) \cup (c,d)$ for all $\varphi \in \mathcal{G}$ and hence F is holomorphic on $(\mathbb{C} \setminus \mathbb{R}) \cup (c,d)$.

To obtain (A.4.17), observe that for $x \in (c,d)$ one has

$$\chi_{\mathbb{R}\setminus(c,d)}(t) \left(\frac{1}{t-\lambda} - \frac{t}{t^2+1} \right) \to \chi_{\mathbb{R}\setminus(c,d)}(t) \left(\frac{1}{t-x} - \frac{t}{t^2+1} \right)$$

when $\lambda \to x$, $\lambda \in \mathbb{C} \setminus \mathbb{R}$, and the functions are uniformly bounded in t. Hence, (A.3.26) and (A.4.18) imply (A.4.17), where the integral in (A.4.17) converges in the strong topology by Proposition A.3.7. Furthermore, the bounded operators $F(x)$, $x \in (c,d)$, are self-adjoint by (A.3.24) and Proposition A.2.9 shows that $x \mapsto (F(x)\varphi, \varphi)$ is nondecreasing on (c,d). $\square$

If F has an analytic continuation to $(c,d) \subset \mathbb{R}$, then it follows from (A.4.17) that F is differentiable on (c,d) and that

$$F'(x) = \beta + \int_{\mathbb{R}} \frac{1}{|t-x|^2} d\Sigma(t), \quad x \in (c,d), \qquad (A.4.19)$$

where the integral exists in the strong sense. In particular, one has for all $\varphi \in \mathcal{G}$:

$$(F'(x)\varphi, \varphi) = (\beta\varphi, \varphi) + \int_{\mathbb{R}} \frac{1}{|t - x|^2} \, d(\Sigma(t)\varphi, \varphi), \quad x \in (c, d);$$

cf. (A.3.25). It is clear that $F'(x)$ is a nonnegative operator in $\mathbf{B}(\mathcal{G})$.

The next observation on isolated singularities of F is useful. If $(c, d) \subset \mathbb{R}$ and F admits an analytic continuation to $(c, d) \setminus \{y\}$ for some $y \in (c, d)$, then Proposition A.4.5 implies

$$F(\lambda) = \alpha + \lambda\beta + \int_{\mathbb{R}\setminus((c,y)\cup(y,d))} \left(\frac{1}{t - \lambda} - \frac{t}{t^2 + 1} \right) d\Sigma(t)$$

for all $\lambda \in (\mathbb{C} \setminus \mathbb{R}) \cup (c, y) \cup (y, d)$. Dominated convergence shows that

$$- \lim_{\lambda \to y} (\lambda - y)(F(\lambda)\varphi, \psi) = (\Sigma(y+)\varphi, \psi) - (\Sigma(y-)\varphi, \psi), \quad \varphi, \psi \in \mathcal{G},$$

whence

$$\lim_{\lambda \to y} \|(\lambda - y)F(\lambda)\| = \|\Sigma(y+) - \Sigma(y-)\|.$$

The following result establishes an important property of operator-valued Nevanlinna functions which will be characteristic for the Weyl functions in this text; see Chapter 4. The proof given here depends on the integral representation of a Nevanlinna function.

Proposition A.4.6. *Let F be a $\mathbf{B}(\mathcal{G})$-valued Nevanlinna function. Then the following statements hold:*

(i) *If $\operatorname{Im} F(\mu)$ is boundedly invertible for some $\mu \in \mathbb{C} \setminus \mathbb{R}$, then $\operatorname{Im} F(\lambda)$ is boundedly invertible for all $\lambda \in \mathbb{C} \setminus \mathbb{R}$.*

(ii) *Assume that F has an analytic continuation to the interval $(c, d) \subset \mathbb{R}$. If either $\operatorname{Im} F(\mu)$ is boundedly invertible for some $\mu \in \mathbb{C} \setminus \mathbb{R}$ or $F'(x_0)$ is boundedly invertible for some $x_0 \in (c, d)$, then $\operatorname{Im} F(\lambda)$ is boundedly invertible for all $\lambda \in \mathbb{C} \setminus \mathbb{R}$ and $F'(x)$ is boundedly invertible for all $x \in (c, d)$.*

Proof. The proof will be given in three steps and involves the nonnegativity of the operators in (A.4.12) and (A.4.19).

Step 1. Assume for some $\mu \in \mathbb{C} \setminus \mathbb{R}$ that $0 \in \sigma_p(\operatorname{Im} F(\mu))$. Then there exists a nontrivial element $\varphi \in \ker(\operatorname{Im} F(\mu))$. Since $\beta \geq 0$, it follows from (A.4.12) that

$$(\beta\varphi, \varphi) = 0 \quad \text{and} \quad \int_{\mathbb{R}} \frac{1}{|t - \mu|^2} \, d(\Sigma(t)\varphi, \varphi) = 0,$$

and therefore $\varphi \in \ker\beta$ and $(\Sigma(t)\varphi, \varphi) = 0$ for all $t \in \mathbb{R}$. Hence, due to (A.4.12) one sees that $(\operatorname{Im} F(\lambda)\varphi, \varphi) = 0$ for all $\lambda \in \mathbb{C} \setminus \mathbb{R}$. For $\lambda \in \mathbb{C}^+$ one concludes by the nonnegativity of $\operatorname{Im} F(\lambda)$ that $\operatorname{Im} F(\lambda)\varphi = 0$. For $\lambda \in \mathbb{C}^-$ one has that $\operatorname{Im} F(\lambda)\varphi = \operatorname{Im} F(\bar{\lambda})\varphi = 0$. Therefore, it follows that $0 \in \sigma_\mathrm{p}(\operatorname{Im} F(\lambda))$ for all $\lambda \in \mathbb{C} \setminus \mathbb{R}$. Moreover, in case there is an analytic continuation to (c,d), (A.4.19) implies that $(F'(x)\varphi, \varphi) = 0$ for all $x \in (c,d)$. For $x \in (c,d)$ one now concludes by the nonnegativity of $F'(x)$ that $F'(x)\varphi = 0$. Hence, it follows that $0 \in \sigma_\mathrm{p}(F'(x))$ for all $x \in (c,d)$.

If F admits an analytic continuation to (c,d), then the above arguments show that the assumption $0 \in \sigma_\mathrm{p}(F'(x_0))$ for some $x_0 \in (c,d)$ leads to $0 \in \sigma_\mathrm{p}(\operatorname{Im} F(\lambda))$ for all $\lambda \in \mathbb{C} \setminus \mathbb{R}$ and $0 \in \sigma_\mathrm{p}(F'(x))$ for all $x \in (c,d)$.

Step 2. Assume for some $\mu \in \mathbb{C} \setminus \mathbb{R}$ that $0 \in \sigma_\mathrm{c}(\operatorname{Im} F(\mu))$. Then there exists a sequence $\varphi_n \in \mathcal{G}$, $\|\varphi_n\| = 1$, such that $\operatorname{Im} F(\mu)\varphi_n \to 0$ for $n \to \infty$. It follows from (A.4.12) that

$$(\beta\varphi_n, \varphi_n) \to 0 \quad \text{and} \quad \int_\mathbb{R} \frac{1}{|t - \mu|^2}\, d(\Sigma(t)\varphi_n, \varphi_n) \to 0, \quad n \to \infty,$$

and hence also $(\operatorname{Im} F(\lambda)\varphi_n, \varphi_n) \to 0$ for any $\lambda \in \mathbb{C} \setminus \mathbb{R}$. Thus, for $\lambda \in \mathbb{C}^+$ one concludes from the nonnegativity of $\operatorname{Im} F(\lambda)$ that $\operatorname{Im} F(\lambda)\varphi_n \to 0$. Hence, one concludes that $0 \in \sigma_\mathrm{c}(\operatorname{Im} F(\lambda))$ for all $\lambda \in \mathbb{C} \setminus \mathbb{R}$. Moreover, if there is an analytic continuation to (c,d), then it follows from (A.4.19) that

$$(\beta\varphi_n, \varphi_n) \to 0 \quad \text{and} \quad \int_\mathbb{R} \frac{1}{|t - x|^2}\, d(\Sigma(t)\varphi_n, \varphi_n) \to 0, \quad n \to \infty,$$

and hence also $(F'(x)\varphi_n, \varphi_n) \to 0$ for any $x \in (c,d)$. Thus, for $x \in (c,d)$ one concludes by the nonnegativity of $F'(x)$ that $F'(x)\varphi_n \to 0$. Hence, $x \in \sigma_\mathrm{c}(F'(x))$ for all $x \in (c,d)$.

Therefore, it has been shown that the assumption $0 \in \sigma_\mathrm{c}(\operatorname{Im} F(\mu))$ for some $\mu \in \mathbb{C} \setminus \mathbb{R}$ leads to

$$0 \in \sigma_\mathrm{c}(\operatorname{Im} F(\lambda)), \quad \lambda \in \mathbb{C} \setminus \mathbb{R}, \quad \text{and} \quad 0 \in \sigma_\mathrm{c}(F'(x)), \quad x \in (c,d).$$

It is clear that if F admits an analytic continuation to (c,d), the above arguments show that the assumption $0 \in \sigma_\mathrm{c}(F'(x_0))$ for some $x_0 \in (c,d)$ leads to $0 \in \sigma_\mathrm{c}(\operatorname{Im} F(\lambda))$ for all $\lambda \in \mathbb{C} \setminus \mathbb{R}$ and $0 \in \sigma_\mathrm{c}(F'(x))$ for all $x \in (c,d)$.

Step 3. Now assume that $\operatorname{Im} F(\mu)$ is boundedly invertible for some $\mu \in \mathbb{C} \setminus \mathbb{R}$ or, if there is an analytic continuation to (c,d), that $F'(x_0)$ is boundedly invertible for some $x_0 \in (c,d)$, in other words, $0 \in \rho(\operatorname{Im} F(\mu))$ or $0 \in \rho(F'(x_0))$. Since $\operatorname{Im} F(\lambda)$, $\lambda \in \mathbb{C} \setminus \mathbb{R}$, and $F'(x)$, $x \in (c,d)$, are self-adjoint operators in $\mathbf{B}(\mathcal{G})$, it follows that $\sigma_\mathrm{r}(\operatorname{Im} F(\lambda)) = \emptyset$ and $\sigma_\mathrm{r}(F'(x)) = \emptyset$. The assertions (i) and (ii) of the proposition now follow from Step 1 and Step 2. $\qquad\square$

Next the notion of a uniformly strict Nevanlinna function will be defined. It can be used in conjunction with Proposition A.4.6.

Definition A.4.7. Let $\mathcal{G}$ be a Hilbert space and let $F : \mathbb{C} \setminus \mathbb{R} \to \mathbf{B}(\mathcal{G})$ be a Nevanlinna function. The function F is said to be *uniformly strict* if its imaginary part $\operatorname{Im} F(\lambda)$ is boundedly invertible for some, and hence for all $\lambda \in \mathbb{C} \setminus \mathbb{R}$.

In this context it is helpful to recall the following facts. Let $A \in \mathbf{B}(\mathcal{G})$ with $\operatorname{Im} A \geq 0$. Then clearly $\operatorname{Im} A$ is boundedly invertible if and only if $\operatorname{Im} A \geq \varepsilon$ for some $\varepsilon > 0$. Note that if $\operatorname{Im} A \geq \varepsilon$ for some $\varepsilon > 0$, then

$$\varepsilon \|\varphi\|^2 \leq (\operatorname{Im} A \, \varphi, \varphi) \leq |(A\varphi, \varphi)| \leq \|A\varphi\| \|\varphi\|, \quad \varphi \in \mathcal{G},$$

yields that $\varepsilon \|\varphi\| \leq \|A\varphi\|$, $\varphi \in \mathcal{G}$. Therefore, if $\operatorname{Im} A$ is boundedly invertible, then so is A itself, i.e., $A^{-1} \in \mathbf{B}(\mathcal{G})$, and furthermore

$$\operatorname{Im} (-A^{-1}) = A^{-1}(\operatorname{Im} A)A^{-*},$$

so that $\operatorname{Im} (-A^{-1}) \geq 0$, and in fact $\operatorname{Im} (-A^{-1}) \geq \varepsilon'$ for some $\varepsilon' > 0$. The next lemma is now clear.

Lemma A.4.8. *Let $\mathcal{G}$ be a Hilbert space and let $F : \mathbb{C} \setminus \mathbb{R} \to \mathbf{B}(\mathcal{G})$ be a Nevanlinna function. If the function F is uniformly strict, then its inverse $-F^{-1}$ is a uniformly strict $\mathbf{B}(\mathcal{G})$-valued Nevanlinna function and*

$$\operatorname{Im} (-F(\lambda)^{-1}) = F(\lambda)^{-1}(\operatorname{Im} F(\lambda))F(\lambda)^{-*}, \quad \lambda \in \mathbb{C} \setminus \mathbb{R}.$$

In general, for an operator-valued Nevanlinna function F with values in $\mathbf{B}(\mathcal{G})$ the values of the inverse $-F^{-1}$ need not be bounded operators.

A.5   Kac functions

The notion of operator-valued Nevanlinna function in Definition A.4.1 gave rise to the integral representation of Nevanlinna functions in Theorem A.4.2. Next special subclasses of Nevanlinna functions with corresponding integral representations will be considered. Whenever one deals with "unbounded measures" the interpretation of the integrals is again via Proposition A.3.7.

Definition A.5.1. Let $\mathcal{G}$ be a Hilbert space and let $F : \mathbb{C} \setminus \mathbb{R} \to \mathbf{B}(\mathcal{G})$ be a Nevanlinna function. Then F is said to belong to the class of *Kac functions* if for all $\varphi \in \mathcal{G}$

$$\int_1^\infty \frac{\operatorname{Im} (F(iy)\varphi, \varphi)}{y} \, dy < \infty. \tag{A.5.1}$$

Note that every $\mathbf{B}(\mathcal{G})$-valued Nevanlinna function F that satisfies

$$\sup_{y>0} y\big(\operatorname{Im} (F(iy)\varphi, \varphi)\big) < \infty, \quad \varphi \in \mathcal{G}, \tag{A.5.2}$$

is a Kac function.

Theorem A.5.2. *Let $\mathcal{G}$ be a Hilbert space and let $F : \mathbb{C} \setminus \mathbb{R} \to \mathbf{B}(\mathcal{G})$ be an operator function. Then the following statements are equivalent:*

(i) *F has an integral representation of the form*

$$F(\lambda) = \gamma + \int_{\mathbb{R}} \frac{1}{t - \lambda} \, d\Sigma(t), \quad \lambda \in \mathbb{C} \setminus \mathbb{R}, \tag{A.5.3}$$

with a self-adjoint operator $\gamma \in \mathbf{B}(\mathcal{G})$ and a nondecreasing self-adjoint operator function $\Sigma : \mathbb{R} \to \mathbf{B}(\mathcal{G})$ such that

$$\int_{\mathbb{R}} \frac{d\Sigma(t)}{|t| + 1} \in \mathbf{B}(\mathcal{G}), \tag{A.5.4}$$

where the integrals in (A.5.3) and (A.5.4) converge in the strong topology.

(ii) *F is a Kac function.*

If either of these equivalent conditions is satisfied, then $\gamma = \lim_{y \to \infty} F(iy)$ in the strong sense.

Proof. It is helpful to recall the identity

$$\int_1^\infty \frac{1}{t^2 + y^2} \, dy = \frac{1}{|t|} \left(\frac{\pi}{2} - \arctan \frac{1}{|t|} \right), \quad t \neq 0, \tag{A.5.5}$$

and the fact that

$$\lim_{|t| \to 0} \frac{1}{|t|} \left(\frac{\pi}{2} - \arctan \frac{1}{|t|} \right) = 1. \tag{A.5.6}$$

(i) $\Rightarrow$ (ii) Write the representation (A.5.3) as

$$F(\lambda) = \gamma + \int_{\mathbb{R}} g_\lambda(t) \frac{d\Sigma(t)}{|t| + 1}, \quad \lambda \in \mathbb{C} \setminus \mathbb{R},$$

where the bounded continuous function $g_\lambda : \mathbb{R} \to \mathbb{C}$ is given by

$$g_\lambda(t) = \frac{|t| + 1}{t - \lambda}.$$

Due to the condition (A.5.4) it follows that $F(\lambda)$ in (A.5.3) is well defined and represents an element in $\mathbf{B}(\mathcal{G})$; cf. Proposition A.3.7. Moreover, (A.5.3) and (A.3.25) imply that for $\varphi \in \mathcal{G}$

$$(F(\lambda)\varphi, \varphi) = (\gamma\varphi, \varphi) + \int_{\mathbb{R}} g_\lambda(t) \frac{d(\Sigma(t)\varphi, \varphi)}{|t| + 1}, \quad \lambda \in \mathbb{C} \setminus \mathbb{R}.$$

Therefore, one sees that the function F is holomorphic on $\mathbb{C} \setminus \mathbb{R}$. Furthermore, it is clear that $F(\lambda)^* = F(\bar{\lambda})$; cf. (A.3.24). Since $(\gamma\varphi, \varphi) \in \mathbb{R}$, one also sees that

$$\frac{\mathrm{Im}\,(F(\lambda)\varphi, \varphi)}{\mathrm{Im}\,\lambda} = \int_{\mathbb{R}} \frac{1}{|t - \lambda|^2} \, d(\Sigma(t)\varphi, \varphi).$$

Thus, F is a Nevanlinna function. Furthermore, integration of this identity shows, after a change of the order of integration, that

$$\int_1^\infty \frac{\operatorname{Im}(F(iy)\varphi, \varphi)}{y}\, dy = \int_{\mathbb{R}} \frac{1}{|t|}\left(\frac{\pi}{2} - \arctan\frac{1}{|t|}\right) d(\Sigma(t)\varphi, \varphi) < \infty.$$

Here the identity (A.5.5) was used in the first equality and (A.5.4), (A.5.6), and $\arctan 1/|t| \to 0$ for $|t| \to \infty$ were used to conclude that the last integral is finite; cf. Proposition A.3.7. This shows that F is a Kac function.

(ii) $\Rightarrow$ (i) Since F is a Nevanlinna function, it follows from the integral representation (A.4.1) and (A.3.25) that for all $\varphi \in \mathcal{G}$

$$\frac{\operatorname{Im}(F(iy)\varphi, \varphi)}{y} = (\beta\varphi, \varphi) + \int_{\mathbb{R}} \frac{1}{t^2 + y^2}\, d(\Sigma(t)\varphi, \varphi).$$

Each of the terms on the right-hand side is nonnegative. Hence, the integrability condition (A.5.1) implies that $\beta = 0$ and furthermore

$$\int_1^\infty \left(\int_{\mathbb{R}} \frac{1}{t^2 + y^2}\, d(\Sigma(t)\varphi, \varphi)\right) dy < \infty.$$

Changing the order of integration and using (A.5.5) gives

$$\int_{\mathbb{R}} \frac{1}{|t|}\left(\frac{\pi}{2} - \arctan\frac{1}{|t|}\right) d(\Sigma(t)\varphi, \varphi) < \infty,$$

and hence $\int_{\mathbb{R}} (1 + |t|)^{-1}\, d(\Sigma(t)\varphi, \varphi) < \infty$. By Proposition A.3.7, this implies that (A.5.4) is satisfied. Observe that for each compact interval $[a, b] \subset \mathbb{R}$ one has the identity

$$\int_a^b \frac{1 + \lambda t}{t - \lambda}\frac{d\Sigma(t)}{t^2 + 1} = \int_a^b \frac{|t| + 1}{t - \lambda}\frac{d\Sigma(t)}{|t| + 1} - \int_a^b \frac{|t| + 1}{t^2 + 1}\frac{d\Sigma(t)}{|t| + 1},$$

and now all integrals have strong limits as $[a, b] \to \mathbb{R}$ by Proposition A.3.7. Hence, one obtains from (A.4.1) with $\beta = 0$ that

$$F(\lambda) = \alpha + \int_{\mathbb{R}} \frac{1 + \lambda t}{t - \lambda}\frac{d\Sigma(t)}{t^2 + 1} = \alpha + \int_{\mathbb{R}} \frac{d\Sigma(t)}{t - \lambda} - \int_{\mathbb{R}} \frac{t}{t^2 + 1}\, d\Sigma(t).$$

Observe that

$$\int_{\mathbb{R}} \frac{t}{t^2 + 1}\, d\Sigma(t) = \int_{\mathbb{R}} \frac{t(|t| + 1)}{t^2 + 1}\frac{d\Sigma(t)}{|t| + 1}$$

is a self-adjoint operator in $\mathbf{B}(\mathcal{G})$. Hence, with γ defined by

$$\gamma = \alpha - \int_{\mathbb{R}} \frac{t}{t^2 + 1}\, d\Sigma(t)$$

one sees that $\gamma \in \mathbf{B}(\mathcal{G})$ is self-adjoint and the assertion (i) follows.

Finally, observe that by means of (A.3.26) one has

$$\lim_{y\to\infty} (F(iy) - \gamma)\varphi = \lim_{y\to\infty} \left(\int_{\mathbb{R}} \frac{1}{t - iy} d\Sigma(t) \right) \varphi = 0$$

for all $\varphi \in \mathcal{G}$. This gives the last assertion. □

A subclass of the Kac functions in Theorem A.5.2 concerns the case with bounded "measures".

Proposition A.5.3. *Let $\mathcal{G}$ be a Hilbert space and let $F : \mathbb{C}\setminus\mathbb{R} \to \mathbf{B}(\mathcal{G})$ be an operator function. Then the following statements are equivalent:*

(i) *F has an integral representation of the form*

$$F(\lambda) = \gamma + \int_{\mathbb{R}} \frac{1}{t - \lambda} d\Sigma(t), \quad \lambda \in \mathbb{C}\setminus\mathbb{R},$$

with a self-adjoint operator $\gamma \in \mathbf{B}(\mathcal{G})$ and a nondecreasing self-adjoint operator function $\Sigma : \mathbb{R} \to \mathbf{B}(\mathcal{G})$ such that $\int_{\mathbb{R}} d\Sigma(t) \in \mathbf{B}(\mathcal{G})$.

(ii) *F is a Kac function satisfying (A.5.2).*

If either of these equivalent conditions is satisfied, then $\gamma = \lim_{y\to\infty} F(iy)$ in the strong sense, and furthermore, for all $\varphi \in \mathcal{G}$

$$\int_{\mathbb{R}} d(\Sigma(t)\varphi, \varphi) = \sup_{y>0} y(\operatorname{Im} F(iy)\varphi, \varphi) < \infty. \tag{A.5.7}$$

Proof. (i) ⇒ (ii) It follows from Theorem A.5.2 that F is a Kac function. Moreover,

$$\sup_{y>0} y\big(\operatorname{Im} F(iy)\varphi, \varphi\big) = \sup_{y>0} \int_{\mathbb{R}} \frac{y^2}{t^2 + y^2} d(\Sigma(t)\varphi, \varphi)$$

$$= \int_{\mathbb{R}} d(\Sigma(t)\varphi, \varphi) - \inf_{y>0} \int_{\mathbb{R}} \frac{t^2}{t^2 + y^2} d(\Sigma(t)\varphi, \varphi) \tag{A.5.8}$$

$$= \int_{\mathbb{R}} d(\Sigma(t)\varphi, \varphi),$$

which shows that the condition (A.5.2) is satisfied.

(ii) ⇒ (i) Since F is a Kac function, the integral representation follows from Theorem A.5.2 with a nondecreasing self-adjoint operator function $\Sigma : \mathbb{R} \to \mathbf{B}(\mathcal{G})$ which satisfies the integrability condition (A.5.4). Now it follows from the assumption (A.5.2) and (A.5.8) that $\int_{\mathbb{R}} d\Sigma(t) \in \mathbf{B}(\mathcal{G})$.

Moreover, the identity $\gamma = \lim_{y\to\infty} F(iy)$ in the strong sense is clear from Theorem A.5.2 and (A.5.7) was shown in (A.5.8). □

The previous proposition has an interesting consequence for a further subclass of the Kac functions which satisfy (A.5.2). The following result is connected with the characterization of generalized resolvents; see Chapter 4, where also the Sz.-Nagy dilation theorem is treated. Note that the conditions $\gamma = \gamma^*$ and $\int_{\mathbb{R}} d\Sigma(t) \in \mathbf{B}(\mathcal{G})$ in Proposition A.5.3 are now specialized to the conditions $\gamma = 0$ and $\| \int_{\mathbb{R}} d\Sigma(t) \| \leq 1$.

Proposition A.5.4. *Let $\mathcal{G}$ be a Hilbert space and let $F : \mathbb{C} \setminus \mathbb{R} \to \mathbf{B}(\mathcal{G})$ be an operator function. Then the following statements are equivalent:*

(i) *F has an integral representation of the form*

$$F(\lambda) = \int_{\mathbb{R}} \frac{1}{t - \lambda}\, d\Sigma(t), \quad \lambda \in \mathbb{C} \setminus \mathbb{R},$$

with a nondecreasing self-adjoint operator function $\Sigma : \mathbb{R} \to \mathbf{B}(\mathcal{G})$ such that $\int_{\mathbb{R}} d\Sigma(t) \in \mathbf{B}(\mathcal{G})$ and $\| \int_{\mathbb{R}} d\Sigma(t) \| \leq 1$.

(ii) *F is a Nevanlinna function which satisfies*

$$\frac{\operatorname{Im} F(\lambda)}{\operatorname{Im} \lambda} - F(\lambda)^* F(\lambda) \geq 0, \quad \lambda \in \mathbb{C} \setminus \mathbb{R}.$$

Proof. (i) $\Rightarrow$ (ii) It is clear from Proposition A.5.3 that F is a Kac function and hence a Nevanlinna function. Moreover, for all $\lambda \in \mathbb{C} \setminus \mathbb{R}$ one has

$$\frac{(\operatorname{Im} F(\lambda)\varphi, \varphi)}{\operatorname{Im} \lambda} = \int_{\mathbb{R}} \frac{1}{|t - \lambda|^2}\, d(\Sigma(t)\varphi, \varphi), \quad \varphi \in \mathcal{G}.$$

Since $\|\Theta(+\infty) - \Theta(-\infty)\| \leq 1$, it follows from Proposition A.3.4 that

$$(F(\lambda)^* F(\lambda)\varphi, \varphi) = \|F(\lambda)\varphi\|^2 = \left\| \left(\int_{\mathbb{R}} \frac{1}{t - \lambda}\, d\Sigma(t) \right) \varphi \right\|^2$$

$$\leq \int_{\mathbb{R}} \frac{1}{|t - \lambda|^2}\, d(\Sigma(t)\varphi, \varphi), \quad \varphi \in \mathcal{G},$$

which gives the desired result.

(ii) $\Rightarrow$ (i) Let F be a Nevanlinna function which satisfies

$$|\operatorname{Im} \lambda|\, \|F(\lambda)\varphi\|^2 \leq |\operatorname{Im} (F(\lambda)\varphi, \varphi)|, \quad \varphi \in \mathcal{G}, \quad \lambda \in \mathbb{C} \setminus \mathbb{R}.$$

Then

$$|\operatorname{Im} \lambda|\, \|F(\lambda)\varphi\| \leq \|\varphi\|, \quad \varphi \in \mathcal{G}, \quad \lambda \in \mathbb{C} \setminus \mathbb{R},$$

which leads to

$$|\operatorname{Im} \lambda|\, |(F(\lambda)\varphi, \varphi)| \leq \|\varphi\|^2, \quad \varphi \in \mathcal{G}, \quad \lambda \in \mathbb{C} \setminus \mathbb{R}.$$

In particular, one has

$$|\operatorname{Im} \lambda|\, |\operatorname{Im} (F(\lambda)\varphi, \varphi)| \le \|\varphi\|^2, \quad \varphi \in \mathcal{G}, \quad \lambda \in \mathbb{C} \setminus \mathbb{R}, \tag{A.5.9}$$

and

$$|\operatorname{Im} \lambda|\, |\operatorname{Re} (F(\lambda)\varphi, \varphi)| \le \|\varphi\|^2, \quad \varphi \in \mathcal{G}, \quad \lambda \in \mathbb{C} \setminus \mathbb{R}. \tag{A.5.10}$$

The inequality (A.5.9) and Proposition A.5.3 imply that F admits the integral representation

$$F(\lambda) = \gamma + \int_{\mathbb{R}} \frac{1}{t - \lambda}\, d\Sigma(t), \quad \lambda \in \mathbb{C} \setminus \mathbb{R},$$

with $\int_{\mathbb{R}} d\Sigma(t) \in \mathbf{B}(\mathcal{G})$, and hence $\int_{\mathbb{R}} d(\Sigma(t)\varphi, \varphi) < \infty$. Moreover, it follows from (A.5.9) and (A.5.7) that

$$\int_{\mathbb{R}} d(\Sigma(t)\varphi, \varphi) = \sup_{y>0} y\big(\operatorname{Im} F(iy)\varphi, \varphi\big) \le \|\varphi\|^2.$$

Hence, one sees that $\| \int_{\mathbb{R}} d\Sigma(t)\| \le 1$. Finally, note that

$$\operatorname{Re} (F(iy)\varphi, \varphi) = (\gamma\varphi, \varphi) + \int_{\mathbb{R}} \frac{t}{t^2 + y^2}\, d(\Sigma(t)\varphi, \varphi), \quad y > 0.$$

Now (A.5.10) and the dominated convergence theorem show that $\gamma = 0$. $\square$

A.6 Stieltjes and inverse Stieltjes functions

Let F be a $\mathbf{B}(\mathcal{G})$-valued Nevanlinna function F with the integral representation (A.4.1) as in Theorem A.4.2. Recall from Proposition A.4.5 that F is holomorphic on $\mathbb{C} \setminus [c, \infty)$ if and only if the function $t \mapsto \Sigma(t)\varphi$ is constant on $(-\infty, c)$ for all $\varphi \in \mathcal{G}$, and in this case one has the integral representation

$$F(\lambda) = \alpha + \lambda\beta + \int_{[c,\infty)} \left(\frac{1}{t - \lambda} - \frac{t}{t^2 + 1} \right) d\Sigma(t), \tag{A.6.1}$$

where the integral converges in the strong topology for all $\lambda \in \mathbb{C} \setminus [c, \infty)$; cf. (A.3.28). Note that (A.6.1) implies

$$\beta\varphi = \lim_{x \downarrow -\infty} \frac{F(x)}{x}\varphi, \quad \varphi \in \mathcal{G}. \tag{A.6.2}$$

This identity can be verified in the same way as in the proof of Lemma A.2.6 using (A.3.26); cf. Lemma A.4.3. Moreover, $x \mapsto F(x)$ is a $\mathbf{B}(\mathcal{G})$-valued operator function on $(-\infty, c)$ with self-adjoint operators as values and one has

$$F(x_1) \le F(x_2), \quad x_1 < x_2 < c, \tag{A.6.3}$$

by Proposition A.4.5. In the present section this class of Nevanlinna functions F will now be further specified by requiring, in addition to holomorphy of F on $\mathbb{C} \setminus [c, \infty)$, a sign condition for the values $F(x)$ for $x \in (-\infty, c)$.

Definition A.6.1. Let $\mathcal{G}$ be a Hilbert space and let $c \in \mathbb{R}$ be fixed. A Nevanlinna function $F : \mathbb{C} \setminus [c, \infty) \to \mathbf{B}(\mathcal{G})$ is said to belong to the class $\mathbf{S}_{\mathcal{G}}(-\infty, c)$ of *Stieltjes functions* if

(i) F is holomorphic on $\mathbb{C} \setminus [c, \infty)$;

(ii) $F(x) \geq 0$ for all $x < c$.

Similarly, a Nevanlinna function $F : \mathbb{C} \setminus [c, \infty) \to \mathbf{B}(\mathcal{G})$ is said to belong to the class $\mathbf{S}_{\mathcal{G}}^{-1}(-\infty, c)$ of *inverse Stieltjes functions* if

(i) F is holomorphic on $\mathbb{C} \setminus [c, \infty)$;

(ii) $F(x) \leq 0$ for all $x < c$.

Thus, if $F \in \mathbf{S}_{\mathcal{G}}(-\infty, c)$, then $x \mapsto F(x)$ is an operator function on $(-\infty, c)$ and

$$0 \leq F(x_1) \leq F(x_2), \quad x_1 < x_2 < c.$$

In particular, $\lim_{x \downarrow -\infty} F(x)$ exists in the strong sense and defines a nonnegative self-adjoint operator in $\mathbf{B}(\mathcal{G})$; cf. Lemma A.3.1. There is also a limit as $x \uparrow c$ in the sense of relations, see for details Chapter 5. Similarly, one sees that if $F \in \mathbf{S}_{\mathcal{G}}^{-1}(-\infty, c)$, then $x \mapsto F(x)$ is an operator function on $(-\infty, c)$ and

$$F(x_1) \leq F(x_2) \leq 0, \quad x_1 < x_2 < c.$$

In particular, $\lim_{x \uparrow c} F(x)$ exists in the strong sense and defines a nonnegative self-adjoint operator in $\mathbf{B}(\mathcal{G})$; cf. Lemma A.3.1. There is also a limit as $x \downarrow -\infty$ in the sense of relations, see for details Chapter 5.

The characterization of Nevanlinna functions in Theorem A.4.2 can be specialized for Stieltjes functions and, in fact, the following result also shows that the Stieltjes functions form a subclass of the Kac functions; cf. Theorem A.5.2.

Theorem A.6.2. *Let $\mathcal{G}$ be a Hilbert space, let $F : \mathbb{C} \setminus \mathbb{R} \to \mathbf{B}(\mathcal{G})$ be an operator function, and let $c \in \mathbb{R}$. Then the following statements are equivalent:*

(i) *F has an integral representation of the form*

$$F(\lambda) = \gamma + \int_{[c,\infty)} \frac{1}{t - \lambda} \, d\Sigma(t), \quad \lambda \in \mathbb{C} \setminus [c, \infty), \quad (A.6.4)$$

with a nonnegative self-adjoint operator $\gamma \in \mathbf{B}(\mathcal{G})$, a nondecreasing self-adjoint operator function $\Sigma : \mathbb{R} \to \mathbf{B}(\mathcal{G})$ such that $t \mapsto \Sigma(t)\varphi$, $\varphi \in \mathcal{G}$, is constant on $(-\infty, c)$, and

$$\int_{[c,\infty)} \frac{d(\Sigma(t)\varphi, \varphi)}{|t| + 1} < \infty, \quad \varphi \in \mathcal{G}, \quad (A.6.5)$$

where the integral in the representation (A.6.4) is interpreted in the weak topology for all $\lambda \in \mathbb{C} \setminus [c, \infty)$.

(ii) *F is a Stieltjes function in $\mathbf{S}_{\mathcal{G}}(-\infty, c)$.*

Proof. (i) $\Rightarrow$ (ii) Assume that the function F has the integral representation
(A.6.4) with $\gamma \in \mathbf{B}(\mathcal{G})$ and $\Sigma : \mathbb{R} \to \mathbf{B}(\mathcal{G})$ as above, such that (A.6.5) holds.
The condition (A.6.5) ensures that the integral in (A.6.4) is well defined in the
weak sense. From the integral representation (A.6.4) one sees that F is holomorphic on $\mathbb{C} \setminus [c, \infty)$,

$$\frac{(\mathrm{Im}\, F(\lambda)\varphi, \varphi)}{\mathrm{Im}\, \lambda} = \int_{[c,\infty)} \frac{1}{|t - \lambda|^2}\, d(\Sigma(t)\varphi, \varphi) \geq 0, \quad \lambda \in \mathbb{C} \setminus \mathbb{R}, \quad \varphi \in \mathcal{G},$$

and $F(\lambda)^* = F(\bar{\lambda})$ for $\lambda \in \mathbb{C} \setminus \mathbb{R}$. Hence, F is a $\mathbf{B}(\mathcal{G})$-valued Nevanlinna function.
Moreover, $\gamma \geq 0$ and

$$(F(x)\varphi, \varphi) = (\gamma\varphi, \varphi) + \int_{[c,\infty)} \frac{1}{t - x}\, d(\Sigma(t)\varphi, \varphi), \quad x \in (-\infty, c), \quad \varphi \in \mathcal{G},$$

imply $F(x) \geq 0$ for all $x < c$. Thus, $F \in \mathbf{S}_\mathcal{G}(-\infty, c)$.

(ii) $\Rightarrow$ (i) Assume that F is a Nevanlinna function in $\mathbf{S}_\mathcal{G}(-\infty, c)$. Then F has the
integral representation (A.6.1), the operators $F(x)$ defined for $-\infty < x < c$ satisfy
(A.6.3), and they are all nonnegative. Hence, the limit

$$\gamma = \lim_{x \downarrow -\infty} F(x) \in \mathbf{B}(\mathcal{G}) \tag{A.6.6}$$

exists in the strong sense and one has $\gamma \geq 0$ by Lemma A.3.1. Note that (A.6.2)
implies $\beta = 0$ in (A.6.1). Therefore, (A.6.1) implies that for all $x < c$:

$$-\int_{[c,\infty)} \frac{1 + tx}{t - x} \frac{d(\Sigma(t)\varphi, \varphi)}{t^2 + 1} \leq (\alpha\varphi, \varphi), \quad \varphi \in \mathcal{G}.$$

Letting $x \downarrow -\infty$ and using the monotone convergence theorem one obtains

$$\int_{[c,\infty)} \frac{t}{t^2 + 1}\, d(\Sigma(t)\varphi, \varphi) \leq (\alpha\varphi, \varphi), \quad \varphi \in \mathcal{G}. \tag{A.6.7}$$

Since

$$\frac{d(\Sigma(t)\varphi, \varphi)}{|t| + 1} = \frac{t^2 + 1}{t(|t| + 1)} \frac{t}{t^2 + 1}\, d(\Sigma(t)\varphi, \varphi)$$

for large t, one concludes from (A.6.7) that (A.6.5) holds. Moreover, one obtains
from (A.6.6) and (A.6.1) that

$$\gamma = \alpha - \int_{[c,\infty)} \frac{t}{t^2 + 1}\, d\Sigma(t).$$

Thus, one may rewrite the integral representation (A.6.1) in the form (A.6.4). $\square$

The inverse Stieltjes functions can also be characterized by integral representations. Here is the analog of Theorem A.6.2.

Theorem A.6.3. *Let $\mathcal{G}$ be a Hilbert space, let $F : \mathbb{C} \setminus \mathbb{R} \to \mathbf{B}(\mathcal{G})$ be an operator function, and let $c \in \mathbb{R}$. Then the following statements are equivalent:*

(i) *F has an integral representation of the form*

$$F(\lambda) = L + \beta(\lambda - c) + \int_{[c,\infty)} \left(\frac{1}{t - \lambda} - \frac{1}{t - c} \right) d\Sigma(t), \quad \lambda \in \mathbb{C} \setminus [c, \infty), \quad \text{(A.6.8)}$$

with $L, \beta \in \mathbf{B}(\mathcal{G})$, $L \leq 0$, $\beta \geq 0$, a nondecreasing self-adjoint operator function $\Sigma : \mathbb{R} \to \mathbf{B}(\mathcal{G})$ such that $t \mapsto \Sigma(t)\varphi$, $\varphi \in \mathcal{G}$, is constant on $(-\infty, c)$, and

$$\int_{[c,\infty)} \frac{d(\Sigma(t)\varphi, \varphi)}{(t - c)(|t| + 1)} < \infty, \qquad \varphi \in \mathcal{G}, \qquad \text{(A.6.9)}$$

where the integral in the representation (A.6.8) is interpreted in the weak topology for all $\lambda \in \mathbb{C} \setminus [c, \infty)$.

(ii) *F is an inverse Stieltjes function in $\mathbf{S}_{\mathcal{G}}^{-1}(-\infty, c)$.*

Proof. (i) $\Rightarrow$ (ii) Assume that the function F has the integral representation (A.6.8) with $L, \beta \in \mathbf{B}(\mathcal{G})$ and $\Sigma : \mathbb{R} \to \mathbf{B}(\mathcal{G})$ as above such that the condition (A.6.9) holds. First observe that

$$\int_{[c,\infty)} \left(\frac{1}{t - \lambda} - \frac{1}{t - c} \right) d(\Sigma(t)\varphi, \varphi) = \int_{[c,\infty)} \frac{(\lambda - c)(|t| + 1)}{t - \lambda} \frac{d(\Sigma(t)\varphi, \varphi)}{(t - c)(|t| + 1)}$$

for $\varphi \in \mathcal{G}$, and hence condition (A.6.9) ensures that the integral in (A.6.8) converges in the weak topology for all $\lambda \in \mathbb{C} \setminus [c, \infty)$. It also follows from (A.6.8) that F is holomorphic on $\mathbb{C} \setminus [c, \infty)$,

$$\frac{(\operatorname{Im} F(\lambda)\varphi, \varphi)}{\operatorname{Im} \lambda} = (\beta\varphi, \varphi) + \int_{[c,\infty)} \frac{1}{|t - \lambda|^2} d(\Sigma(t)\varphi, \varphi) \geq 0$$

for $\lambda \in \mathbb{C} \setminus \mathbb{R}$ and $\varphi \in \mathcal{G}$, and $F(\lambda)^* = F(\bar{\lambda})$ for $\operatorname{Im} \lambda \neq 0$. This shows that F is a $\mathbf{B}(\mathcal{G})$-valued Nevanlinna function. Now if $x < c$, then $\beta(x - c) \leq 0$ and the integrand in (A.6.8) is nonpositive. Since also $L \leq 0$, one concludes that $F(x) \leq 0$ for all $x < c$. Thus, $F \in \mathbf{S}_{\mathcal{G}}^{-1}(-\infty, c)$.

(ii) $\Rightarrow$ (i) Assume that the function F is a Nevanlinna function in $\mathbf{S}_{\mathcal{G}}^{-1}(-\infty, c)$. Then F has the integral representation (A.6.1) and the operators $F(x)$ defined for $-\infty < x < c$ satisfy (A.6.3) and they are all nonpositive. Hence, the limit

$$L = \lim_{x \uparrow c} F(x) \in \mathbf{B}(\mathcal{G})$$

exists in the strong sense and one has $L \leq 0$ by Lemma A.3.1. Furthermore, for $\varphi \in \mathcal{G}$ and $-\infty < x < c$ one has

$$(F(x)\varphi, \varphi) = (\alpha\varphi, \varphi) + x(\beta\varphi, \varphi) + \int_{[c,\infty)} \left(\frac{1}{t-x} - \frac{t}{t^2+1} \right) d(\Sigma(t)\varphi, \varphi),$$

so that for $x \uparrow c$ the monotone convergence theorem gives

$$(L\varphi, \varphi) = (\alpha\varphi, \varphi) + c(\beta\varphi, \varphi) + \int_{[c,\infty)} \left(\frac{1}{t-c} - \frac{t}{t^2+1} \right) d(\Sigma(t)\varphi, \varphi). \quad \text{(A.6.10)}$$

In particular, the integral on the right-hand side of (A.6.10) exists for all $\varphi \in \mathcal{G}$. From

$$\frac{d(\Sigma(t)\varphi, \varphi)}{(t-c)(|t|+1)} = \frac{t^2+1}{(1+tc)(|t|+1)} \left(\frac{1}{t-c} - \frac{t}{t^2+1} \right) d(\Sigma(t)\varphi, \varphi)$$

one then concludes that (A.6.9) holds. Using (A.6.10) the integral representation (A.6.1) can be rewritten as

$$(F(\lambda)\varphi, \varphi) = (L\varphi, \varphi) + (\lambda - c)(\beta\varphi, \varphi) + \int_{[c,\infty)} \left(\frac{1}{t-\lambda} - \frac{1}{t-c} \right) d(\Sigma(t)\varphi, \varphi)$$

for $\lambda \in \mathbb{C} \setminus [c, \infty)$ and $\varphi \in \mathcal{G}$. This proves (A.6.8). $\qquad\square$

Remark A.6.4. The integral representations in (A.6.4) and (A.6.8) are understood in the weak sense. Using Proposition A.3.7 one can verify that the integral representation for Stieltjes functions in (A.6.4) remains valid in the strong sense and that the integrability condition (A.6.5) can be replaced by the condition

$$\int_{[c,\infty)} \frac{d\Sigma(t)}{|t|+1} \in \mathbf{B}(\mathcal{G}),$$

where the integral exists in the strong sense. Within the theory of operator-valued integrals developed in Section A.3, the integral representation (A.6.8) for inverse Stieltjes functions and the integrability condition (A.6.9) cannot be interpreted directly in the strong sense.

Let F be a Nevanlinna function and let $c \in \mathbb{R}$. Define the functions F_c and F_c^- by

$$F_c(\lambda) = (\lambda - c)F(\lambda), \quad \lambda \in \mathbb{C} \setminus \mathbb{R},$$

and

$$F_c^-(\lambda) = (\lambda - c)^{-1}F(\lambda), \quad \lambda \in \mathbb{C} \setminus \mathbb{R}.$$

The class of Nevanlinna functions is not stable under either of the mappings

$$F \mapsto F_c \quad \text{or} \quad F \mapsto F_c^-.$$

In fact, the next results show that the set of Nevanlinna functions that is stable under these mappings coincides with $\mathbf{S}_\mathcal{G}(-\infty, c)$ or $\mathbf{S}_\mathcal{G}^{-1}(-\infty, c)$, respectively.

Proposition A.6.5. *Let $\mathcal{G}$ be a Hilbert space, let F be a $\mathbf{B}(\mathcal{G})$-valued Nevanlinna function F, and let $c \in \mathbb{R}$. Then the following statements are equivalent:*

(i) *F belongs to the class $\mathbf{S}_{\mathcal{G}}(-\infty, c)$;*

(ii) *F_c is a Nevanlinna function.*

In fact, if $F \in \mathbf{S}_{\mathcal{G}}(-\infty, c)$, then $F_c \in \mathbf{S}_{\mathcal{G}}^{-1}(-\infty, c)$.

Proof. (i) $\Rightarrow$ (ii) Assume that F belongs to the class $\mathbf{S}_{\mathcal{G}}(-\infty, c)$. Then, by Theorem A.6.2, the function F has the integral representation (A.6.4) interpreted in the weak sense. This implies that F_c has the representation

$$F_c(\lambda) = \gamma(\lambda - c) + \int_{[c,\infty)} \frac{\lambda - c}{t - \lambda} \, d\Sigma(t), \quad \lambda \in \mathbb{C} \setminus [c, \infty), \qquad (A.6.11)$$

in the weak sense. It follows that F_c is holomorphic on $\mathbb{C} \setminus [c, \infty)$,

$$\frac{(\operatorname{Im} F_c(\lambda)\varphi, \varphi)}{\operatorname{Im} \lambda} = (\gamma\varphi, \varphi) + \int_{[c,\infty)} \frac{t - c}{|t - \lambda|^2} \, d(\Sigma(t)\varphi, \varphi) \geq 0$$

for $\lambda \in \mathbb{C} \setminus \mathbb{R}$ and $\varphi \in \mathcal{G}$, and $F_c(\lambda)^* = F_c(\bar{\lambda})$ for $\lambda \in \mathbb{C} \setminus \mathbb{R}$. Therefore, F_c is a $\mathbf{B}(\mathcal{G})$-valued Nevanlinna function. It is also clear from (A.6.11) that $F_c(x) \leq 0$ for $x < c$, and hence $F_c \in \mathbf{S}_{\mathcal{G}}^{-1}(-\infty, c)$.

(ii) $\Rightarrow$ (i) Assume that with F also F_c is a Nevanlinna function. Let $\varphi \in \mathcal{G}$ and express the scalar Nevanlinna function $f_\varphi(\lambda) = (F(\lambda)\varphi, \varphi)$ with its integral representation

$$f_\varphi(\lambda) = \alpha_\varphi + \beta_\varphi \lambda + \int_{\mathbb{R}} \left(\frac{1}{t - \lambda} - \frac{t}{t^2 + 1} \right) d\sigma_\varphi(t), \quad \lambda \in \mathbb{C} \setminus \mathbb{R},$$

in Theorem A.2.5. The function $g(\lambda) = \lambda - c$, $c \in \mathbb{R}$, is entire and real on $\mathbb{R}$. According to the Stieltjes inversion formula in Lemma A.2.7, for any compact subinterval $[a, b] \subset (-\infty, c)$ one obtains

$$\lim_{\varepsilon \downarrow 0} \frac{1}{2\pi i} \int_a^b \left[(gf_\varphi)(s + i\varepsilon) - (gf_\varphi)(s - i\varepsilon) \right] ds$$
$$= \frac{1}{2} \int_{\{a\}} (t - c) \, d\sigma_\varphi(t) + \int_{a+}^{b-} (t - c) \, d\sigma_\varphi(t) + \frac{1}{2} \int_{\{b\}} (t - c) \, d\sigma_\varphi(t). \qquad (A.6.12)$$

Since the function $g(\lambda)f_\varphi(\lambda) = (F_c(\lambda)\varphi, \varphi)$ is also a Nevanlinna function, the limit in (A.6.12) is nonnegative. However, since $t - c < 0$ for all $t \in [a, b]$ one concludes that the right-hand side in (A.6.12) is nonpositive and consequently $\sigma(t)$ must take a constant value on the whole interval $[a, b]$. Proposition A.2.9 implies that f_φ is holomorphic on $\mathbb{C} \setminus [c, \infty)$ for all $\varphi \in \mathcal{G}$. This shows that F is holomorphic on $\mathbb{C} \setminus [c, \infty)$.

Finally, to see that F takes nonnegative values on the interval $(-\infty, c)$ Proposition A.2.9 will be applied. Again consider the two functions $f_\varphi(\lambda)$ and $(\lambda - c)f_\varphi(\lambda)$. Both of them are differentiable and nondecreasing on the interval $(-\infty, c)$. In particular, for all $x < c$ one has $f'_\varphi(x) \geq 0$ and

$$f_\varphi(x) + (x - c)f'_\varphi(x) = \frac{d}{dx}\left((x - c)f_\varphi(x)\right) \geq 0.$$

Since $x < c$, this implies that

$$f_\varphi(x) \geq (c - x)f'_\varphi(x) \geq 0,$$

and hence $(F(x)\varphi, \varphi) \geq 0$ for all $\varphi \in \mathcal{G}$. Therefore, $F(x) \geq 0$ for all $x < c$ and the claim $F \in \mathbf{S}_\mathcal{G}(-\infty, c)$ is proved. $\square$

The next proposition is a variant of Proposition A.6.5 for inverse Stieltjes functions.

Proposition A.6.6. *Let $\mathcal{G}$ be a Hilbert space, let F be a $\mathbf{B}(\mathcal{G})$-valued Nevanlinna function F, and let $c \in \mathbb{R}$. Then the following statements are equivalent:*

(i) *F belongs to the class $\mathbf{S}_\mathcal{G}^{-1}(-\infty, c)$;*

(ii) *F_c^- is a Nevanlinna function.*

In fact, if $F \in \mathbf{S}_\mathcal{G}^{-1}(-\infty, c)$, then $F_c^- \in \mathbf{S}_\mathcal{G}(-\infty, c)$.

Proof. (i) $\Rightarrow$ (ii) Assume that F belongs to the class $\mathbf{S}_\mathcal{G}^{-1}(-\infty, c)$. Then, by Theorem A.6.3, the function F has the integral representation (A.6.8) interpreted in the weak sense. This implies that F_c^- has the representation

$$F_c^-(\lambda) = \frac{-L}{c - \lambda} + \beta + \int_{(c,\infty)} \frac{d\Sigma(t)}{(t - c)(t - \lambda)}, \quad \lambda \in \mathbb{C} \setminus [c, \infty), \qquad (A.6.13)$$

in the weak sense. It follows that F_c^- is holomorphic on $\mathbb{C} \setminus [c, \infty)$,

$$\frac{(\operatorname{Im} F_c^-(\lambda)\varphi, \varphi)}{\operatorname{Im} \lambda} = \frac{(-L\varphi, \varphi)}{|c - \lambda|^2} + \int_{[c,\infty)} \frac{1}{(t - c)|t - \lambda|^2} \, d(\Sigma(t)\varphi, \varphi) \geq 0$$

for $\lambda \in \mathbb{C} \setminus \mathbb{R}$ and $\varphi \in \mathcal{G}$, and $F_c^-(\lambda)^* = F_c^-(\bar{\lambda})$ for $\lambda \in \mathbb{C} \setminus \mathbb{R}$. Therefore, F_c^- is a $\mathbf{B}(\mathcal{G})$-valued Nevanlinna function. It is also clear from (A.6.13) that $F_c^-(x) \geq 0$ for $x < c$, and hence $F_c^- \in \mathbf{S}_\mathcal{G}(-\infty, c)$.

(ii) $\Rightarrow$ (i) Assume that with F also F_c^- is a Nevanlinna function. Let $\varphi \in \mathcal{G}$ and express the scalar Nevanlinna function $f_\varphi(\lambda) = (F(\lambda)\varphi, \varphi)$ with its integral representation in Theorem A.2.5. Since the function $g(\lambda) = (\lambda - c)^{-1}$ is holomorphic when $\lambda \neq c$ and real on $\mathbb{R} \setminus \{c\}$, the same argument as in the proof of Theorem A.6.5 shows that f_φ is holomorphic on $\mathbb{C} \setminus [c, \infty)$ for all $\varphi \in \mathcal{G}$, and hence F is holomorphic on $\mathbb{C} \setminus [c, \infty)$.

To see that F takes nonnegative values on the interval $(-\infty, c)$ one applies Proposition A.2.9. Consider the functions $f_\varphi(\lambda)$ and $(\lambda - c)^{-1} f_\varphi(\lambda)$. Both of them are differentiable and nondecreasing on the interval $(-\infty, c)$. In particular, for all $x < c$ one has $f'_\varphi(x) \geq 0$ and

$$\frac{f'_\varphi(x)}{x - c} - \frac{f_\varphi(x)}{(x - c)^2} = \frac{d}{dx}\left((x - c)^{-1} f_\varphi(x)\right) \geq 0.$$

Since $x < c$, this implies that

$$f_\varphi(x) \leq (x - c) f'_\varphi(x) \leq 0$$

and hence $(F(x)\varphi, \varphi) \leq 0$ for all $\varphi \in \mathcal{G}$. Therefore, $F(x) \leq 0$ for all $x < c$ and the claim $F \in \mathbf{S}_\mathcal{G}^{-1}(-\infty, c)$ is proved. $\qquad\square$

The classes of Stieltjes functions and inverse Stieltjes functions are also connected to each other by inversion. For an operator-valued Nevanlinna function F with values in $\mathbf{B}(\mathcal{G})$ the values of the inverse $-F^{-1}$ need not be bounded operators. For simplicity, it is assumed here that the relevant functions are uniformly strict, see Definition A.4.7. Recall that if F is a uniformly strict $\mathbf{B}(\mathcal{G})$-valued Nevanlinna function, then so is the function $-F^{-1}$; cf. Lemma A.4.8.

Proposition A.6.7. *Let $\mathcal{G}$ be a Hilbert space, let F be a $\mathbf{B}(\mathcal{G})$-valued Nevanlinna function, and assume that F is uniformly strict. Then the following statements hold:*

(i) *if $F \in \mathbf{S}_\mathcal{G}(-\infty, c)$, then $-F^{-1} \in \mathbf{S}_\mathcal{G}^{-1}(-\infty, c)$;*

(ii) *if $F \in \mathbf{S}_\mathcal{G}^{-1}(-\infty, c)$, then $-F^{-1} \in \mathbf{S}_\mathcal{G}(-\infty, c)$.*

Proof. (i) Let $F \in \mathbf{S}_\mathcal{G}(-\infty, c)$ and note first that in the integral representation (A.6.1) one has $\beta = 0$, as follows from (A.6.2); cf. the proof of Theorem A.6.2. Hence, one concludes from (A.6.1) that

$$\operatorname{Im} F(i) = \int_{[c,\infty)} \frac{d\Sigma(t)}{t^2 + 1},$$

which by assumption is a nonnegative boundedly invertible operator. Recall that F has the integral representation (A.6.4) with the same Σ as in (A.6.1). It is straightforward to see that for every $x < c$ there exists a constant $C_x > 0$ such that

$$\frac{t - x}{t^2 + 1} \leq C_x \quad \text{for all} \quad t \geq c.$$

Thus, one concludes that

$$(F(x)\varphi, \varphi) = (\gamma\varphi, \varphi) + \int_{[c,\infty)} \frac{1}{t - x}\, d(\Sigma(t)\varphi, \varphi)$$

$$\geq (\gamma\varphi, \varphi) + \frac{1}{C_x} \int_{[c,\infty)} \frac{d(\Sigma(t)\varphi, \varphi)}{t^2 + 1}$$

for all $\varphi \in \mathcal{G}$. It follows that $(F(x)\varphi, \varphi) \geq \delta_x \|\varphi\|^2$, $\varphi \in \mathcal{G}$, for some $\delta_x > 0$. Therefore, $F(x)^{-1} \in \mathbf{B}(\mathcal{G})$ for all $x < c$. Thus, the function $-F^{-1}$ is holomorphic in the region $\mathbb{C} \setminus [c, \infty)$. Since $F(x) \geq 0$, it is clear that $-F^{-1}(x) \leq 0$ for all $x < c$. Therefore, $-F^{-1} \in \mathbf{S}_{\mathcal{G}}^{-1}(-\infty, c)$.

(ii) Let $F \in \mathbf{S}_{\mathcal{G}}^{-1}(-\infty, c)$. Then it follows from Theorem A.6.3 that for all $x < c$

$$F(x) = L + (x - c)\left(\beta + \int_{[c,\infty)} \frac{d\Sigma(t)}{(t-x)(t-c)}\right) \leq 0, \qquad (A.6.14)$$

where the integral is interpreted in the weak sense. Since F is uniformly strict, Proposition A.4.6 asserts that for all $x < c$

$$F'(x) = \beta + \int_{[c,\infty)} \frac{d\Sigma(t)}{(t-x)^2}$$

is a nonnegative boundedly invertible operator. It follows that

$$(\beta\varphi, \varphi) + \int_{[c,\infty)} \frac{d(\Sigma(t)\varphi, \varphi)}{(t-x)(t-c)} \geq (\beta\varphi, \varphi) + \int_{[c,\infty)} \frac{d(\Sigma(t)\varphi, \varphi)}{(t-x)^2} \geq \delta_x \|\varphi\|^2$$

for all $\varphi \in \mathcal{G}$ and for some $\delta_x > 0$, $x < c$. This implies that $F(x)$ in (A.6.14) has a bounded inverse for every $x < c$. Thus, $-F^{-1}$ is holomorphic in the region $\mathbb{C} \setminus [c, \infty)$. Since $F(x) \leq 0$, it is clear that $0 \leq -F(x)^{-1}$ for all $x < c$. Therefore, $-F^{-1} \in \mathbf{S}_{\mathcal{G}}(-\infty, c)$. $\qquad\square$

Appendix B

Self-Adjoint Operators

Let A be a self-adjoint operator in the Hilbert space $\mathfrak{H}$ and let $E(\cdot)$ be the corresponding spectral measure. The operator A will be diagonalized by means of a self-adjoint operator Q in a Hilbert space $L^2_{d\rho}(\mathbb{R})$, where ρ is a nondecreasing function on $\mathbb{R}$. Here Q stands for multiplication by the independent variable in $L^2_{d\rho}(\mathbb{R})$. By means of the spectral measure one constructs an integral transform $\mathcal{F}$ which maps $\mathfrak{H}$ unitarily onto $L^2_{d\rho}(\mathbb{R})$ such that $A = \mathcal{F}^* Q \mathcal{F}$. This transform shares the diagonalization property with the classical Fourier transform and hence, for convenience, it will be referred to as Fourier transform in the following. There are two cases of interest in the present text.

The first (scalar) case is where $\mathfrak{H} = L^2_r(a, b)$ and r is a locally integrable function that is positive almost everywhere, and where A is a self-adjoint operator in $\mathfrak{H}$ with spectral measure $E(\cdot)$. The Fourier transform is initially defined on the compactly supported scalar functions in $L^2_r(a, b)$ by

$$\widehat{f}(x) = \int_a^b \omega(t, x) f(t) \, r(t) \, dt,$$

where $(t, x) \mapsto \omega(t, x)$ is a continuous real function defined on the square $(a, b) \times \mathbb{R}$. The spectral measure of A and the Fourier transform are assumed to be connected by

$$(E(\delta) f, f) = \int_\delta \widehat{f}(x) \overline{\widehat{f}(x)} \, d\rho(x), \quad \delta \subset \mathbb{R}, \tag{B.0.1}$$

for all $f \in L^2_r(a, b)$ with compact support. Due to this condition the Fourier transform can be extended to an isometry on all of $L^2_r(a, b)$. Under an additional assumption on ω it is shown that the Fourier transform is a unitary map.

The second (vector) case is where $\mathfrak{H} = L^2_\Delta(a, b)$ and Δ is a nonnegative 2×2 matrix function with locally integrable coefficients and where A is a self-adjoint relation. Then A_{op} is a self-adjoint operator in $\mathfrak{H} \ominus \mathrm{mul}\, A$ with the spectral measure $E(\cdot)$. The Fourier transform is initially defined on the compactly supported 2×1 vector functions in $L^2_\Delta(a, b)$ by

$$\widehat{f}(x) = \int_a^b \omega(t, x)\Delta(t)f(t)\, dt,$$

where $(t, x) \mapsto \omega(t, x)$ is a continuous 1×2 matrix function defined on the square $(a, b) \times \mathbb{R}$. The spectral measure and the Fourier transform are assumed to be connected by (B.0.1) for all $f \in L^2_\Delta(a, b)$ with compact support. Due to this condition, the Fourier transform can be extended to a partial isometry on all of $L^2_\Delta(a, b)$. Under an additional assumption on ω it is shown that the Fourier transform is onto.

For the scalar case the theory will be developed in detail; the results in the vector case will be only briefly explained, as the details are very much the same.

B.1   The scalar case

Let (a, b) be an open interval and let r be a locally integrable function that is positive almost everywhere. Let A be a self-adjoint operator in the Hilbert space $L^2_r(a, b)$ and let $E(\cdot)$ be the spectral measure of A. The basic ingredient is a continuous real function $(t, x) \mapsto \omega(t, x)$ on $(a, b) \times \mathbb{R}$. Hence, the *Fourier transform* $\widehat{f}$ of a compactly supported function $f \in L^2_r(a, b)$ defined by

$$\widehat{f}(x) = \int_a^b \omega(t, x)f(t)r(t)\, dt, \quad x \in \mathbb{R}, \tag{B.1.1}$$

is a well-defined complex function which is continuous on $\mathbb{R}$. The main assumption is that there exists a nondecreasing function ρ on $\mathbb{R}$ such the identity

$$(E(\delta)f, g) = \int_\delta \widehat{f}(x)\overline{\widehat{g}(x)}\, d\rho(x), \quad \delta \subset \mathbb{R}, \tag{B.1.2}$$

holds for all $f, g \in L^2_r(a, b)$ with compact support and all bounded open intervals $\delta \subset \mathbb{R}$, whose endpoints are not eigenvalues of A. Since the eigenvalues of A are discrete, an approximation argument and the dominated convergence theorem show that the identity (B.1.2) is in fact true for all $f, g \in L^2_r(a, b)$ with compact support and all bounded open intervals $\delta \subset \mathbb{R}$, regardless of whether their endpoints being eigenvalues or not.

It follows from the assumption (B.1.2) that for every function $f \in L^2_r(a, b)$ with compact support the continuous function $\widehat{f}$ belongs to $L^2_{d\rho}(\mathbb{R})$. In fact, the following result is valid.

Lemma B.1.1. *The Fourier transform $f \to \widehat{f}$ in (B.1.1) extends by continuity from the compactly supported functions in $L_r^2(a,b)$ to an isometric mapping*

$$\mathcal{F} : L_r^2(a,b) \to L_{d\rho}^2(\mathbb{R})$$

such that for all $f \in L_r^2(a,b)$

$$\lim_{\alpha \to a, \beta \to b} \int_{\mathbb{R}} \left| (\mathcal{F}f)(x) - \int_\alpha^\beta \omega(t,x) f(t) r(t)\, dt \right|^2 d\rho(x) = 0. \tag{B.1.3}$$

The Fourier transform and the spectral measure $E(\cdot)$ are related via

$$(E(\delta)f, g) = \int_\delta (\mathcal{F}f)(x)\overline{(\mathcal{F}g)(x)}\, d\rho(x) \tag{B.1.4}$$

for all $f, g \in L_r^2(a,b)$, where $\delta \subset \mathbb{R}$ is any bounded open interval.

Proof. Step 1. The mapping $f \mapsto \widehat{f}$ is a contraction on the functions in $L_r^2(a,b)$ that have compact support. To see this, let $f \in L_r^2(a,b)$ have compact support. Then it follows from the assumption (B.1.2) that

$$\int_\delta \widehat{f}(x)\overline{\widehat{f}(x)}\, d\rho(x) = (E(\delta)f, f) \leq (f, f),$$

where δ is an arbitrary bounded open interval. The monotone convergence theorem shows that $\widehat{f}$ belongs to $L_{d\rho}^2(\mathbb{R})$ and

$$(\widehat{f}, \widehat{f})_\rho \leq (f, f),$$

when $f \in L_r^2(a,b)$ has compact support.

Step 2. The mapping $f \mapsto \widehat{f}$, defined on the functions in $L_r^2(a,b)$ that have compact support, can be extended as a contraction to all of $L_r^2(a,b)$. For this, let $f \in L_r^2(a,b)$ and approximate f in $L_r^2(a,b)$ by functions $f_n \in L_r^2(a,b)$ with compact support. Then $f_n - f_m$ is a Cauchy sequence in $L_r^2(a,b)$ and since

$$\|\widehat{f_n} - \widehat{f_m}\|_\rho \leq \|f_n - f_m\|,$$

there is an element $\varphi \in L_{d\rho}^2(\mathbb{R})$ such that $\widehat{f_n} \to \varphi$ in $L_{d\rho}^2(\mathbb{R})$. It follows from $\|\widehat{f_n}\|_\rho \leq \|f_n\|$ that, in fact,

$$\|\varphi\|_\rho \leq \|f\|.$$

In particular, the mapping $f \mapsto \varphi$ from $L_r^2(a,b)$ to $L_{d\rho}^2(\mathbb{R})$ is a contraction. Hence, the operator $\mathcal{F}$ given by $\mathcal{F}f = \varphi$ is well defined and takes $L_r^2(a,b)$ contractively to $L_{d\rho}^2(\mathbb{R})$. The assertion in (B.1.3) is clear by multiplying $f \in L_r^2(a,b)$ with appropriately chosen characteristic functions.

Step 3. The operator $\mathcal{F}$ is an isometry. To see this, let $f, g \in L_r^2(a, b)$ and approximate them by compactly supported functions $f_n, g_n \in L_r^2(a, b)$. Then by the assumption (B.1.2) one obtains

$$(E(\delta)f_n, g_n) = \int_\delta (\mathcal{F}f_n)(x)\overline{(\mathcal{F}g_n)(x)}\, d\rho(x). \tag{B.1.5}$$

Taking limits in (B.1.5) yields (B.1.4). It follows from (B.1.4) and the dominated convergence theorem that

$$(f, g) = \int_\delta (\mathcal{F}f)(x)\overline{(\mathcal{F}g)(x)}\, d\rho(x)$$

holds for all functions $f, g \in L_r^2(a, b)$. Therefore, $\mathcal{F}$ is an isometry. $\qquad\square$

The following simple observation is a direct consequence of Lemma B.1.1.

Corollary B.1.2. *For any bounded open interval $\delta \subset \mathbb{R}$ one has*

$$\mathcal{F}(E(\delta)f) = \chi_\delta\, \mathcal{F}f, \quad f \in L_r^2(a, b).$$

Proof. It follows from (B.1.4) that

$$(E(\delta)f, g) = \int_\delta (\mathcal{F}f)(x)\overline{(\mathcal{F}g)(x)}d\rho(x) = (\chi_\delta\mathcal{F}f, \mathcal{F}g)_\rho$$

for all $f, g \in L_r^2(a, b)$. Consequently,

$$\begin{aligned}
\|E(\delta)f - g\|^2 &= (E(\delta)f - g, E(\delta)f - g)\\
&= (E(\delta)f, f) - (E(\delta)f, g) - (g, E(\delta)f) + (g, g)\\
&= (\chi_\delta\mathcal{F}f, \mathcal{F}f)_\rho - (\chi_\delta\mathcal{F}f, \mathcal{F}g)_\rho\\
&\qquad\qquad - (\mathcal{F}g, \chi_\delta\mathcal{F}f)_\rho + (\mathcal{F}g, \mathcal{F}g)_\rho\\
&= \|\chi_\delta\mathcal{F}f - \mathcal{F}g\|_\rho^2.
\end{aligned}$$

Setting $g = E(\delta)f$, the desired result follows. $\qquad\square$

Let $\delta \subset \mathbb{R}$ be a bounded open interval. The identity (B.1.4) is now written in the equivalent form

$$\int_\mathbb{R} \chi_\delta(t)d(E(t)f, g) = \int_\mathbb{R} \chi_\delta(x)(\mathcal{F}f)(x)\overline{(\mathcal{F}g)(x)}\, d\rho(x)$$

for all $f, g \in L_r^2(a, b)$. Let Ξ be a bounded Borel measurable function on $\mathbb{R}$. An approximation argument involving characteristic functions shows that then also

$$\int_\mathbb{R} \Xi(t)d(E(t)f, g) = \int_\mathbb{R} \Xi(x)(\mathcal{F}f)(x)\overline{(\mathcal{F}g)(x)}\, d\rho(x) \tag{B.1.6}$$

for all $f, g \in L_r^2(a, b)$. As a particular case of (B.1.6) one has

$$((A - \lambda)^{-1} f, g) = \int_{\mathbb{R}} \frac{\mathcal{F}f(x)\overline{\mathcal{F}g(x)}}{x - \lambda} \, d\rho(x) \tag{B.1.7}$$

for $f, g \in L_r^2(a, b)$ and $\lambda \in \rho(A)$. Moreover, if $g \in L_r^2(a, b)$ has compact support one obtains

$$\int_a^b ((A - \lambda)^{-1} f)(t)\overline{g(t)} \, r(t) \, dt = \int_{\mathbb{R}} \frac{\mathcal{F}f(x)}{x - \lambda} \left(\int_a^b w(t, x)\overline{g(t)} \, r(t) \, dt \right) d\rho(x)$$

$$= \int_a^b \left(\int_{\mathbb{R}} \frac{\mathcal{F}f(x)}{x - \lambda} w(t, x) \, d\rho(x) \right) \overline{g(t)} \, r(t) \, dt.$$

The Fubini theorem now implies that

$$((A - \lambda)^{-1} f)(t) = \int_{\mathbb{R}} \frac{w(t, x)}{x - \lambda} \mathcal{F}f(x) \, d\rho(x) \tag{B.1.8}$$

for almost all $t \in (a, b)$ and, in particular, the integrand on the right-hand side is integrable for almost all $t \in (a, b)$.

Parallel to the Fourier transform one may also introduce a reverse Fourier transform acting on the space $L_{d\rho}^2(\mathbb{R})$. Let $\varphi \in L_{d\rho}^2(\mathbb{R})$ have compact support and define the *reverse Fourier transform* $\check{\varphi}$ by

$$\check{\varphi}(t) = \int_{\mathbb{R}} w(t, x)\varphi(x) \, d\rho(x), \quad t \in (a, b). \tag{B.1.9}$$

Then $\check{\varphi}$ is a well-defined function which is continuous on (a, b). By means of Lemma B.1.1 one can now prove the following result.

Lemma B.1.3. *The reverse Fourier transform $\varphi \to \check{\varphi}$ in (B.1.9) extends by continuity from the compactly supported functions in $L_{d\rho}^2(\mathbb{R})$ to a contractive mapping $\mathcal{G} : L_{d\rho}^2(\mathbb{R}) \to L_r^2(a, b)$ such that for all $\varphi \in L_{d\rho}^2(\mathbb{R})$*

$$\lim_{\eta \to \mathbb{R}} \int_a^b \left| \mathcal{G}\varphi(t) - \int_\eta w(t, x)\varphi(x) \, d\rho(x) \right|^2 r(t) dt = 0. \tag{B.1.10}$$

In fact, the extension $\mathcal{G}$ satisfies $\mathcal{G}\mathcal{F}f = f$ for all $f \in L_r^2(a, b)$.

Proof. Step 1. The mapping $\varphi \mapsto \check{\varphi}$ takes the compactly supported functions in $L_{d\rho}^2(\mathbb{R})$ contractively into $L_r^2(a, b)$. To see this, let $\varphi \in L_{d\rho}^2(\mathbb{R})$ have compact support and let $[\alpha, \beta] \subset (a, b)$ be a compact interval. Then

$$\int_\alpha^\beta |\check\varphi(t)|^2 r(t)\, dt = \int_\alpha^\beta \check\varphi(t) \left(\int_\mathbb{R} \omega(t, x)\overline{\varphi(x)}\, d\rho(x) \right) r(t)\, dt$$

$$= \int_\mathbb{R} \left(\int_\alpha^\beta \omega(t, x)\check\varphi(t) r(t)\, dt \right) \overline{\varphi(x)}\, d\rho(x)$$

$$= \int_\mathbb{R} (\mathcal{F}(\chi_{[\alpha,\beta]}\check\varphi))(x)\overline{\varphi(x)}\, d\rho(x)$$

$$\leq \|\mathcal{F}(\chi_{[\alpha,\beta]}\check\varphi)\|_\rho \|\varphi\|_\rho = \|\chi_{[\alpha,\beta]}\check\varphi\| \|\varphi\|_\rho,$$

where Lemma B.1.1 was used in the last step. This estimate gives that for any compact interval $[\alpha, \beta] \subset (a, b)$,

$$\sqrt{\int_\alpha^\beta |\check\varphi(t)|^2 r(t)\, dt} \leq \|\varphi\|_\rho.$$

By the monotone convergence theorem this leads to the inequality

$$\|\check\varphi\| \leq \|\varphi\|_\rho.$$

Hence, the mapping $\varphi \mapsto \check\varphi$ takes the compactly supported functions in $L^2_{d\rho}(\mathbb{R})$ contractively into $L^2_r(a, b)$.

Step 2. The mapping $\varphi \mapsto \check\varphi$ defined on the functions in $L^2_{d\rho}(\mathbb{R})$ that have compact support can be contractively extended to all of $L^2_{d\rho}(\mathbb{R})$. For this, let $\varphi \in L^2_{d\rho}(\mathbb{R})$ and approximate φ in $L^2_{d\rho}(\mathbb{R})$ by functions $\varphi_n \in L^2_{d\rho}(\mathbb{R})$ with compact support. Then $\varphi_n - \varphi_m$ is a Cauchy sequence in $L^2_{d\rho}(\mathbb{R})$ and since

$$\|\check\varphi_n - \check\varphi_m\| \leq \|\varphi_n - \varphi_m\|_\rho,$$

there is an element $f \in L^2_r(a, b)$ such that $\check\varphi_n \to f$ in $L^2_r(a, b)$. It follows from $\|\check\varphi_n\| \leq \|\varphi_n\|_\rho$ that, in fact,

$$\|f\| \leq \|\varphi\|_\rho.$$

In particular, the mapping $\varphi \mapsto f$ from $L^2_{d\rho}(\mathbb{R})$ to $L^2_r(a, b)$ is a contraction. Hence, the operator $\mathcal{G}$ given by $\mathcal{G}\varphi = f$ is well defined and takes $L^2_{d\rho}(\mathbb{R})$ contractively to $L^2_r(a, b)$. The assertion in (B.1.10) is clear by multiplying $\varphi \in L^2_{d\rho}(\mathbb{R})$ by appropriately chosen characteristic functions.

Step 3. The extended mapping $\mathcal{G}$ is a left inverse of the Fourier transform $\mathcal{F}$. For this, observe that

$$(f, g) = \int_\mathbb{R} (\mathcal{F}f)(x)\overline{(\mathcal{F}g)(x)}\, d\rho(x)$$

for all $f, g \in L_r^2(a, b)$, since $\mathcal{F}$ is isometric by Lemma B.1.1. Thus, by means of the Fubini theorem one concludes that for $f, g \in L_r^2(a, b)$ and g having compact support,

$$
\begin{aligned}
(f, g) &= \lim_{\eta \to \mathbb{R}} \int_\eta (\mathcal{F}f)(x) \left(\int_a^b \omega(t, x) \overline{g(t)} r(t) dt \right) d\rho(x) \\
&= \lim_{\eta \to \mathbb{R}} \int_a^b \left(\int_\eta \omega(t, x) (\mathcal{F}f)(x) \, d\rho(x) \right) \overline{g(t)} r(t) dt \\
&= (\mathcal{G}\mathcal{F}f, g),
\end{aligned}
$$

where, in the last step, (B.1.10) and the continuity of the inner product have been used. Since the functions with compact support are dense in $L_r^2(a, b)$, one obtains $\mathcal{G}\mathcal{F}f = f$ for all $f \in L_r^2(a, b)$. $\qquad\square$

Recall that multiplication by the independent variable in the Hilbert space $L_{d\rho}^2(\mathbb{R})$ generates a self-adjoint operator Q whose resolvent is given by

$$
(Q - \lambda)^{-1} = \frac{1}{x - \lambda}, \qquad \lambda \in \mathbb{C} \setminus \mathbb{R}.
$$

Hence, by (B.1.7), one sees that

$$
((A - \lambda)^{-1} f, g) = ((Q - \lambda)^{-1} \mathcal{F}f, \mathcal{F}g)_\rho = (\mathcal{F}^*(Q - \lambda)^{-1} \mathcal{F}f, g)
$$

for all $f, g \in L_r^2(a, b)$, which leads to

$$
(A - \lambda)^{-1} = \mathcal{F}^*(Q - \lambda)^{-1} \mathcal{F}, \qquad \lambda \in \mathbb{C} \setminus \mathbb{R}. \tag{B.1.11}
$$

In general, here the Fourier transform $\mathcal{F}$ is an isometry, which is not necessarily onto.

The above results have been proved under the assumption that ω is a continuous function on $(a, b) \times \mathbb{R}$ which satisfies (B.1.2). For the following theorem one needs the additional condition:

$$
\begin{array}{cc}
\text{for each } x_0 \in \mathbb{R} \text{ there exists a compactly} & \\
\text{supported function } f \in L_r^2(a, b) \text{ such that } (\mathcal{F}f)(x_0) \neq 0. & \tag{B.1.12}
\end{array}
$$

In the present situation this condition is equivalent to the following:

$$
\text{for each } x_0 \in (a, b) \text{ there exists } t_0 \in \mathbb{R} \text{ such that } \omega(t_0, x_0) \neq 0. \tag{B.1.13}
$$

Theorem B.1.4. *Let ω be continuous on $(a, b) \times \mathbb{R}$ and assume that (B.1.2) and one of the equivalent conditions (B.1.12) or (B.1.13) are satisfied. Then the Fourier transform*

$$
f \mapsto \widehat{f}, \quad \widehat{f}(x) = \int_a^b \omega(t, x) f(t) \, r(t) \, dt, \quad x \in \mathbb{R},
$$

extends by continuity from the compactly supported functions $f \in L^2_r(a,b)$ *to a unitary mapping* $\mathcal{F} : L^2_r(a,b) \to L^2_{d\rho}(\mathbb{R})$. *Moreover, the self-adjoint operator* A *in* $L^2_r(a,b)$ *is unitarily equivalent to the multiplication operator* Q *by the independent variable in* $L^2_{d\rho}(\mathbb{R})$ *via the Fourier transform* $\mathcal{F}$:

$$A = \mathcal{F}^* Q \mathcal{F}. \tag{B.1.14}$$

Proof. It is clear from Lemma B.1.1 that $\mathcal{F} : L^2_r(a,b) \to L^2_{d\rho}(\mathbb{R})$ is isometric and hence it remains to show for the first part of the theorem that $\mathcal{F}$ is surjective.

Assume that $\psi \in L^2_{d\rho}(\mathbb{R})$, $\psi \neq 0$, is orthogonal to $\operatorname{ran} \mathcal{F}$. Note that

$$(\psi, \mathcal{F}f)_\rho = 0 \quad \Rightarrow \quad (\overline{\psi}, \mathcal{F}\overline{f})_\rho = 0$$

and thus it follows that $\operatorname{Re}\psi$ and $\operatorname{Im}\psi$ are also orthogonal to $\operatorname{ran} \mathcal{F}$; hence it is no restriction to assume that $\psi \in L^2_{d\rho}(\mathbb{R})$ is real. Since $\psi \neq 0$, there exist an interval $I = [\alpha, \beta]$ and a Borel set $B \subset I$ with positive ρ-measure, such that $\psi(x) > 0$ (or $\psi(x) < 0$) for all $x \in B$. By the condition (B.1.12), there exists for each $x_0 \in I$ a compactly supported function $f_{x_0} \in L^2_r(a,b)$ such that $(\mathcal{F}f_{x_0})(x_0) > 0$. Since $\mathcal{F}f_{x_0}$ is continuous, there exists an open interval I_{x_0} containing x_0 such that $(\mathcal{F}f_{x_0})(x) > 0$ for all $x \in I_{x_0}$. As I is compact there exist finitely many points $x_1, \ldots, x_n \in I$ such that

$$I \subset \bigcup_{i=1}^n I_{x_i}.$$

Hence, $B \cap I_{x_j}$ has positive ρ-measure for some $j \in \{1, \ldots, n\}$ and therefore

$$(\chi_{B \cap I_{x_j}} \psi, \mathcal{F}f_{x_j})_\rho = \int_{\mathbb{R}} \chi_{B \cap I_{x_j}}(x)\psi(x)(\mathcal{F}f_{x_j})(x)\, d\rho(x) \neq 0. \tag{B.1.15}$$

On the other hand, since $\psi \perp \operatorname{ran} \mathcal{F}$, Corollary B.1.2 implies that

$$(\chi_\delta \psi, \mathcal{F}f)_\rho = (\psi, \chi_\delta \mathcal{F}f)_\rho = (\psi, \mathcal{F}E(\delta)f)_\rho = 0$$

for all $f \in L^2_r(a,b)$ and all bounded open intervals $\delta \subset \mathbb{R}$. Hence, by the regularity of the Borel measure ρ, also $(\chi_{B \cap I_{x_j}} \psi, \mathcal{F}f)_\rho = 0$ for all $f \in L^2_r(a,b)$; this contradicts (B.1.15). Therefore, $\operatorname{ran} \mathcal{F}$ is dense in $L^2_{d\rho}(\mathbb{R})$, and since $\mathcal{F} : L^2_r(a,b) \to L^2_{d\rho}(\mathbb{R})$ is isometric, one obtains that $\mathcal{F}$ is surjective.

The identity (B.1.14) follows from (B.1.11), the fact that $\mathcal{F}$ is unitary, and Lemma 1.3.8. $\qquad\qquad \square$

In the situation of Theorem B.1.4 the inverse of the Fourier transform $\mathcal{F}$ is actually given by the reverse Fourier transform $\mathcal{G}$ in Lemma B.1.3, which is now a unitary mapping from $L^2_{d\rho}(\mathbb{R})$ to $L^2_r(a,b)$. Thus, for all $\varphi \in L^2_{d\rho}(\mathbb{R})$ one has

$$\lim_{\eta \to \mathbb{R}} \int_a^b \left| \mathcal{F}^{-1}\varphi(t) - \int_\eta \omega(t,x)\varphi(x)\, d\rho(x) \right|^2 r(t)dt = 0.$$

B.2 The vector case

Let (a, b) be an open interval and let Δ be a measurable nonnegative 2×2 matrix function on (a, b). Let A be a self-adjoint relation in the corresponding Hilbert space $L^2_\Delta(a, b)$ and assume that its multivalued part is at most finite-dimensional. Let $E(\cdot)$ be the spectral measure of A_{op}, where A_{op} is the orthogonal operator part of A: it is a self-adjoint operator in $L^2_\Delta(a, b) \ominus \operatorname{mul} A$. In this case the basic ingredient is a continuous 1×2 matrix function $(t, x) \mapsto \omega(t, x)$ on $(a, b) \times \mathbb{R}$ whose entries are real. Hence, the *Fourier transform* $\widehat{f}$ of a compactly supported function $f \in L^2_\Delta(a, b)$ given by

$$\widehat{f}(x) = \int_a^b \omega(t, x) \Delta(t) f(t)\, dt, \quad x \in \mathbb{R}, \tag{B.2.1}$$

is a well-defined complex function that is continuous on $\mathbb{R}$. The main assumption is that there exists a nondecreasing function ρ on $\mathbb{R}$ such the identity

$$(E(\delta)f, g) = \int_\delta \widehat{f}(x)\overline{\widehat{g}(x)}\, d\rho(x), \quad \delta \subset \mathbb{R}, \tag{B.2.2}$$

holds for all $f, g \in L^2_\Delta(a, b)$ with compact support and all bounded open intervals $\delta \subset \mathbb{R}$, whose endpoints are not eigenvalues of A_{op}. An approximation argument shows that (B.2.2) remains valid for all $f, g \in L^2_\Delta(a, b)$ with compact support and all bounded open intervals $\delta \subset \mathbb{R}$.

It follows from the assumption (B.2.2) that for every function $f \in L^2_\Delta(a, b)$ with compact support the continuous function $\widehat{f}$ belongs to $L^2_{d\rho}(\mathbb{R})$. In the present situation Lemma B.1.1 remains valid in a slightly modified form; here $\mathcal{F}$ is a partial isometry with $\ker \mathcal{F} = \operatorname{mul} A$.

Lemma B.2.1. *The Fourier transform $f \mapsto \widehat{f}$ in (B.2.1) extends by continuity from the compactly supported functions in $L^2_\Delta(a, b)$ to a partial isometry*

$$\mathcal{F} : L^2_\Delta(a, b) \to L^2_{d\rho}(\mathbb{R})$$

with $\ker \mathcal{F} = \operatorname{mul} A$ such that for all $f \in L^2_\Delta(a, b)$

$$\lim_{\alpha \to a, \beta \to b} \int_\mathbb{R} \left| (\mathcal{F}f)(x) - \int_\alpha^\beta \omega(t, x)\Delta(t)f(t)\, dt \right|^2 d\rho(x) = 0.$$

The Fourier transform $\mathcal{F}$ and the spectral measure $E(\cdot)$ are related via

$$(E(\delta)f, g) = \int_\delta (\mathcal{F}f)(x)\overline{(\mathcal{F}g)(x)}\, d\rho(x) \tag{B.2.3}$$

for all $f, g \in L^2_\Delta(a, b)$, where $\delta \subset \mathbb{R}$ is any bounded open interval.

Next one verifies as in the proof of Corollary B.1.2 that the identity

$$\mathcal{F}(E(\delta)f) = \chi_\delta\,\mathcal{F}f, \quad f \in L^2_\Delta(a,b),$$

holds for bounded open intervals $\delta \subset \mathbb{R}$. Further, (B.2.3) and an approximation argument using characteristic functions show that

$$((A-\lambda)^{-1}f, g) = \int_\mathbb{R} \frac{\mathcal{F}f(x)\overline{\mathcal{F}g(x)}}{x-\lambda}\, d\rho(x) \tag{B.2.4}$$

for all $f, g \in L^2_\Delta(a,b)$ and $\lambda \in \rho(A)$. Moreover,

$$((A-\lambda)^{-1}f)(t) = \int_\mathbb{R} \frac{\omega(t,x)^*}{x-\lambda}\,\mathcal{F}f(x)\,d\rho(x), \quad \lambda \in \mathbb{C}\setminus\mathbb{R}, \tag{B.2.5}$$

for almost all $t \in (a,b)$ and, in particular, the integrand on the right-hand side is integrable for almost all $t \in (a,b)$.

The *reverse Fourier transform* of a function $\varphi \in L^2_{d\rho}(\mathbb{R})$ with compact support is defined by

$$\check{\varphi}(t) = \int_\mathbb{R} \omega(t,x)^*\varphi(x)\,d\rho(x), \quad t \in (a,b). \tag{B.2.6}$$

Then $\check{\varphi}$ is a well-defined 2×1 matrix function which is continuous on (a,b). By means of Lemma B.2.1 one now obtains the following result, which is similar to Lemma B.1.3. There are some slight differences in the proof for the vector case, which is provided here for completeness.

Lemma B.2.2. *The reverse Fourier transform* $\varphi \mapsto \check{\varphi}$ *in* (B.2.6) *extends by continuity from the compactly supported functions in* $L^2_{d\rho}(\mathbb{R})$ *to a contractive mapping* $\mathcal{G}: L^2_{d\rho}(\mathbb{R}) \to L^2_\Delta(a,b)$ *such that for all* $\varphi \in L^2_{d\rho}(\mathbb{R})$

$$\lim_{\eta\to\mathbb{R}} \int_a^b \left| \mathcal{G}\varphi(t) - \int_\eta \omega(t,x)^*\varphi(x)\,d\rho(x) \right|^2 dt = 0. \tag{B.2.7}$$

In fact, the extension $\mathcal{G}$ *satisfies* $\mathcal{G}\mathcal{F}f = f$ *for all* $f \in L^2_\Delta(a,b) \ominus \operatorname{mul} A$.

Proof. Step 1. The mapping $\varphi \mapsto \check{\varphi}$ takes the compactly supported functions in $L^2_{d\rho}(\mathbb{R})$ contractively into $L^2_\Delta(a,b)$. For this, let $\varphi \in L^2_{d\rho}(\mathbb{R})$ have compact support and let $[\alpha, \beta] \subset (a,b)$ be a compact interval. First observe that $\chi_{[\alpha,\beta]}\check{\varphi} \in L^2_\Delta(a,b)$, so that indeed

$$\int_\alpha^\beta \omega(t,x)\Delta(t)\check{\varphi}(t)\,dt = (\mathcal{F}(\chi_{[\alpha,\beta]}\check{\varphi}))(x).$$

Therefore, one obtains

$$\int_\alpha^\beta \check{\varphi}(t)^* \Delta(t) \check{\varphi}(t)\, dt = \int_\alpha^\beta \left(\int_{\mathbb{R}} \omega(t,x)^* \varphi(x)\, d\rho(x) \right)^* \Delta(t) \check{\varphi}(t)\, dt$$

$$= \int_{\mathbb{R}} \left(\int_\alpha^\beta \omega(t,x) \Delta(t) \check{\varphi}(t)\, dt \right) \overline{\varphi(x)}\, d\rho(x)$$

$$= \int_{\mathbb{R}} (\mathcal{F}(\chi_{[\alpha,\beta]} \check{\varphi}))(x) \overline{\varphi(x)}\, d\rho(x)$$

$$\leq \|\mathcal{F}(\chi_{[\alpha,\beta]} \check{\varphi})\|_\rho \|\varphi\|_\rho \leq \|\chi_{[\alpha,\beta]} \check{\varphi}\| \|\varphi\|_\rho,$$

where Lemma B.2.1 was used in the last step. The above estimate gives for any compact interval $[\alpha, \beta] \subset (a, b)$

$$\sqrt{\int_\alpha^\beta \check{\varphi}(t)^* \Delta(t) \check{\varphi}(t)\, dt} \leq \|\varphi\|_\rho.$$

By the monotone convergence theorem this leads to the inequality

$$\|\check{\varphi}\| \leq \|\varphi\|_\rho.$$

Hence, the mapping $\varphi \mapsto \check{\varphi}$ takes the compactly supported functions in $L^2_{d\rho}(\mathbb{R})$ contractively into $L^2_\Delta(a, b)$.

Step 2. The mapping $\varphi \mapsto \check{\varphi}$ defined on the functions in $L^2_{d\rho}(\mathbb{R})$ that have compact support can be contractively extended to all of $L^2_{d\rho}(\mathbb{R})$. To see this, let $\varphi \in L^2_{d\rho}(\mathbb{R})$ and approximate φ in $L^2_{d\rho}(\mathbb{R})$ by functions $\varphi_n \in L^2_{d\rho}(\mathbb{R})$ with compact support. Then $\varphi_n - \varphi_m$ is a Cauchy sequence in $L^2_{d\rho}(\mathbb{R})$ and since

$$\|\check{\varphi}_n - \check{\varphi}_m\| \leq \|\varphi_n - \varphi_m\|_\rho,$$

there is an element $f \in L^2_\Delta(a, b)$ such that $\check{\varphi}_n \to f$ in $L^2_\Delta(a, b)$. It follows from $\|\check{\varphi}_n\| \leq \|\varphi_n\|_\rho$ that in fact

$$\|f\| \leq \|\varphi\|_\rho.$$

In particular, the mapping $\varphi \mapsto f$ from $L^2_{d\rho}(\mathbb{R})$ to $L^2_\Delta(a, b)$ is a contraction. Hence, the operator $\mathcal{G}$ given by $\mathcal{G}\varphi = f$ is well defined and takes $L^2_{d\rho}(\mathbb{R})$ contractively to $L^2_\Delta(a, b)$. The assertion in (B.2.7) is clear by multiplying $\varphi \in L^2_{d\rho}(\mathbb{R})$ by appropriately chosen characteristic functions.

Step 3. The extended mapping $\mathcal{G}$ is a left inverse of the Fourier transform $\mathcal{F}$. For this, first note that it follows from (B.2.3) and dominated convergence that

$$(f, g) = \int_{\mathbb{R}} (\mathcal{F}f)(x) \overline{(\mathcal{F}g)(x)}\, d\rho(x)$$

for all $f \in L^2_\Delta(a,b) \ominus \operatorname{mul} A$ and $g \in L^2_\Delta(a,b)$. Thus, by means of the Fubini theorem, one concludes for $f \in L^2_\Delta(a,b) \ominus \operatorname{mul} A$ and $g \in L^2_\Delta(a,b)$, each having compact support, that

$$
\begin{aligned}
(f,g) &= \lim_{\eta \to \mathbb{R}} \int_\eta (\mathcal{F}f)(x) \left(\int_a^b g(t)^* \Delta(t) \omega(t,x)^* \, dt \right) d\rho(x) \\
&= \lim_{\eta \to \mathbb{R}} \int_a^b g(t)^* \Delta(t) \left(\int_\eta \omega(t,x)^* (\mathcal{F}f)(x) \, d\rho(x) \right) dt \\
&= (\mathcal{G}\mathcal{F}f, g),
\end{aligned}
$$

where, in the last step, (B.2.7) and the continuity of the inner product have been used. This implies

$$
\mathcal{G}\mathcal{F}f = f
$$

for all $f \in L^2_\Delta(a,b) \ominus \operatorname{mul} A$ with compact support. Since the functions in $L^2_\Delta(a,b)$ with compact support are dense in $L^2_\Delta(a,b)$ and $\operatorname{mul} A$ is finite-dimensional by assumption, it follows that the functions with compact support are also dense in $L^2_\Delta(a,b) \ominus \operatorname{mul} A$. Therefore, one obtains from the contractivity of $\mathcal{F}$ and $\mathcal{G}$ that $\mathcal{G}\mathcal{F}f = f$ for all $f \in L^2_\Delta(a,b) \ominus \operatorname{mul} A$. $\qquad\square$

Recall that multiplication by the independent variable in the Hilbert space $L^2_{d\rho}(\mathbb{R})$ generates a self-adjoint operator Q whose resolvent is given by

$$
(Q - \lambda)^{-1} = \frac{1}{x - \lambda}, \qquad \lambda \in \mathbb{C} \setminus \mathbb{R}.
$$

Hence, by (B.2.4), one sees that

$$
\left((A - \lambda)^{-1} f, g \right) = \left((Q - \lambda)^{-1} \mathcal{F}f, \mathcal{F}g \right)_\rho = \left(\mathcal{F}^* (Q - \lambda)^{-1} \mathcal{F}f, g \right)
$$

for all $f \in L^2_\Delta(a,b)$ and $g \in L^2_\Delta(a,b)$, which leads to

$$
(A - \lambda)^{-1} = \mathcal{F}^* (Q - \lambda)^{-1} \mathcal{F}, \qquad \lambda \in \mathbb{C} \setminus \mathbb{R}. \tag{B.2.8}
$$

Consider the restriction $\mathcal{F}_{\text{op}} : L^2_\Delta(a,b) \ominus \operatorname{mul} A \to L^2_{d\rho}(\mathbb{R})$ of the partial isometry $\mathcal{F}$ onto $L^2_\Delta(a,b) \ominus \operatorname{mul} A$ and recall that

$$
\ker \mathcal{F} = \operatorname{mul} A = \ker (A - \lambda)^{-1}, \qquad \lambda \in \mathbb{C} \setminus \mathbb{R}.
$$

Then $\mathcal{F}_{\text{op}}$ is an isometry and (B.2.8) leads to

$$
(A_{\text{op}} - \lambda)^{-1} = \mathcal{F}^*_{\text{op}} (Q - \lambda)^{-1} \mathcal{F}_{\text{op}}, \qquad \lambda \in \mathbb{C} \setminus \mathbb{R}.
$$

In general, here the restricted Fourier transform $\mathcal{F}_{\text{op}}$ is not necessarily onto.

In order to prove that $\mathcal{F}_{\text{op}}$ is surjective one needs an additional condition:

$$
\begin{array}{c}
\text{for each } x_0 \in \mathbb{R} \text{ there exists a compactly} \\
\text{supported function } f \in L^2_\Delta(a,b) \text{ such that } (\mathcal{F}_{\text{op}}f)(x_0) \neq 0.
\end{array} \tag{B.2.9}
$$

For the following result, recall that it is assumed that mul A is finite-dimensional.

Theorem B.2.3. *Let* ω *be a continuous* 1×2 *matrix function* $(t, x) \mapsto \omega(t, x)$ *on* $(a, b) \times \mathbb{R}$ *whose entries are real, and assume that the conditions* (B.2.2) *and* (B.2.9) *are satisfied. Then the Fourier transform*

$$f \mapsto \widehat{f}, \quad \widehat{f}(x) = \int_a^b \omega(t, x) \Delta(t) f(t) \, dt, \quad x \in \mathbb{R},$$

extends by continuity from the compactly supported functions $f \in L^2_\Delta(a, b)$ *to a surjective partial isometry* $\mathcal{F}$ *from* $L^2_\Delta(a, b)$ *to* $L^2_{d\rho}(\mathbb{R})$ *with* $\ker \mathcal{F} = \operatorname{mul} A$, *that is, the restriction*

$$\mathcal{F}_{\mathrm{op}} : L^2_\Delta(a, b) \ominus \operatorname{mul} A \to L^2_{d\rho}(\mathbb{R})$$

is a unitary mapping. Moreover, the self-adjoint operator A_{op} *in* $L^2_\Delta(a, b) \ominus \operatorname{mul} A$ *is unitarily equivalent to the multiplication operator* Q *by the independent variable in* $L^2_{d\rho}(\mathbb{R})$ *via the restricted Fourier transform* $\mathcal{F}_{\mathrm{op}}$:

$$A_{\mathrm{op}} = \mathcal{F}_{\mathrm{op}}^* Q \mathcal{F}_{\mathrm{op}}. \tag{B.2.10}$$

Proof. Is is clear from Lemma B.2.1 that $\mathcal{F} : L^2_\Delta(a, b) \to L^2_{d\rho}(\mathbb{R})$ is a partial isometry with $\ker \mathcal{F} = \operatorname{mul} A$. One verifies in the same way as in the proof of Theorem B.1.4 that $\mathcal{F}$ is surjective, and hence $\mathcal{F}_{\mathrm{op}}$ is unitary. The identity (B.2.10) follows from (B.2.8). $\square$

In the situation of Theorem B.2.3 the inverse of the Fourier transform $\mathcal{F}_{\mathrm{op}}$ is actually given by the reverse Fourier transform $\mathcal{G}$ in Lemma B.2.2, which is now a unitary mapping from $L^2_{d\rho}(\mathbb{R})$ to $L^2_\Delta(a, b) \ominus \operatorname{mul} A$. Thus, for all $\varphi \in L^2_{d\rho}(\mathbb{R})$ one has

$$\lim_{\eta \to \mathbb{R}} \int_a^b \left| \Delta(t)^{\frac{1}{2}} \left(\mathcal{F}_{\mathrm{op}}^{-1} \varphi(t) - \int_\eta \omega(t, x)^* \varphi(x) \, d\rho(x) \right) \right|^2 dt = 0.$$

Appendix C

Measures of Closed Subspaces of a Hilbert Space

In this appendix the sum of closed (linear) subspaces in a Hilbert space is discussed and, in particular, conditions are given so that sums of closed subspaces are closed. There is also a brief review of the opening and gap of closed subspaces.

In the following $\mathfrak{M}$ and $\mathfrak{N}$ will be closed subspaces of a Hilbert space $\mathfrak{H}$. The first lemma on the sum of closed subspaces is preliminary.

Lemma C.1. *Let $\mathfrak{M}$ and $\mathfrak{N}$ be closed subspaces of $\mathfrak{H}$. Then the following statements are equivalent:*

(i) *$\mathfrak{M} + \mathfrak{N}$ is closed and $\mathfrak{M} \cap \mathfrak{N} = \{0\}$;*

(ii) *there exists $0 < \rho < 1$ such that*

$$\rho\sqrt{\|f\|^2 + \|g\|^2} \leq \|f + g\|, \quad f \in \mathfrak{M}, \ g \in \mathfrak{N}. \tag{C.1}$$

Proof. (i) $\Rightarrow$ (ii) Assume that $\mathfrak{M} + \mathfrak{N}$ is closed and $\mathfrak{M} \cap \mathfrak{N} = \{0\}$. The projection from the Hilbert space $\mathfrak{M} + \mathfrak{N}$ onto $\mathfrak{M}$ parallel to $\mathfrak{N}$ is a closed, everywhere defined operator. Hence, by the closed graph theorem, the projection is bounded and there exists $C > 0$ such that

$$\|f\| \leq C\|f + g\|, \quad f \in \mathfrak{M}, \ g \in \mathfrak{N}.$$

Likewise, there exists $D > 0$ such that

$$\|g\| \leq D\|f + g\|, \quad f \in \mathfrak{M}, \ g \in \mathfrak{N}.$$

A combination of these inequalities leads to

$$\|f\|^2 + \|g\|^2 \leq (C^2 + D^2)\|f + g\|^2, \quad f \in \mathfrak{M}, \ g \in \mathfrak{N}.$$

By eventually enlarging $C^2 + D^2$ the inequality (C.1) follows with $0 < \rho < 1$.

(ii) $\Rightarrow$ (i) Let (h_n) be a sequence in $\mathfrak{M} + \mathfrak{N}$ converging to $h \in \mathfrak{H}$. Decompose each element h_n as

$$h_n = f_n + g_n, \quad f_n \in \mathfrak{M}, \quad g_n \in \mathfrak{N}.$$

Since (h_n) is a Cauchy sequence in $\mathfrak{H}$ it follows from (C.1) that (f_n) and (g_n) are Cauchy sequences in $\mathfrak{M}$ and $\mathfrak{N}$, respectively. Therefore, there exist elements $f \in \mathfrak{M}$ and $g \in \mathfrak{N}$, so that $f_n \to f$ in $\mathfrak{M}$ and $g_n \to g$ in $\mathfrak{N}$. Hence, $h = f + g \in \mathfrak{M} + \mathfrak{N}$. Thus, $\mathfrak{M} + \mathfrak{N}$ is closed. To see that the sum is direct, assume that $h \in \mathfrak{M} \cap \mathfrak{N}$. Then, in particular, $h \in \mathfrak{M}$ and $-h \in \mathfrak{N}$ and the inequality (C.1) with $f = h$ and $g = -h$ implies $h = 0$. □

Let $\mathfrak{M}$ and $\mathfrak{N}$ be closed subspaces of $\mathfrak{H}$. Then the intersection $\mathfrak{M} \cap \mathfrak{N}$ is a closed subspace which generates an orthogonal decomposition of the Hilbert space:

$$\mathfrak{H} = (\mathfrak{M} \cap \mathfrak{N})^\perp \oplus (\mathfrak{M} \cap \mathfrak{N}).$$

In order to study properties of the sums $\mathfrak{M} + \mathfrak{N}$ and $\mathfrak{M}^\perp + \mathfrak{N}^\perp$ it is sometimes useful to reduce to a direct sum.

Lemma C.2. *Let $\mathfrak{M}$ and $\mathfrak{N}$ be closed subspaces of $\mathfrak{H}$. Then the subspaces*

$$\mathfrak{M} \cap (\mathfrak{M} \cap \mathfrak{N})^\perp \quad and \quad \mathfrak{N} \cap (\mathfrak{M} \cap \mathfrak{N})^\perp$$

have a trivial intersection and due to

$$(\mathfrak{M} \cap \mathfrak{N})^\perp = [\mathfrak{M} \cap (\mathfrak{M} \cap \mathfrak{N})^\perp] \oplus \mathfrak{M}^\perp = [\mathfrak{N} \cap (\mathfrak{M} \cap \mathfrak{N})^\perp] \oplus \mathfrak{N}^\perp \qquad (C.2)$$

their orthogonal complements in the subspace $(\mathfrak{M} \cap \mathfrak{N})^\perp$ coincide with $\mathfrak{M}^\perp$ and $\mathfrak{N}^\perp$, respectively. Moreover,

$$\begin{aligned}
\mathfrak{M} &= [\mathfrak{M} \cap (\mathfrak{M} \cap \mathfrak{N})^\perp] \oplus (\mathfrak{M} \cap \mathfrak{N}), \\
\mathfrak{N} &= [\mathfrak{N} \cap (\mathfrak{M} \cap \mathfrak{N})^\perp] \oplus (\mathfrak{M} \cap \mathfrak{N}),
\end{aligned} \qquad (C.3)$$

and, consequently, $\mathfrak{M} + \mathfrak{N}$ has the decomposition

$$\mathfrak{M} + \mathfrak{N} = [\mathfrak{M} \cap (\mathfrak{M} \cap \mathfrak{N})^\perp + \mathfrak{N} \cap (\mathfrak{M} \cap \mathfrak{N})^\perp] \oplus (\mathfrak{M} \cap \mathfrak{N}). \qquad (C.4)$$

Proof. It is clear that $\mathfrak{M} \cap (\mathfrak{M} \cap \mathfrak{N})^\perp$ and $\mathfrak{N} \cap (\mathfrak{M} \cap \mathfrak{N})^\perp$ have a trivial intersection. To see the first identity in (C.2) note that

$$\mathfrak{M} \cap (\mathfrak{M} \cap \mathfrak{N})^\perp \subset (\mathfrak{M} \cap \mathfrak{N})^\perp \quad and \quad \mathfrak{M}^\perp \subset (\mathfrak{M} \cap \mathfrak{N})^\perp,$$

so that the right-hand side is contained in the left-hand side. To see the other inclusion, decompose $\mathfrak{H}$ as

$$\mathfrak{H} = \mathfrak{M} \oplus \mathfrak{M}^\perp.$$

Let $f \in (\mathfrak{M} \cap \mathfrak{N})^\perp$. Then according to this decomposition one has

$$f = g + h, \quad g \in \mathfrak{M}, \quad h \in \mathfrak{M}^\perp.$$

Since $\mathfrak{M}^\perp \subset (\mathfrak{M} \cap \mathfrak{N})^\perp$ one sees that $g \in \mathfrak{M} \cap (\mathfrak{M} \cap \mathfrak{N})^\perp$. Hence, the left-hand side is contained in the right-hand side. The second identity in (C.2) follows by interchanging $\mathfrak{M}$ and $\mathfrak{N}$.

Since $\mathfrak{M} \cap \mathfrak{N}$ is a closed subspace, it is clear that $\mathfrak{H}$ has the following orthogonal decomposition

$$\mathfrak{H} = (\mathfrak{M} \cap \mathfrak{N})^\perp \oplus (\mathfrak{M} \cap \mathfrak{N}).$$

Hence, in an analogous way as above, the decompositions (C.3) are clear. Thus, (C.4) follows. □

The above reduction process will play a role in the following theorem.

Theorem C.3. *Let $\mathfrak{M}$ and $\mathfrak{N}$ be closed subspaces of $\mathfrak{H}$. Then the following statements are equivalent:*

(i) *$\mathfrak{M} + \mathfrak{N}$ is closed;*

(ii) *$\mathfrak{M}^\perp + \mathfrak{N}^\perp$ is closed.*

Furthermore, the following statements are equivalent:

(iii) *$\mathfrak{M} + \mathfrak{N}$ is closed and $\mathfrak{M} \cap \mathfrak{N} = \{0\}$;*

(iv) *$\mathfrak{M}^\perp + \mathfrak{N}^\perp = \mathfrak{H}$,*

and the following statements are equivalent:

(v) *$\mathfrak{M} + \mathfrak{N} = \mathfrak{H}$ and $\mathfrak{M} \cap \mathfrak{N} = \{0\}$;*

(vi) *$\mathfrak{M}^\perp + \mathfrak{N}^\perp = \mathfrak{H}$ and $\mathfrak{M}^\perp \cap \mathfrak{N}^\perp = \{0\}$.*

Proof. (iii) $\Rightarrow$ (iv) Assume that $\mathfrak{M} \cap \mathfrak{N} = \{0\}$. It will be shown that $\mathfrak{H} = \mathfrak{M}^\perp + \mathfrak{N}^\perp$ or, equivalently, that $\mathfrak{H} \subset \mathfrak{M}^\perp + \mathfrak{N}^\perp$. To see this, choose $h \in \mathfrak{H}$. The element $h \in \mathfrak{H}$ induces two linear functionals F and G on $\mathfrak{M} + \mathfrak{N}$ by

$$F(\gamma) = (\alpha, h), \quad G(\gamma) = (\beta, h), \quad \gamma = \alpha + \beta, \quad \alpha \in \mathfrak{M}, \ \beta \in \mathfrak{N}. \tag{C.5}$$

Since $\mathfrak{M} + \mathfrak{N}$ is closed and $\mathfrak{M} \cap \mathfrak{N} = \{0\}$, it follows from the inequality (C.1) that the functionals F and G are bounded on $\mathfrak{M} + \mathfrak{N}$. Extend the functionals F and G trivially to bounded linear functionals on all of $\mathfrak{H}$, and denote the extensions by F and G, respectively. By the Riesz representation theorem there exist unique elements $f, g \in \mathfrak{H}$ such that

$$F(\gamma) = (\gamma, f), \quad G(\gamma) = (\gamma, g), \quad \gamma \in \mathfrak{H}. \tag{C.6}$$

The definition in (C.5) implies that $F(\gamma) = 0$, $\gamma \in \mathfrak{N}$, and $G(\gamma) = 0$, $\gamma \in \mathfrak{M}$. For the corresponding elements f and g in (C.6) this means that $f \in \mathfrak{N}^\perp$ and $g \in \mathfrak{M}^\perp$. Now for $\gamma = \alpha + \beta$, $\alpha \in \mathfrak{M}$, $\beta \in \mathfrak{N}$, it follows from (C.5) and (C.6) that

$$(\gamma, h) = (\alpha, h) + (\beta, h) = F(\gamma) + G(\gamma) = (\gamma, f) + (\gamma, g).$$

This implies that $k = h - f - g \in (\mathfrak{M} + \mathfrak{N})^{\perp} \subset \mathfrak{M}^{\perp}$ and hence $h = f + g + k$ with $f \in \mathfrak{N}^{\perp}$ and $g + k \in \mathfrak{M}^{\perp}$. Therefore, $\mathfrak{H} \subset \mathfrak{M}^{\perp} + \mathfrak{N}^{\perp}$, which completes the proof.

(i) $\Rightarrow$ (ii) Use the reduction process from Lemma C.2. Since $\mathfrak{M} + \mathfrak{N}$ is assumed to be closed it follows from (C.4) that

$$\mathfrak{M} \cap (\mathfrak{M} \cap \mathfrak{N})^{\perp} + \mathfrak{N} \cap (\mathfrak{M} \cap \mathfrak{N})^{\perp}$$

is closed in $(\mathfrak{M} \cap \mathfrak{N})^{\perp}$. Since this is a direct sum, the sum of their orthogonal complements is closed in $(\mathfrak{M} \cap \mathfrak{N})^{\perp}$ by the implication (iii) $\Rightarrow$ (iv). Recall that the sum of their orthogonal complements coincides with $\mathfrak{M}^{\perp} + \mathfrak{N}^{\perp}$.

(ii) $\Rightarrow$ (i) This follows from the previous implication by symmetry.

(iv) $\Rightarrow$ (iii) Apply (ii) $\Rightarrow$ (i) to conclude that $\mathfrak{M} + \mathfrak{N}$ is closed. This sum is direct since $\{0\} = (\mathfrak{M}^{\perp} + \mathfrak{N}^{\perp})^{\perp}$ by assumption.

(v) $\Leftrightarrow$ (vi) It suffices to show (v) $\Rightarrow$ (vi). Note that it follows from (iii) $\Rightarrow$ (iv) that $\mathfrak{M}^{\perp} + \mathfrak{N}^{\perp} = \mathfrak{H}$. Moreover, $\mathfrak{M}^{\perp} \cap \mathfrak{N}^{\perp} = (\mathfrak{M} + \mathfrak{N})^{\perp} = \{0\}$ by assumption, which completes the argument. $\qquad\square$

Let again $\mathfrak{M}$ and $\mathfrak{N}$ be closed subspaces of $\mathfrak{H}$ and assume that $\mathfrak{M} + \mathfrak{N}$ is closed. It follows from (C.3) that $\mathfrak{M} + \mathfrak{N}$ can be written as

$$\begin{aligned} \mathfrak{M} + \mathfrak{N} &= \big([\mathfrak{M} \cap (\mathfrak{M} \cap \mathfrak{N})^{\perp}] \oplus (\mathfrak{M} \cap \mathfrak{N})\big) + \mathfrak{N} \\ &= [\mathfrak{M} \cap (\mathfrak{M} \cap \mathfrak{N})^{\perp}] + \mathfrak{N}. \end{aligned} \tag{C.7}$$

Since $\mathfrak{M} + \mathfrak{N}$ is closed, the orthogonal decomposition $\mathfrak{H} = (\mathfrak{M} + \mathfrak{N})^{\perp} \oplus (\mathfrak{M} + \mathfrak{N})$ and (C.7) show that

$$\begin{aligned} \mathfrak{H} &= (\mathfrak{M} + \mathfrak{N})^{\perp} \oplus \big([\mathfrak{M} \cap (\mathfrak{M} \cap \mathfrak{N})^{\perp}] + \mathfrak{N}\big) \\ &= \big((\mathfrak{M} + \mathfrak{N})^{\perp} \oplus [\mathfrak{M} \cap (\mathfrak{M} \cap \mathfrak{N})^{\perp}]\big) + \mathfrak{N}. \end{aligned} \tag{C.8}$$

The sum in the identity (C.8) is direct. To see this, assume that

$$f + g + \varphi = 0, \quad f \in (\mathfrak{M} + \mathfrak{N})^{\perp}, \quad g \in \mathfrak{M} \cap (\mathfrak{M} \cap \mathfrak{N})^{\perp}, \quad \varphi \in \mathfrak{N}.$$

Then $f = -g - \varphi$, where $-g - \varphi \in \mathfrak{M} + \mathfrak{N}$. Hence, $f = 0$ and $g = -\varphi$ implies that $g = 0$ and $\varphi = 0$.

The decompositions (C.7) and (C.8) will be used in the proof of the next lemma.

Lemma C.4. *Let $B \in \mathbf{B}(\mathfrak{H}, \mathfrak{K})$ have a closed range and let $\mathfrak{M}$ be a closed subspace of $\mathfrak{H}$. Then $\mathfrak{M} + \ker B$ is closed if and only if $B(\mathfrak{M})$ is closed.*

Proof. Assume that $\mathfrak{M} + \ker B$ is closed and set $\mathfrak{N} = \ker B$. It follows from the direct sum decomposition (C.8) that B maps

$$(\mathfrak{M} + \mathfrak{N})^{\perp} \oplus [\mathfrak{M} \cap (\mathfrak{M} \cap \mathfrak{N})^{\perp}]$$

bijectively onto the closed subspace $\operatorname{ran} B$ and hence B also provides a bijection between the closed subspaces of $(\mathfrak{M} + \mathfrak{N})^{\perp} \oplus [\mathfrak{M} \cap (\mathfrak{M} \cap \mathfrak{N})^{\perp}]$ and those of $\operatorname{ran} B$. In particular, it follows from this observation and (C.7) that

$$B(\mathfrak{M}) = B\big(\mathfrak{M} \cap (\mathfrak{M} \cap \mathfrak{N})^{\perp}\big)$$

is closed in $\mathfrak{K}$.

For the converse statement assume that $B(\mathfrak{M})$ is closed. In order to show that $\mathfrak{M} + \ker B$ is closed, let $f_n \in \mathfrak{M}$ and $\varphi_n \in \ker B$ have the property that

$$f_n + \varphi_n \to \chi$$

for some $\chi \in \mathfrak{H}$. In particular, this shows that $B f_n \to B\chi$. The assumption that $B(\mathfrak{M})$ is closed implies that $B f_n \to B f$ for some $f \in \mathfrak{M}$. Hence,

$$\chi - f = \varphi \in \ker B \quad \text{or} \quad \chi = f + \varphi \in \mathfrak{M} + \ker B.$$

It follows that $\mathfrak{M} + \ker B$ is closed. $\qquad\square$

There is another way to approach the topic of sums of closed subspaces, namely via various notions of opening or gap between closed subspaces.

Definition C.5. Let $\mathfrak{M}$ and $\mathfrak{N}$ be closed subspaces of $\mathfrak{H}$ with corresponding orthogonal projections $P_{\mathfrak{M}}$ and $P_{\mathfrak{N}}$, respectively. The *opening* $\omega(\mathfrak{M}, \mathfrak{N})$ between $\mathfrak{M}$ and $\mathfrak{N}$ is defined as

$$\omega(\mathfrak{M}, \mathfrak{N}) = \|P_{\mathfrak{M}} P_{\mathfrak{N}}\|.$$

It is clear that $0 \leq \omega(\mathfrak{M}, \mathfrak{N}) \leq 1$ and that $\omega(\mathfrak{M}, \mathfrak{N}) = \omega(\mathfrak{N}, \mathfrak{M})$, since one has $\|A\| = \|A^*\|$ for any $A \in \mathbf{B}(\mathfrak{H})$.

Proposition C.6. *Let $\mathfrak{M}$ and $\mathfrak{N}$ be closed subspaces of $\mathfrak{H}$. Then the following statements are equivalent:*

(i) $\omega(\mathfrak{M}, \mathfrak{N}) < 1$;

(ii) $\mathfrak{M} + \mathfrak{N}$ *is closed and* $\mathfrak{M} \cap \mathfrak{N} = \{0\}$.

Proof. (i) $\Rightarrow$ (ii) Assume that $\omega(\mathfrak{M}, \mathfrak{N}) < 1$. Observe that for $f \in \mathfrak{M}$ and $g \in \mathfrak{N}$ one has

$$|(f, g)| = |(P_{\mathfrak{M}} f, P_{\mathfrak{N}} g)| \leq \omega(\mathfrak{M}, \mathfrak{N}) \|f\| \|g\|.$$

It follows that for all $f \in \mathfrak{M}$ and $g \in \mathfrak{N}$

$$\begin{aligned}
\|f\|^2 + \|g\|^2 &= \|f + g\|^2 - 2\operatorname{Re}(f, g) \\
&\leq \|f + g\|^2 + 2|(f, g)| \\
&\leq \|f + g\|^2 + 2\omega(\mathfrak{M}, \mathfrak{N}) \|f\| \|g\| \\
&\leq \|f + g\|^2 + \omega(\mathfrak{M}, \mathfrak{N})\big(\|f\|^2 + \|g\|^2\big).
\end{aligned}$$

In particular, this shows that

$$(1 - \omega(\mathfrak{M}, \mathfrak{N}))\left(\|f\|^2 + \|g\|^2\right) \leq \|f + g\|^2, \quad f \in \mathfrak{M}, \quad g \in \mathfrak{N}.$$

Hence, Lemma C.1 implies that $\mathfrak{M} \cap \mathfrak{N} = \{0\}$ and that $\mathfrak{M} + \mathfrak{N}$ is closed, which gives (ii).

(ii) $\Rightarrow$ (i) Assume that $\mathfrak{M} + \mathfrak{N}$ is closed and $\mathfrak{M} \cap \mathfrak{N} = \{0\}$. By Lemma C.1, the inequality (C.1) holds for some $0 < \rho < 1$, hence for all $f \in \mathfrak{M}$ and $g \in \mathfrak{N}$

$$\rho^2 \left(\|f\|^2 + \|g\|^2\right) \leq \|f\|^2 + 2\operatorname{Re}(f, g) + \|g\|^2$$

or, equivalently,

$$- 2\operatorname{Re}(f, g) \leq (1 - \rho^2)\left(\|f\|^2 + \|g\|^2\right). \tag{C.9}$$

Now let $h, k \in \mathfrak{H}$ with $\|h\| \leq 1$ and $\|k\| \leq 1$ and choose $\theta \in \mathbb{R}$ such that

$$e^{i\theta}(P_{\mathfrak{M}}h, P_{\mathfrak{N}}k) = -|(P_{\mathfrak{M}}h, P_{\mathfrak{N}}k)|.$$

Then $(e^{i\theta}P_{\mathfrak{M}}h, P_{\mathfrak{N}}k) \in \mathbb{R}$ and (C.9) with $f = e^{i\theta}P_{\mathfrak{M}}h$ and $g = P_{\mathfrak{N}}k$ yields

$$\begin{aligned}
|(h, P_{\mathfrak{M}}P_{\mathfrak{N}}k)| &= |(P_{\mathfrak{M}}h, P_{\mathfrak{N}}k)| \\
&= -\operatorname{Re}(e^{i\theta}P_{\mathfrak{M}}h, P_{\mathfrak{N}}k) \\
&\leq \frac{1 - \rho^2}{2}\left(\|e^{i\theta}P_{\mathfrak{M}}h\|^2 + \|P_{\mathfrak{N}}k\|^2\right) \\
&\leq 1 - \rho^2.
\end{aligned}$$

This implies $\|P_{\mathfrak{M}}P_{\mathfrak{N}}\| \leq 1 - \rho^2 < 1$ and hence (i) holds. $\qquad\square$

Corollary C.7. *Let $\mathfrak{M}$ and $\mathfrak{N}$ be closed subspaces of $\mathfrak{H}$ such that $\mathfrak{M} + \mathfrak{N}$ is closed and $\mathfrak{M} \cap \mathfrak{N} = \{0\}$. Then for closed subspaces $\mathfrak{M}_1 \subset \mathfrak{M}$ and $\mathfrak{N}_1 \subset \mathfrak{N}$ also the subspace $\mathfrak{M}_1 + \mathfrak{N}_1$ is closed.*

Proof. This statement follows immediately from Proposition C.6 by noting that $\omega(\mathfrak{M}_1, \mathfrak{N}_1) \leq \omega(\mathfrak{M}, \mathfrak{N}) < 1$. $\qquad\square$

Proposition C.6 and the reduction of closed subspaces in Lemma C.2 leads to the following corollary.

Corollary C.8. *Let $\mathfrak{M}$ and $\mathfrak{N}$ be closed subspaces of $\mathfrak{H}$. Then the following statements are equivalent:*

(i) $\omega\left(\mathfrak{M} \cap (\mathfrak{M} \cap \mathfrak{N})^\perp, \mathfrak{N} \cap (\mathfrak{M} \cap \mathfrak{N})^\perp\right) < 1$;

(ii) $\mathfrak{M} + \mathfrak{N}$ *is closed.*

Proof. Since the closed subspaces $\mathfrak{M} \cap (\mathfrak{M} \cap \mathfrak{N})^\perp$ and $\mathfrak{N} \cap (\mathfrak{M} \cap \mathfrak{N})^\perp$ have trivial intersection, it follows from Proposition C.6 that (i) is equivalent to

$$\mathfrak{M} \cap (\mathfrak{M} \cap \mathfrak{N})^\perp + \mathfrak{N} \cap (\mathfrak{M} \cap \mathfrak{N})^\perp \qquad (C.10)$$

is closed. Furthermore, the decomposition (C.4) shows that the space in (C.10) is closed if and only if $\mathfrak{M} + \mathfrak{N}$ is closed. □

The notion of opening is now supplemented with the notion of gap between closed subspaces.

Definition C.9. Let $\mathfrak{M}$ and $\mathfrak{N}$ be closed subspaces of $\mathfrak{H}$ with corresponding orthogonal projections $P_{\mathfrak{M}}$ and $P_{\mathfrak{N}}$, respectively. The *gap* $g(\mathfrak{M}, \mathfrak{N})$ between $\mathfrak{M}$ and $\mathfrak{N}$ is defined as

$$g(\mathfrak{M}, \mathfrak{N}) = \| P_{\mathfrak{M}} - P_{\mathfrak{N}} \|.$$

Note that it follows directly from the definition that

$$g(\mathfrak{M}, \mathfrak{N}) = g(\mathfrak{N}, \mathfrak{M}) \quad \text{and} \quad g(\mathfrak{M}^\perp, \mathfrak{N}^\perp) = g(\mathfrak{M}, \mathfrak{N}). \qquad (C.11)$$

The connection between the gap and the opening between closed subspaces is contained in the following proposition.

Proposition C.10. *Let $\mathfrak{M}$ and $\mathfrak{N}$ be closed subspaces in $\mathfrak{H}$. Then*

$$g(\mathfrak{M}, \mathfrak{N}) = \max \left\{ \omega(\mathfrak{M}, \mathfrak{N}^\perp), \omega(\mathfrak{M}^\perp, \mathfrak{N}) \right\} \qquad (C.12)$$

and, in particular, $g(\mathfrak{M}, \mathfrak{N}) \leq 1$.

Proof. First observe that

$$P_{\mathfrak{N}^\perp} P_{\mathfrak{M}} = (I - P_{\mathfrak{N}}) P_{\mathfrak{M}} = (P_{\mathfrak{M}} - P_{\mathfrak{N}}) P_{\mathfrak{M}},$$

so that $\omega(\mathfrak{M}, \mathfrak{N}^\perp) = \omega(\mathfrak{N}^\perp, \mathfrak{M}) \leq g(\mathfrak{M}, \mathfrak{N}) \| P_{\mathfrak{M}} \| \leq g(\mathfrak{M}, \mathfrak{N})$. Therefore, also

$$\omega(\mathfrak{M}^\perp, \mathfrak{N}) \leq g(\mathfrak{M}^\perp, \mathfrak{N}^\perp) = g(\mathfrak{M}, \mathfrak{N})$$

and hence

$$\max \left\{ \omega(\mathfrak{M}, \mathfrak{N}^\perp), \omega(\mathfrak{M}^\perp, \mathfrak{N}) \right\} \leq g(\mathfrak{M}, \mathfrak{N}).$$

Furthermore, observe that for all $h \in \mathfrak{H}$

$$\begin{aligned}
\| (P_{\mathfrak{M}} - P_{\mathfrak{N}}) h \|^2 &= \| (I - P_{\mathfrak{N}}) P_{\mathfrak{M}} h - P_{\mathfrak{N}} (I - P_{\mathfrak{M}}) h \|^2 \\
&= \| P_{\mathfrak{N}^\perp} P_{\mathfrak{M}} h \|^2 + \| P_{\mathfrak{N}} P_{\mathfrak{M}^\perp} h \|^2 \\
&= \| P_{\mathfrak{N}^\perp} P_{\mathfrak{M}} P_{\mathfrak{M}} h \|^2 + \| P_{\mathfrak{N}} P_{\mathfrak{M}^\perp} P_{\mathfrak{M}^\perp} h \|^2 \\
&\leq \omega(\mathfrak{M}, \mathfrak{N}^\perp)^2 \| P_{\mathfrak{M}} h \|^2 + \omega(\mathfrak{M}^\perp, \mathfrak{N})^2 \| P_{\mathfrak{M}^\perp} h \|^2,
\end{aligned}$$

and the last term is majorized by

$$\max\left\{\omega(\mathfrak{M},\mathfrak{N}^{\perp})^2,\omega(\mathfrak{M}^{\perp},\mathfrak{N})^2\right\}\left(\|P_{\mathfrak{M}}h\|^2+\|P_{\mathfrak{M}^{\perp}}h\|^2\right).$$

It follows that

$$g(\mathfrak{M},\mathfrak{N})\leq\max\left\{\omega(\mathfrak{M},\mathfrak{N}^{\perp}),\omega(\mathfrak{M}^{\perp},\mathfrak{N})\right\}.$$

Therefore, (C.12) has been established. □

Proposition C.11. *Let $\mathfrak{M}$ and $\mathfrak{N}$ be closed subspaces of $\mathfrak{H}$. Then*

$$g(\mathfrak{M},\mathfrak{N})<1 \quad\Rightarrow\quad \dim\mathfrak{M}=\dim\mathfrak{N}.$$

Proof. If $g(\mathfrak{M},\mathfrak{N})=\|P_{\mathfrak{M}}-P_{\mathfrak{N}}\|<1$, then the bounded operator

$$I-(P_{\mathfrak{M}}-P_{\mathfrak{N}})$$

maps $\mathfrak{H}$ onto itself with bounded inverse. Hence,

$$P_{\mathfrak{M}}P_{\mathfrak{N}}\,\mathfrak{H}=P_{\mathfrak{M}}\left[I-(P_{\mathfrak{M}}-P_{\mathfrak{N}})\right]\mathfrak{H}=P_{\mathfrak{M}}\,\mathfrak{H},$$

so that $P_{\mathfrak{M}}$ maps $\operatorname{ran}P_{\mathfrak{N}}$ onto $\operatorname{ran}P_{\mathfrak{M}}$. Observe that for $h\in\operatorname{ran}P_{\mathfrak{N}}$

$$\|P_{\mathfrak{M}}h\|=\|h+(P_{\mathfrak{M}}-P_{\mathfrak{N}})h\|\geq\left(1-\|P_{\mathfrak{M}}-P_{\mathfrak{N}}\|\right)\|h\|$$

and hence $P_{\mathfrak{M}}$ maps $\operatorname{ran}P_{\mathfrak{N}}$ boundedly and boundedly invertible onto $\operatorname{ran}P_{\mathfrak{M}}$. This implies that the dimensions of $\mathfrak{M}=\operatorname{ran}P_{\mathfrak{M}}$ and $\mathfrak{N}=\operatorname{ran}P_{\mathfrak{N}}$ coincide. □

Theorem C.12. *Let $\mathfrak{M}$ and $\mathfrak{N}$ be closed subspaces of $\mathfrak{H}$. Then the following statements are equivalent:*

(i) $g(\mathfrak{M},\mathfrak{N}^{\perp})<1$;

(ii) $\mathfrak{M}+\mathfrak{N}=\mathfrak{H}$ *and* $\mathfrak{M}\cap\mathfrak{N}=\{0\}$,

and in this case $\dim\mathfrak{M}=\dim\mathfrak{N}^{\perp}$ *and* $\dim\mathfrak{M}^{\perp}=\dim\mathfrak{N}$.

Proof. (i) $\Rightarrow$ (ii) Proposition C.10 implies $\omega(\mathfrak{M},\mathfrak{N})<1$ and $\omega(\mathfrak{M}^{\perp},\mathfrak{N}^{\perp})<1$. By Proposition C.6, the condition $\omega(\mathfrak{M},\mathfrak{N})<1$ implies that $\mathfrak{M}+\mathfrak{N}$ is closed and that $\mathfrak{M}\cap\mathfrak{N}=\{0\}$, and in the same way $\omega(\mathfrak{M}^{\perp},\mathfrak{N}^{\perp})<1$ yields that $\mathfrak{M}^{\perp}+\mathfrak{N}^{\perp}$ is closed and that $\mathfrak{M}^{\perp}\cap\mathfrak{N}^{\perp}=\{0\}$. The identity $\mathfrak{M}^{\perp}\cap\mathfrak{N}^{\perp}=\{0\}$ implies that $\mathfrak{M}+\mathfrak{N}$ is dense in $\mathfrak{H}$. Therefore, $\mathfrak{M}+\mathfrak{N}=\mathfrak{H}$.

(ii) $\Rightarrow$ (i) It follows from Proposition C.6 that $\omega(\mathfrak{M},\mathfrak{N})<1$. Moreover, the equivalence of (v) and (vi) in Theorem C.3 shows $\mathfrak{M}^{\perp}+\mathfrak{N}^{\perp}=\mathfrak{H}$ and $\mathfrak{M}^{\perp}\cap\mathfrak{N}^{\perp}=\{0\}$, and therefore $\omega(\mathfrak{M}^{\perp},\mathfrak{N}^{\perp})<1$ by Proposition C.6. Now Proposition C.10 implies that $g(\mathfrak{M},\mathfrak{N}^{\perp})<1$.

It is clear from (i) and Proposition C.11 that $\dim\mathfrak{M}=\dim\mathfrak{N}^{\perp}$. Furthermore, as $g(\mathfrak{M}^{\perp},\mathfrak{N})=g(\mathfrak{M},\mathfrak{N}^{\perp})$ by (C.11), the same argument gives $\dim\mathfrak{M}^{\perp}=\dim\mathfrak{N}$. □

Appendix D

Bounded Linear Operators

This appendix contains a number of results pertaining to the factorization of bounded linear operators based on range inclusions or norm inequalities. These results will be useful in conjunction with range inclusions or norm inequalities for relations.

Let $\mathfrak{H}$ and $\mathfrak{K}$ be Hilbert spaces and let $H \in \mathbf{B}(\mathfrak{H}, \mathfrak{K})$. The restriction of H to $(\ker H)^{\perp} \subset \mathfrak{H}$,

$$H : (\ker H)^{\perp} \to \operatorname{ran} H,$$

is a bijective mapping between $(\ker H)^{\perp}$ and $\operatorname{ran} H$. The inverse of this restriction is a (linear) operator from $\operatorname{ran} H$ onto $(\ker H)^{\perp}$ and is called the *Moore–Penrose inverse* of the operator H, which in general is an unbounded operator. Since $\mathfrak{H} = \ker H \oplus (\ker H)^{\perp}$, one sees immediately that

$$H^{(-1)}H = P_{(\ker H)^{\perp}}. \tag{D.1}$$

Moreover, the closed graph theorem shows that $\operatorname{ran} H$ is closed if and only if $H^{(-1)}$ takes $\operatorname{ran} H$ boundedly into $(\ker H)^{\perp}$. Hence, if $\operatorname{ran} H$ is closed, then

$$H^{(-1)} \in \mathbf{B}\big(\operatorname{ran} H, (\ker H)^{\perp}\big) \quad \text{and} \quad (H^{(-1)})^{\times} \in \mathbf{B}\big((\ker H)^{\perp}, \operatorname{ran} H\big), \tag{D.2}$$

where $^{\times}$ denotes the adjoint with respect to the scalar products in $\operatorname{ran} H$ and $(\ker H)^{\perp}$.

Lemma D.1. *Let $\mathfrak{H}$ and $\mathfrak{K}$ be Hilbert spaces and let $H \in \mathbf{B}(\mathfrak{H}, \mathfrak{K})$. Then $\operatorname{ran} H$ is closed if and only if $\operatorname{ran} H^{*}$ is closed.*

Proof. Since $H \in \mathbf{B}(\mathfrak{H}, \mathfrak{K})$, it suffices to show that if $\operatorname{ran} H$ is closed in $\mathfrak{K}$, then $\operatorname{ran} H^{*}$ is closed in $\mathfrak{H}$. Since $\overline{\operatorname{ran}} H^{*} = (\ker H)^{\perp}$, one only needs to show

$$(\ker H)^{\perp} \subset \operatorname{ran} H^{*}.$$

For this let $h \in (\ker H)^{\perp}$ and $f \in \mathfrak{H}$. Then it follows from (D.1) and (D.2) that with $(H^{(-1)})^{\times} \in \mathbf{B}((\ker H)^{\perp}, \operatorname{ran} H)$ in (D.2) one has

$$
\begin{aligned}
(h, f)_{\mathfrak{H}} &= (h, P_{(\ker H)^{\perp}} f)_{(\ker H)^{\perp}} \\
&= (h, H^{(-1)} H f)_{(\ker H)^{\perp}} \\
&= ((H^{(-1)})^{\times} h, H f)_{\operatorname{ran} H} \\
&= ((H^{(-1)})^{\times} h, H f)_{\mathfrak{K}} \\
&= (H^* (H^{(-1)})^{\times} h, f)_{\mathfrak{H}}
\end{aligned}
$$

and it follows that $h = H^* (H^{(-1)})^{\times} h$. Hence, $h \in \operatorname{ran} H^*$ and therefore $\operatorname{ran} H^*$ is closed. $\qquad\square$

Lemma D.1 has a direct consequence, which is useful.

Lemma D.2. *For $H \in \mathbf{B}(\mathfrak{H}, \mathfrak{K})$ there are the inclusions*

$$
\operatorname{ran} H H^* \subset \operatorname{ran} H \subset \overline{\operatorname{ran}} H = \overline{\operatorname{ran}} H H^*. \tag{D.3}
$$

Moreover, the following statements are equivalent:

(i) $\operatorname{ran} H H^*$ *is closed;*

(ii) $\operatorname{ran} H$ *is closed;*

(iii) $\operatorname{ran} H H^* = \operatorname{ran} H$,

in which case all inclusions in (D.3) *are identities.*

Proof. The chain of inclusions in (D.3) is clear; the last equality is a consequence of the identity $\ker H^* = \ker H H^*$.

(i) $\Rightarrow$ (iii) Assume that $\operatorname{ran} H H^*$ is closed. Then all inclusions in (D.3) are equalities and, in particular, (iii) follows.

(iii) $\Rightarrow$ (ii) Assume that $\operatorname{ran} H H^* = \operatorname{ran} H$. It will be sufficient to show that $\overline{\operatorname{ran}} H^* \subset \operatorname{ran} H^*$, which implies that $\operatorname{ran} H^*$ is closed and hence also $\operatorname{ran} H$ is closed by Lemma D.1. Assume that $h \in \overline{\operatorname{ran}} H^* = (\ker H)^{\perp}$. By the assumption, it follows that $Hh = H H^* k$ for some k, which gives $H(h - H^* k) = 0$. Note that $h \in (\ker H)^{\perp}$ and $H^* k \in \operatorname{ran} H^* \subset \overline{\operatorname{ran}} H^* = (\ker H)^{\perp}$, and hence $h = H^* k$. Thus, $\overline{\operatorname{ran}} H^* \subset \operatorname{ran} H^*$ which implies that $\operatorname{ran} H^*$ is closed, so that also $\operatorname{ran} H$ is closed by Lemma D.1.

(ii) $\Rightarrow$ (i) Assume that $\operatorname{ran} H$ is closed. Then it follows from (D.3) that

$$
\overline{\operatorname{ran}} H H^* = \operatorname{ran} H. \tag{D.4}
$$

It will suffice to show that $\overline{\operatorname{ran}} H H^* \subset \operatorname{ran} H H^*$. Assume that $k \in \overline{\operatorname{ran}} H H^*$. Then $k = Hh$ for some $h \in (\ker H)^{\perp}$ by (D.4). As $(\ker H)^{\perp} = \overline{\operatorname{ran}} H^* = \operatorname{ran} H^*$ by the assumption (ii) and Lemma D.1, one has $h \in \operatorname{ran} H^*$ and therefore $k \in \operatorname{ran} H H^*$. This shows that $\overline{\operatorname{ran}} H H^* \subset \operatorname{ran} H H^*$ which implies that $\operatorname{ran} H H^*$ is closed. $\qquad\square$

Let $\mathfrak{F}$, $\mathfrak{G}$, and $\mathfrak{H}$ be Hilbert spaces, let $A \in \mathbf{B}(\mathfrak{F},\mathfrak{H})$, $B \in \mathbf{B}(\mathfrak{G},\mathfrak{H})$, and $C \in \mathbf{B}(\mathfrak{F},\mathfrak{G})$, and assume that the following factorization holds:

$$A = BC. \tag{D.5}$$

Then it follows from (D.5) that $\ker C \subset \ker A$. Note that the orthogonal decomposition $\mathfrak{G} = \ker B \oplus \overline{\operatorname{ran}} B^*$ allows one to write $A = BPC$, where P is the orthogonal projection in $\mathfrak{G}$ onto $\overline{\operatorname{ran}} B^*$. Hence, one may always assume that $\operatorname{ran} C \subset \overline{\operatorname{ran}} B^*$. In this case $\ker C = \ker A$ and hence C maps $(\ker A)^\perp = \overline{\operatorname{ran}} A^*$ into $\overline{\operatorname{ran}} B^*$ injectively. Moreover, in this case the operator C in (D.5) is uniquely determined. The following proposition is a version of the well-known *Douglas lemma*.

Proposition D.3. *Assume that $A \in \mathbf{B}(\mathfrak{F},\mathfrak{H})$ and $B \in \mathbf{B}(\mathfrak{G},\mathfrak{H})$. Let $\rho > 0$, then the following statements are equivalent:*

(i) $A = BC$ *for some* $C \in \mathbf{B}(\mathfrak{F},\mathfrak{G})$ *with* $\operatorname{ran} C \subset \overline{\operatorname{ran}} B^*$ *and* $\|C\| \leq \rho$;

(ii) $\operatorname{ran} A \subset \operatorname{ran} B$ *and* $\|B^{(-1)}\varphi\| \leq \rho \|A^{(-1)}\varphi\|$, $\varphi \in \operatorname{ran} A$;

(iii) $AA^* \leq \rho^2 BB^*$.

Moreover, if $\operatorname{ran} A \subset \operatorname{ran} B$, then $\|B^{(-1)}\varphi\| \leq \rho \|A^{(-1)}\varphi\|$, $\varphi \in \operatorname{ran} A$, for some $\rho > 0$.

Proof. (i) $\Rightarrow$ (ii) It is clear that $\operatorname{ran} A \subset \operatorname{ran} B$ and therefore also

$$B^{(-1)}A = B^{(-1)}BC = P_{(\ker B)^\perp}C = P_{\overline{\operatorname{ran}} B^*}C = C.$$

From the last identity one sees that

$$\|B^{(-1)}A\psi\| \leq \|C\| \|\psi\|, \quad \psi \in \mathfrak{F}.$$

Now let $\varphi \in \operatorname{ran} A$. Then there exists $\psi \perp \ker A$ such that $\varphi = A\psi$. Therefore, $\psi = A^{(-1)}\varphi$ and it follows immediately that

$$\|B^{(-1)}\varphi\| \leq \|C\| \|A^{(-1)}\varphi\|, \quad \varphi \in \operatorname{ran} A.$$

Hence, (ii) has been shown.

(ii) $\Rightarrow$ (i) It follows from (ii) that for each $h \in \mathfrak{F}$ there exists a uniquely determined element $k \in (\ker B)^\perp = \overline{\operatorname{ran}} B^*$ such that $Ah = Bk$. Hence, the mapping $h \mapsto k$ from $\mathfrak{F}$ to $\overline{\operatorname{ran}} B^* \subset \mathfrak{G}$ defines a linear operator C with $\operatorname{dom} C = \mathfrak{F}$ and one has $A = BC$. To show that the operator C is closed, assume that

$$h_n \to h \quad \text{and} \quad k_n = Ch_n \to k,$$

where $k_n \in \overline{\operatorname{ran}} B^*$. Since A and B are bounded linear operators, it follows from from $Ah_n = Bk_n$ that $Ah = Bk$. Furthermore, $\overline{\operatorname{ran}} B^*$ is closed, which implies that $k \in \overline{\operatorname{ran}} B^*$. Hence, $k = Ch$ and thus C is closed. By the closed graph theorem, it follows that $C \in \mathbf{B}(\mathfrak{F}, \mathfrak{G})$. The property $\operatorname{ran} C \subset \overline{\operatorname{ran}} B^*$ holds by construction. Furthermore, one sees that for $\psi \in \mathfrak{F}$

$$
\begin{aligned}
\|C\psi\| &= \|P_{\overline{\operatorname{ran}} B^*} C\psi\| \\
&= \|P_{(\ker B)^\perp} C\psi\| \\
&= \|B^{(-1)} BC\psi\| \\
&\leq \rho \|A^{(-1)} A\psi\| \\
&= \rho \|P_{(\ker A)^\perp} \psi\| \\
&\leq \rho \|\psi\|,
\end{aligned}
$$

and so $\|C\| \leq \rho$.

(i) $\Rightarrow$ (iii) It is clear that $AA^* = BCC^*B^*$. For $\psi \in \mathfrak{F}$ one has

$$
\left(AA^*\psi, \psi\right) = \|C^*B^*\psi\|^2 \leq \|C^*\|^2 \|B^*\psi\|^2 = \|C^*\|^2 \left(BB^*\psi, \psi\right)
$$

and hence $AA^* \leq \|C^*\|^2 BB^*$. Since $\|C^*\| = \|C\| \leq \rho$ this leads to (iii).

(iii) $\Rightarrow$ (i) The mapping D from $\operatorname{ran} B^*$ to $\operatorname{ran} A^*$ given by

$$
B^*h \mapsto A^*h, \quad h \in \mathfrak{H},
$$

is well defined and linear. To see that it is well defined observe that $B^*h = 0$ implies $AA^*h = 0$ and hence $A^*h = 0$. Then $DB^*h = A^*h$ for $h \in \mathfrak{H}$ and one has $D \in \mathbf{B}(\operatorname{ran} B^*, \operatorname{ran} A^*)$ with $\|DB^*h\| \leq \rho^2 \|B^*h\|$. Due to the boundedness, this operator has a unique extension in $\mathbf{B}(\overline{\operatorname{ran}} B^*, \overline{\operatorname{ran}} A^*)$, denoted again by D, and $\|D\| \leq \rho^2$. Finally, this mapping has a trivial extension $D \in \mathbf{B}(\mathfrak{G}, \mathfrak{F})$, satisfying $DB^*h = A^*h$ for $h \in \mathfrak{H}$ and again $\|D\| \leq \rho^2$. With $C = D^*$ one obtains that $C \in \mathbf{B}(\mathfrak{F}, \mathfrak{G})$, $A = BC$, and $\|C\| \leq \rho^2$. In view of the construction of D one has $\ker B = (\overline{\operatorname{ran}} B^*)^\perp \subset \ker D$, so that $\operatorname{ran} C = \operatorname{ran} D^* \subset \overline{\operatorname{ran}} B^*$.

For the last statement it suffices to observe that the inclusion $\operatorname{ran} A \subset \operatorname{ran} B$ implies $A = BC$ for some $C \in \mathbf{B}(\mathfrak{F}, \mathfrak{G})$ (see (ii) $\Rightarrow$ (i)). However, this implies $AA^* \leq \rho^2 BB^*$ for some $\rho > 0$ (see (i) $\Rightarrow$ (iii)). $\square$

The next result contains a strengthening of Proposition D.3. Recall that $\operatorname{ran} C \subset \overline{\operatorname{ran}} B^*$ implies that $\ker C = \ker A$. Hence, if C maps $\overline{\operatorname{ran}} A^*$ onto $\overline{\operatorname{ran}} B^*$, then C maps $\overline{\operatorname{ran}} A^*$ onto $\operatorname{ran} C = \overline{\operatorname{ran}} B^*$ bijectively.

Corollary D.4. *Assume that $A \in \mathbf{B}(\mathfrak{F}, \mathfrak{H})$ and $B \in \mathbf{B}(\mathfrak{G}, \mathfrak{H})$. Let $\rho > 0$. Then the following statements are equivalent:*

(a) $A = BC$ *for some* $C \in \mathbf{B}(\mathfrak{F}, \mathfrak{G})$ *with* $\operatorname{ran} C = \overline{\operatorname{ran}} B^*$ *and* $\|C\| \leq \rho$;

(b) $\operatorname{ran} A = \operatorname{ran} B$ *and* $\|B^{(-1)}\varphi\| \leq \rho \|A^{(-1)}\varphi\|$, $\varphi \in \operatorname{ran} A$;

(c) $\varepsilon^2 BB^* \leq AA^* \leq \rho^2 BB^*$ *for some* $0 < \varepsilon < \rho$.

Moreover, if $\operatorname{ran} A = \operatorname{ran} B$, *then* $\|B^{(-1)}\varphi\| \leq \rho \|A^{(-1)}\varphi\|$, $\varphi \in \operatorname{ran} A$, *for some* $\rho > 0$.

Proof. The assertions in Proposition D.3 will be freely used in the arguments below.

(a) $\Rightarrow$ (c) Recall that C maps $\overline{\mathrm{ran}}\, A^*$ bijectively onto $\overline{\mathrm{ran}}\, B^*$. Since

$$\overline{\mathrm{ran}}\, C^* = \overline{\mathrm{ran}}\, A^* \quad \text{and} \quad \ker C^* = \ker B,$$

one sees that C^* maps $\overline{\mathrm{ran}}\, B^*$ bijectively onto $\overline{\mathrm{ran}}\, A^*$, as $\mathrm{ran}\, C^*$ is closed by Lemma D.1. Thus, CC^* is a bijective mapping from $\overline{\mathrm{ran}}\, B^*$ onto itself and hence

$$(AA^*h, h) = (CC^*B^*h, B^*h) \geq \varepsilon^2(B^*h, B^*h) = \varepsilon^2(BB^*h, h), \quad h \in \mathfrak{H}.$$

Therefore, (c) follows.

(c) $\Rightarrow$ (b) The inequality $BB^* \leq \varepsilon^{-2}AA^*$ implies $\mathrm{ran}\, B \subset \mathrm{ran}\, A$.

(b) $\Rightarrow$ (a) It suffices to show that $\overline{\mathrm{ran}}\, B^* \subset \mathrm{ran}\, C$. Let $k \in \overline{\mathrm{ran}}\, B^*$. The assumption $\mathrm{ran}\, B \subset \mathrm{ran}\, A$ implies that $Bk = Ah$ for some $h \in \mathfrak{F}$. Since $A = BC$, one sees that $Bk = BCh$, and hence $k - Ch \in \ker B$. On the other hand, $k \in \overline{\mathrm{ran}}\, B^*$ and $Ch \in \mathrm{ran}\, C \subset \overline{\mathrm{ran}}\, B^*$ yield $k - Ch \in \overline{\mathrm{ran}}\, B^* = (\ker B)^\perp$. Therefore, $k = Ch$ and it follows that $\overline{\mathrm{ran}}\, B^* \subset \mathrm{ran}\, C$ holds. $\square$

An operator $T \in \mathbf{B}(\mathfrak{F}, \mathfrak{G})$ is said to be a *partial isometry* if the restriction of T to $(\ker T)^\perp$ is an isometry. The *initial space* is $(\ker T)^\perp$ and the *final space* is $\mathrm{ran}\, T$, which is automatically closed since $(\ker T)^\perp$ is closed and the restriction of T is isometric. Let $T \in \mathbf{B}(\mathfrak{F}, \mathfrak{G})$, then the following statements are equivalent:

(i) T is a partial isometry with initial space $\mathfrak{M}$ and final space $\mathfrak{N}$;

(ii) T^*T is an orthogonal projection in $\mathfrak{F}$ onto $\mathfrak{M}$;

(iii) T^* is a partial isometry with initial space $\mathfrak{N}$ and final space $\mathfrak{M}$;

(vi) TT^* is an orthogonal projection in $\mathfrak{G}$ onto $\mathfrak{N}$.

The next result can be considered to be a special case of Proposition D.3 and Corollary D.4.

Corollary D.5. *Assume that $A \in \mathbf{B}(\mathfrak{F}, \mathfrak{H})$ and $B \in \mathbf{B}(\mathfrak{G}, \mathfrak{H})$. Then the following statements are equivalent:*

(i) *$A = BC$ for some $C \in \mathbf{B}(\mathfrak{F}, \mathfrak{G})$, where C is a partial isometry with initial space $\overline{\mathrm{ran}}\, A^*$ and final space $\overline{\mathrm{ran}}\, B^*$;*

(ii) *$AA^* = BB^*$.*

Proof. (i) $\Rightarrow$ (ii) Observe that $AA^* = BCC^*B^*$. By assumption, C is a partial isometry with initial space $\overline{\mathrm{ran}}\, A^*$ and final space $\overline{\mathrm{ran}}\, B^*$. Therefore, C^* is a partial isometry with initial space $\overline{\mathrm{ran}}\, B^*$ and final space $\overline{\mathrm{ran}}\, A^*$, and it follows that CC^* is the orthogonal projection onto the final space $\overline{\mathrm{ran}}\, B^*$ of C. Hence, $AA^* = BB^*$.

(ii) $\Rightarrow$ (i) It follows from Corollary D.4 that $A = BC$ for some $C \in \mathbf{B}(\mathfrak{F}, \mathfrak{G})$ with $\mathrm{ran}\, C = \overline{\mathrm{ran}}\, B^*$ and $\|C\| \leq 1$. Moreover

$$\|C^*B^*h\| = \|A^*h\| = \|B^*h\|, \quad h \in \mathfrak{H},$$

shows that C^* is a partial isometry with initial space $\overline{\mathrm{ran}}\, B^*$ and final space $\overline{\mathrm{ran}}\, A^*$. Therefore, C is a partial isometry with initial space $\overline{\mathrm{ran}}\, A^*$ and final space $\overline{\mathrm{ran}}\, B^*$. $\qquad\qquad\qquad\qquad\qquad\qquad\qquad\qquad\qquad\qquad\qquad\qquad\square$

The following *polar decomposition* of $H \in \mathbf{B}(\mathfrak{F}, \mathfrak{H})$ can be seen as a consequence of the previous corollaries. Recall that $|H^*| \in \mathbf{B}(\mathfrak{H})$ is a nonnegative operator in $\mathfrak{H}$ defined by $|H^*| = (HH^*)^{\frac{1}{2}}$.

Corollary D.6. *Let $H \in \mathbf{B}(\mathfrak{F}, \mathfrak{H})$. Then $H = |H^*|C$ and $|H^*| = HC^*$ for some partial isometry $C \in \mathbf{B}(\mathfrak{F}, \mathfrak{H})$ with initial space $\overline{\mathrm{ran}}\, H^*$ and final space $\overline{\mathrm{ran}}\, |H^*|$. In addition, one has $\mathrm{ran}\, H = \mathrm{ran}\, |H^*|$.*

The following corollary is a direct consequence of Lemma D.2.

Corollary D.7. *Let $H \in \mathbf{B}(\mathfrak{H})$ be self-adjoint and nonnegative and let $H^{\frac{1}{2}}$ be its square root. Then there are the inclusions*

$$\mathrm{ran}\, H \subset \mathrm{ran}\, H^{\frac{1}{2}} \subset \overline{\mathrm{ran}}\, H^{\frac{1}{2}} = \overline{\mathrm{ran}}\, H. \tag{D.6}$$

Moreover, the following statements are equivalent:

(i) $\mathrm{ran}\, H$ *is closed;*

(ii) $\mathrm{ran}\, H^{\frac{1}{2}}$ *is closed;*

(iii) $\mathrm{ran}\, H = \mathrm{ran}\, H^{\frac{1}{2}}$,

in which case all the inclusions in (D.6) *are identities.*

Recall that the Moore–Penrose inverse of $H^{\frac{1}{2}}$ is the uniquely defined inverse of the restriction

$$H^{\frac{1}{2}} : (\ker H^{\frac{1}{2}})^{\perp} = (\ker H)^{\perp} \to \mathrm{ran}\, H^{\frac{1}{2}}.$$

In the sequel the Moore–Penrose inverse of $H^{\frac{1}{2}}$ will be denoted by $H^{(-\frac{1}{2})}$, i.e.,

$$H^{(-\frac{1}{2})} := (H^{\frac{1}{2}})^{(-1)}.$$

It is clear that

$$H^{(-\frac{1}{2})} \in \mathbf{B}(\mathrm{ran}\, H^{\frac{1}{2}}, (\ker H)^{\perp}) \quad \Leftrightarrow \quad \mathrm{ran}\, H^{\frac{1}{2}} \text{ is closed.}$$

Here is the version of Proposition D.3 for nonnegative operators.

Proposition D.8. *Let $A, B \in \mathbf{B}(\mathfrak{H})$ be self-adjoint and nonnegative, and let $\rho > 0$. Then the following statements are equivalent:*

(i) $A^{\frac{1}{2}} = B^{\frac{1}{2}}C$ *for some $C \in \mathbf{B}(\mathfrak{H})$ with $\mathrm{ran}\, C \subset \overline{\mathrm{ran}}\, B^{\frac{1}{2}}$ and $\|C\| \leq \rho$;*

(ii) $\mathrm{ran}\, A^{\frac{1}{2}} \subset \mathrm{ran}\, B^{\frac{1}{2}}$ *and $\|B^{(-\frac{1}{2})}\varphi\| \leq \rho\,\|A^{(-\frac{1}{2})}\varphi\|$, $\varphi \in \mathrm{ran}\, A^{\frac{1}{2}}$;*

(iii) $A \leq \rho^2 B$.

Morover, if $\mathrm{ran}\, A^{\frac{1}{2}} \subset \mathrm{ran}\, B^{\frac{1}{2}}$, then $\|B^{(-\frac{1}{2})}\varphi\| \leq \rho\,\|A^{(-\frac{1}{2})}\varphi\|$, $\varphi \in \mathrm{ran}\, A^{\frac{1}{2}}$, for some $\rho > 0$.

Notes

These pages contain a number of comments about the various results in this monograph and some further developments. We also give some historical remarks without any claim of completeness; the history of this area is complicated, also because of the limited exchange of ideas that has existed between East and West. The main setting of the book is that of operators and relations in Hilbert spaces. For an introduction and comprehensive treatment of linear operators in Hilbert spaces and related topics we refer the reader to the standard textbooks [2, 133, 141, 272, 276, 341, 348, 598, 649, 650, 651, 652, 673, 691, 723, 738, 743, 752, 756, 757]. We shall occasionally address topics that fall outside the scope of the monograph itself, e.g., the role of operators and relations in spaces with an indefinite inner product is indicated with proper references.

Notes on Chapter 1

An Overview of Linear Relations go back at least to Arens [42]; they attracted attention because of their usefulness in the extension theory of not necessarily densely defined operators; see for instance [82, 128, 202, 218, 266, 297, 523, 619, 638]. Linear relations also appear in a natural way in the description of boundary conditions; cf. [2, 660]. Related material can be found in [74, 298, 406, 512, 513, 514, 515, 516, 576, 683, 685, 686]. The description of self-adjoint, maximal dissi-pative, or accumulative extensions of a symmetric operator (see Sections 1.4, 1.5, and 1.6) in terms of operators between the defect spaces in Theorem 1.7.12 and Theorem 1.7.14 goes back to von Neumann and ˇStraus [610, 731] in the densely defined case and was worked out in the general case by Coddington [202, 203]; see also [266, 490, 733]. These descriptions may be seen as building stones for the extension theory via boundary triplets in Chapter 2.

Sections 1.1, 1.2, and 1.3 present elementary facts. The identity (1.1.10) goes back to Štraus [726]. The notions of spectrum and points of regular type for a linear relation in Definition 1.2.1 and Definition 1.2.3 are introduced as in the operator case [268, 269, 404], while the adjoint of a linear relation in Definition 1.3.1 is introduced as in [611], see also [42, 659] and [396, 599]. Note that in Proposition 1.2.9 one sees a "pseudo-resolvent" which already shows that there is a relation in the background; cf. [421]. In Theorem 1.3.14 we present the notion of the operator part of a relation and the corresponding Hilbert space decomposition in the spirit of the Lebesgue decomposition of a relation [397, 398]; see also [439, 621, 622, 623] for the operator case. Operator parts have a connection with generalized inverses (see for instance [127, 609]), such as the Moore–Penrose inverse in Definition 1.3.17. Note that the more general definition $P_{\ker \overline{H}} H^{-1}$ gives a Moore–Penrose inverse that is a closable operator. The special classes of relations and their transforms are developed in the usual way; see [660, 705, 706]. The presentation is influenced by personal communication with McKelvey (see also [572]) and lecture notes by Kaltenbäck [451] and Woracek [775, 776]; see also [365, 556, 662] for more references. Although we assume a working knowledge of the spectral theory of self-adjoint operators in Hilbert spaces, we highlight in Section 1.5 a couple of useful facts, with special attention paid to the semibounded case. Lemma 1.5.7 goes back to [210], Proposition 1.5.11 is a consequence of the Douglas lemma in Appendix D, and Lemma 1.5.12 is taken from [95]. The extension theory in the sense of von Neumann can be found in Section 1.7. Note that extensions of a symmetric relation may be disjoint or transversal. Our present characterization of these notions in Theorem 1.7.3 seems to be new; see also [391].

For the development of boundary triplets we only need a few basic facts concerning spaces with an indefinite inner product, see Section 1.8. A typical result in this respect is the automatic boundedness property in Lemma 1.8.1; see [73, 77, 235, 236]. The transform in Definition 1.8.4 goes back to Shmuljan [506, 705, 706]. The notions of strong graph convergence and strong resolvent convergence of relations are treated in Section 1.9 (see [650] for the case of self-adjoint operators). Here we prove the equivalence of the two notions when there is a uniform bound. For Corollary 1.9.5 see [755]. The parametric representation of linear relations in Section 1.10 is closely connected with the theory of operator ranges. Here we only consider such representations from the point of view of boundary value problems; see also [2, 660] and Chapter 2. However, these parametric representations also play an important role in the above mentioned Lebesgue decompositions of relations and, more generally, in the Lebesgue type decompositions of relations; cf. [397]. Most of the material in the beginning of Section 1.11 consists of straightforward generalizations of properties of the resolvent. Lemma 1.11.4 is well known for the usual resolvent and seems to be new in the generalized context; it is used in Section 2.8. Lemma 1.11.5 was inspired by [523]. Nevanlinna families and pairs are generalizations of operator-valued Nevanlinna functions; cf. Appendix A. The notion of Nevanlinna pair goes back to [617]. The notion of Nevanlinna family appears already in [523] (and in a different guise in [266]) and was later more systematically studied in [20, 21, 22, 89, 94, 95, 234, 246].

Although in this monograph we will restrict ourselves mainly to Hilbert spaces, we will make some comments in the notes concerning the setting of indefinite inner product spaces; see for instance [31, 77, 79, 142, 430, 452] and also [775, 776] for Pontryagin spaces, almost Pontryagin spaces, Kreĭn spaces, and almost Kreĭn spaces. The spectral theory of operators and relations in such spaces has been studied extensively; in Kreĭn spaces the operators are often required to have (locally) finitely many negative squares or to be (locally) definitizable, that is, roughly speaking some polynomial (or rational function) of the operator or relation under consideration is (locally) nonnegative in the Kreĭn space sense. Here we only mention work related to the present topics [75, 78, 79, 99, 220, 222, 253, 260, 264, 268, 269, 274, 275, 434, 435, 436, 437, 497, 498, 499, 500, 501, 502, 503, 519, 522, 631, 632, 633, 634, 674, 717, 718, 719, 720, 721, 722, 764, 765].

Notes on Chapter 2

Boundary triplets were originally introduced in the works of Bruk [176, 177, 178] and Kochubeĭ [466] and were used in the study of operator differential equations in the monograph [346]; see also [509]. The notion of boundary triplet appeared implicitly in a different form already much earlier in the work of Calkin [186, 187] (see also [276, Chapter XII] and [410]). The Weyl function was introduced by Derkach and Malamud in [243, 244]; it can also be interpreted as the Q-function from the works of Kreĭn [491, 492] (see also [679]) and, e.g., the papers [499, 500, 501, 502, 503, 504] of Kreĭn and Langer, and [523] of Langer and Textorius. We also mention the more recent monographs [359, 691], where boundary triplets and Weyl functions are briefly discussed, as well as the very recent monograph by Derkach and Malamud [247] (in Russian), in which a more detailed exposition and more than 400 references (also many older papers published in Russian) can be found. From our point of view the most influential general papers on boundary triplets and Weyl functions area are [245, 246] and [184], the latter also has an introduction to the basic notions.

While a large part of the material in this chapter can be found in the literature, the presentation here is given in a unified form for general symmetric relations. The connection between the abstract Green identity and the appropriate indefinite inner products (see Section 1.8) is used in Section 2.1 when deriving some of the basic properties of boundary triplets and their transforms. The inverse result for boundary triplets in Theorem 2.1.9 is taken from [103, 104]. The discussion in Section 2.2 on parametric representations of boundary conditions is given to establish a connection between abstract and concrete boundary value problems. The definitions and the central properties of γ-fields and Weyl functions are presented in Section 2.3 and these results can be found in [245, 246]. In particular, Proposition 2.3.2 (part (ii)) and Proposition 2.3.6 (parts (iii), (v)) show the connection with the notion of Q-function (see [184, 245, 246]), while the formula in part (vi) of Proposition 2.3.6 is taken from [265]. Without specification of the γ-field and Weyl function, constructions of boundary triplets appear already in the early literature; see [176, 466, 577] and the monographs [346, 509]. Here the constructions are based on decompositions that are closely related to the von Neumann formulas in Section 1.7. The Weyl function in Theorem 2.4.1 coincides with the abstract Donoghue type M-function that was studied, for instance, in [317, 320, 327]. Transformations of boundary triplets and the corresponding

γ-fields and Weyl functions have been treated in [234, 237, 245, 246]. In particular, the description of boundary triplets in Theorem 2.5.1 is taken from [246]. We make the precise connection with Q-functions in Corollary 2.5.8. The present notion of unitary equivalence of boundary triplets in Definition 2.5.14 is slightly more general than the definition used in [382].

Section 2.6 is devoted to Kreĭn's formula for intermediate extensions. The Kreĭn type formula (for intermediate extensions) phrased in terms of boundary triplets in Theorem 2.6.1 is adapted to the present setting from [233]. This yields new simple proofs for the description of various parts of the spectra of intermediate extensions in Theorem 2.6.2. This description appears in a similar form in [243, 244, 245, 246] and is treated for dissipative extensions in [346, 468] by means of characteristic functions. The same remark applies to the proof of Theorem 2.6.5, which can be found in a different form in [184]. These results indicate the importance of Kreĭn's formula for the spectral analysis of self-adjoint extensions of symmetric operators. Closely related is the completion problem, briefly touched upon in Remark 2.4.4, which is investigated in [383], where also an analogous boundary triplet appears for the nonnegative case. This case of extensions of a bounded symmetric operator was studied earlier in, for instance, [735, 736].

Section 2.7 is devoted to the Kreĭn–Naĭmark formula, which is our terminology for the Kreĭn formula for compressed resolvents of self-adjoint exit space extensions, in short, Kreĭn's formula for exit space extensions. Actually, it was Naĭmark [605, 606, 607] who investigated different types of extensions (with exit) and a proper interpretation then yields the corresponding resolvent formula. For the notion of Štraus family and for Kreĭn's formula for exit space extensions we refer to [491, 492, 679] and to [726, 727, 728, 729, 730, 731, 732, 733]; some other related papers are [51, 121, 126, 219, 234, 237, 254, 266, 320, 497, 523, 554, 595, 596, 624]. The Štraus families are restrictions of the adjoint in terms of "λ-dependent boundary conditions" given by a Nevanlinna family or corresponding Nevanlinna pair. When the exit space is a Pontryagin space, the same mechanism is in force and the boundary conditions are now given by a generalized Nevanlinna family or corresponding Nevanlinna pair (generalized in the sense that the family or pair has negative squares); we mention [76, 100, 111, 253, 257, 258, 259, 260, 261, 263, 290, 521, 523] and [664] for the special case of a finite-dimensional exit space. Boundary triplets for symmetric relations in Pontryagin and even in Kreĭn spaces were introduced by Derkach (see [101, 227, 228, 229, 230]) and for isometric operators in Pontryagin spaces see [80]; cf. Chapter 6 for concrete λ-dependent boundary conditions.

The perturbation problems in Section 2.8 and Kreĭn's formula are closely related. The $\mathfrak{S}_p$-perturbation result in Theorem 2.8.3 appears in a similar form in [245]. Standard (additive) perturbations of an unbounded self-adjoint operator yield an analogous situation, where the symmetric operator is maximally nondensely defined [405].

Rigged Hilbert spaces offer a framework where extension theory of unbounded symmetric operators can be developed in a somewhat analogous manner as in the bounded case; see [371, 400, 401, 746]. A closely related approach to extensions of symmetric relations relies on the concept of graph perturbations studied in [203, 204, 205, 206, 238, 264, 267, 270, 399, 403]. There has been a considerable interest in the related concept of singular perturbations, where perturbation elements belong to rigged Hilbert spaces with negative indices; see [10], which contains an extensive list of earlier literature, and see also the notes to Chapter 6 and Chapter 8 in the context of differential operators with singular potentials. General operator models for highly singular perturbations involve lifting of operators in Pontryagin spaces [117, 239, 240, 241, 252, 256, 640, 641, 707].

The interest in boundary triplets and Weyl functions has substantially grown in the last decade, so a complete list of references is beyond the scope of these notes. However, for a selection of papers in which boundary triplet techniques were applied to differential operators and other related problems we refer to [28, 50, 86, 87, 97, 112, 113, 123, 143, 144, 155, 169, 170, 173, 174, 184, 185, 241, 288, 307, 342, 343, 344, 377, 461, 475, 477, 478, 483, 511, 529, 551, 562, 563, 564, 565, 589, 590, 591, 626, 627, 642, 644, 677, 750, 751]. For boundary triplets and similar techniques in the analysis of quantum graphs see [110, 134, 172, 183, 196, 287, 289, 294, 536, 625, 637, 645, 646].

Boundary triplets and their Weyl functions for symmetric operators have been further generalized in [103, 105, 109, 110, 115, 233, 236, 237, 246] by relaxing some of the conditions in the definition of a boundary triplet. Moreover, in the setting of dual pairs of operators (see [525]) boundary triplets have been introduced in [559, 560] and applied, e.g., in [170, 173, 300, 302, 381]; they were specialized to the case of isometric operators in [561]. Boundary triplets and their extensions also occur naturally in a system-theoretic environment, where the underlying operator is often isometric, contractive, or skew-symmetric; cf. [65, 66, 67, 102, 394, 750, 751].

By applying a transform of the boundary triplet, resulting in a Cayley transform of the Weyl function, one gets a connection with the notion of characteristic function, which was used to elaborate an alternative analytic tool for studying extension theory for symmetric operators, for some literature in this direction, see, e.g., [165, 166, 246, 346, 468, 509, 603, 728].

Notes on Chapter 3

Most of the material in Section 3.1 and Section 3.2 is working knowledge from measure theory. Typical classical textbooks providing a more detailed exposition on Borel measures are [272, 335, 416, 676, 682] and the papers [61, 60, 271] are listed here for symmetric derivatives and the limit behavior of the Borel transform (see Appendix A). We also recommend the more recent monograph [738]. Section 3.3 is a brief exposition of the notions of absolutely continuous and singular spectrum of self-adjoint operators (see [649, 691, 738, 757]) in the context of self-adjoint relations; cf. Section 1.5 and Theorem 1.5.1.

Simplicity of symmetric operators and the decomposition of a symmetric operator into a self-adjoint and a simple symmetric part in Theorem 3.4.4 go back to [496]; cf. [523] for a version of this result for symmetric relations. The local version of simplicity in Definition 3.4.9 appears in papers of Jonas (see, e.g., [437]) and was more recently considered in [119].

The characterization of the spectrum via the limit properties of the Weyl function in Section 3.5 and Section 3.6 is well known for the special case of singular Sturm–Liouville differential operators and the Titchmarsh–Weyl m-function; cf. Chapter 6 and the classics [209, 740, 741] and, e.g., [60, 97, 175, 224, 321, 328, 335, 333, 334, 426, 479, 711, 712] and also [195]. This description was extended in [155] to the abstract setting of boundary triplets and Weyl functions in the case where the underlying symmetric operator is densely defined and simple (on $\mathbb{R}$). These ideas where further generalized in [119, 120] and applied to elliptic partial differential operators. The present treatment in the context of linear relations seems to be new; cf. [265] for Theorem 3.6.1 (i).

Our presentation of the material in Section 3.7 is strongly inspired by the contribution [523] of Langer and Textorius. Due to the limit properties in this section there are important connections between analytic properties of the Weyl functions and the geometric properties of the associated self-adjoint extensions. The characterizations stated in Proposition 3.7.1 and Proposition 3.7.4 have been established initially in [371, 374, 379], where they were used to introduce the notions of generalized Friedrichs and Kreĭn–von Neumann extensions for non-semibounded symmetric operators or relations; see the notes on Chapter 5.

Finally, the translation of the results in Section 3.5 and Section 3.6 to arbitrary self-adjoint extensions in Section 3.8 based on the corresponding transformation of the Weyl function (see Section 2.5) is immediate.

Notes on Chapter 4

One of the central themes in this chapter is the construction of a Hilbert space via a nonnegative reproducing kernel; such a construction has been studied intensively after Aronszjan [59]. For some introductory texts on this topic we refer the reader to [32, 272, 277, 278, 680, 737]. It should be mentioned that there are different approaches avoiding the mechanism of reproducing kernel Hilbert spaces; see for instance Brodskiĭ [165] and for another method Sz.-Nagy and Korányi [604] (sometimes called the ε-method).

Section 4.1 gives a short review of all the facts about reproducing kernel Hilbert spaces that are needed in the monograph. We remind the reader that special reproducing kernels were considered by de Branges and Rovnyak for the setting of Nevanlinna functions or Schur functions [146, 147, 148, 149, 150, 151, 152, 153]; see also [24, 25, 33].

In Section 4.2 the case of operator-valued Nevanlinna functions is taken up first. Theorem 4.2.1 states that the Nevanlinna kernel of a Nevanlinna function M is nonnegative; the proof can be found in [450]. In this case the reproducing kernel approach leads to the realization of the function $-(M(\lambda) + \lambda)^{-1}$ in terms of compressed resolvents in Theorem 4.2.2; see [231, Theorem 2.5 and Remark 2.6] and also [409]. The unitary equivalence of different models in Theorem 4.2.3 is based on a standard argument. If the Nevanlinna function is uniformly strict, then it can be regarded as a Weyl function; see Theorem 4.2.4 as follows by an inspection of the proof of Theorem 4.2.2. The uniqueness of the model in Theorem 4.2.6 is obtained by a further specialization of the proof of Theorem 4.2.3.

Section 4.3 presents a reproducing kernel approach for scalar Nevanlinna functions via their integral representation (see Appendix A) as can be found in [246]; cf. [19, 379, 530]. In the present monograph only the case of scalar Nevanlinna functions is treated in this way; the matrix-valued case is a straightforward generalization. For a rigorous treatment involving operator-valued Nevanlinna functions one should apply [558].

Parallel to the above treatment of operator-valued Nevanlinna functions the case of Nevanlinna families or Nevanlinna pairs is taken up in Section 4.4. The nonnegativity of the Nevanlinna kernel for a Nevanlinna family or Nevanlinna pair in Theorem 4.4.1 is proved via a simple reduction argument. The representation model for Nevanlinna pairs in Theorem 4.4.2 is again given along the lines of Derkach's work [231]; see also [265]. In fact, the proof can be carried forward to show that any Nevanlinna family is the Weyl family of a boundary relation (see the notes for Chapter 2). Another reproducing kernel approach was followed in [95, Theorem 6.1] by a reduction to the corresponding case of Schur functions; cf. [152, 153]. Closely connected is the notion of generalized resolvents given in Definition 4.4.5. It was formalized by McKelvey [572], see also [207]. Its representation in Corollary 4.4.7 is equivalent to the representation of Nevanlinna pairs in Theorem 4.4.2. In Corollary 4.4.9 we present Naĭmark's dilation result [606, 607] as a straightforward consequence of the characterization of generalized resolvents in Theorem 4.4.8; cf. [659].

Our construction of the exit space extension in Theorem 4.5.2 uses the above representation of generalized resolvents; cf. [497]. The identity in (4.5.3) gives the precise connection between the exit space and the Nevanlinna pair leading to the exit space. Closely related is an interpretation via the coupling method in Section 4.6. By taking an "orthogonal sum of two boundary triplets" as in Proposition 4.6.1 one obtains an exit space extension for one of the original symmetric relations whose compressed resolvent is described in Corollary 4.6.3. This leads to an interpretation of the Kreĭn formula when the parameter family coincides with a uniformly strict Nevanlinna function; see [373, 375, 376, 630]. However, a similar interpretation exists when the parameter family is a general Nevanlinna family, once it is interpreted as the Weyl family of a boundary relation; see [236, 237, 238]. It has already been explained in the notes on Chapter 2 that it is quite natural to have exit spaces which may be indefinite, in which case the Nevanlinna family must be replaced by a more general object.

And, indeed, more general reproducing kernel spaces can be constructed. For the Pontryagin space situation and reproducing kernels with finitely many negative squares one may consult the monograph [23] and the papers [89, 232, 409, 497, 499, 500]; see also [454, 455, 456, 457, 458, 459]. For the setting of almost Pontryagin spaces, see [452, 775, 777]. Finally, for Kreĭn spaces we refer to [15, 16, 17, 262, 434, 435, 437, 778].

Notes on Chapter 5

The approach to semibounded forms in Section 5.1 follows the lines of Kato's work [462], where densely defined semibounded and sectorial forms are treated; see also [2, 49, 246, 346, 359, 552, 649, 650, 690, 691]. An arbitrary form is not necessarily closable, but there is a Lebesgue decomposition into the sum of a closable form and a singular form; see [395, 397, 404, 476, 710]. Versions of the representation theorems for nondensely defined closed forms appear in [14, 44, 393, 661, 710]; see Theorem 5.1.18 and Theorem 5.1.23. Nondensely defined forms appear, for instance, when treating sums of densely defined forms, leading to a representation problem for the form sum; see [295, 296, 389, 390, 392]. The ordering of semibounded forms and semibounded self-adjoint relations in Section 5.2 results in Proposition 5.2.7; the present simple treatment via the Douglas lemma extends [390]; see Proposition 1.5.11. The characterization in terms of the corresponding resolvent operators goes back to Rellich [654]; see also [210, 390, 412, 462]. For the monotonicity principle (Theorem 5.2.11) see [96], and for similar statements [83, 245, 246, 618, 661, 709, 710, 753].

The main properties of the Friedrichs extension, namely Theorem 5.3.3 and Proposition 5.3.6, follow from the present approach via forms; cf. [202, 370, 734]. The square root decomposition approach involving resolvents to describe those semibounded self-adjoint extensions which are transversal with the Friedrichs extension extends [391] (for the nonnegative case). The transversality criterion in Theorem 5.3.8 is an extension of Malamud's earlier result [553]. The introduction of Kreĭn type extensions in Definition 5.4.2 is inspired by [210]. The minimality of the Kreĭn type extension is inherited from the maximality of the Friedrichs extension and this yields a characterization of all semibounded extensions whose lower bounds belong to a certain prescribed interval in Theorem 5.4.6. This is the semibounded analog of a characterization of nonnegative self-adjoint extensions in the nonnegative case [210]. The interpretation of the Friedrichs and Kreĭn type extensions as strong resolvent limits in Theorem 5.4.10 goes back in the operator case to [34] and in the general case to [383]. The present treatment is based on the general monotonicity principle in Theorem 5.2.11. The concept of a positively closable operator (see (iii) of Corollary 5.4.8) was introduced in [34] to detect the nonnegative self-adjoint operator extensions. This notion is closely connected with [490], where self-adjoint operator extensions were determined. Closely related to the Friedrichs and the Kreĭn–von Neumann extensions is the study of all extremal extensions of a nonnegative or a sectorial operator in [43, 44, 45, 46, 47, 53, 55, 383, 389, 391, 392, 393, 510, 648, 702, 703].

If a symmetric relation is semibounded, then a boundary triplet can be chosen so that A_0 is the Friedrichs extension and A_1 is a semibounded self-adjoint extension (for instance, a Kreĭn type extension, when transversality is satisfied). This type of boundary triplet serves as an extension of the notions of positive boundary triplets introduced in [43, 467]. The main part of the results in Section 5.5 when specialized to the nonnegative case, reduce to the results in [245, 246]. A result similar to Proposition 5.5.8 (ii) can be found in [349, 355], see also the more recent contribution [362] that deals with elliptic partial differential operators on exterior domains. We remark that the sufficient condition in Lemma 5.5.7 is also necessary for the extension A_Θ to be semibounded for every semibounded self-adjoint relation Θ in $\mathcal{G}$; see [245, 246]. For an example where Θ is a nonzero bounded operator with arbitrary small operator norm $\|\Theta\|$, while A_Θ is not semibounded from below, see [378]. In the context of elliptic partial differential operators on bounded domains the boundary triplet in Example 5.5.13 appears implicitly already in Grubb [352]; see also [169, 359, 557] and the notes to Chapter 8 for more details.

The first Green identity appearing in Theorem 5.5.14 establishes a link with the notion of boundary pairs for semibounded operators and the corresponding semibounded forms in Section 5.6. Definition 5.6.1 of a boundary pair for a semibounded relation S is adapted from Arlinskiĭ [44], see also [50] for the context of generalized boundary triplets. This notion makes it possible to describe the closed forms associated with all semibounded self-adjoint extensions of a given semibounded relation S by means of closed semibounded forms in the parameter space. In particular, Theorem 5.6.11 contains the description of all nonnegative closed forms generated by the nonnegative self-adjoint extensions in [53]; for a similar result in the case of maximal sectorial extensions see [44]. By connecting boundary pairs with compatible boundary triplets (see Theorem 5.6.6 and Theorem 5.6.10) one obtains an explicit description of the forms associated with all semibounded self-adjoint extensions by means of the boundary conditions, see Theorem 5.6.13.

We remind the reader of the study of nonnegative self-adjoint extensions of a densely defined nonnegative operator as initiated by von Neumann [610]; see [280, 309, 310, 724, 773] and the papers by Kreĭn [491, 492, 493, 494]. Kreĭn's analysis was complemented by Vishik [747] and Birman [139]; see also [2, 14, 58, 157, 217, 295, 346, 347, 352, 354, 372, 464, 714]. For an operator matrix completion approach see [383, 472], see also [81] for an extension in a Pontryagin space setting. This approach has its origin in the famous paper of Shmuljan [704]. The notion of boundary pairs for forms can also be traced back to Kreĭn [493, 494], Vishik [747], and Birman [139]. Some further developments can be found, e.g., in [36, 43, 44, 47, 48, 54, 56, 57, 248, 301, 357, 359, 363, 553, 555, 556], where also accretive and sectorial extensions, and associated closed forms are treated, and see also [552, 580, 581, 582, 636, 725]. A related concept of boundary pairs designed for elliptic partial differential operators and corresponding quadratic forms was recently proposed by Post [647].

A study of symmetric forms which are not semibounded and concepts of generalized Friedrichs and Kreĭn–von Neumann extensions goes beyond the scope of the present text. However, an interested reader may consult in this direction, e.g., [125, 215, 216, 303, 306, 364, 371, 374, 379, 570, 571].

Notes on Chapter 6

The main textbooks in this area are [2, 26, 72, 208, 281, 284, 414, 420, 438, 489, 542, 574, 608, 663, 724, 740, 741, 754, 757, 782]. The Sturm–Liouville differential expressions that we consider here have coefficients which are assumed to satisfy some weak integrability conditions giving rise to quasiderivatives (where under stronger smoothness conditions ordinary derivatives would suffice); see for instance [724] where already such quasiderivatives were used. Most of the material concerning singular Sturm–Liouville equations has been stimulated by the work of Weyl [758, 759, 760]; see also [761]. For later work on operators of higher order see, for instance, [200, 201, 465, 469, 470].

Section 6.1 and Section 6.2 contain standard material, where we follow parts of [414, 757]; for the quasiderivatives in the limit-circle case we refer to [312]. The treatment of the regular and limit-circle cases in Section 6.3 is straightforward; see [312] and also [12] for a boundary triplet in the limit-circle case. For an alternative description of boundary conditions and Weyl functions in the regular case we mention the recent papers [197, 199, 332] using the concept of boundary data maps. The limit-point case is treated in Section 6.4. In the proof of Proposition 6.4.4 concerning the simplicity of the minimal operator we follow the arguments of R.C. Gilbert [336]; cf. [263]. The treatment of the Fourier transform in Lemma 6.4.6 and Theorem 6.4.7 uses the theory in Appendix B. The surjectivity is direct in this case. The statement in Lemma 6.4.8 that the Fourier transform provides a model for the Weyl function uses an argument provided in [281].

In general, interface conditions are a tool to paste together several boundary value problems. The interface conditions which we consider in Section 6.5 make it possible in the case of two endpoints in the limit-point case to consider minimal operators [339, 740] and to apply the coupling of boundary triplets as explained in Section 4.6. This automatically leads to some kind of exit space extensions. As this goes beyond the present text, we only refer to [336, 338, 441, 713].

The present theory of subordinate solutions in Section 6.7 goes back to D.J. Gilbert and D.B. Pearson [335]; see also [333, 334, 463]. Further contributions can be found in [433, 533, 655, 656, 657]; see also [738]. Our presentation is modelled on [388].

If the minimal operator is semibounded one can apply the form methods from Chapter 5. In the regular case the treatment of the forms associated with the Sturm–Liouville expression in Section 6.8 is based on the inequality in Lemma 6.8.2 and Theorem 5.1.16; see [212, 493, 494, 757]. This makes it possible to introduce boundary pairs which are compatible with the given boundary triplet. Section 6.9 and Section 6.10 contain preparatory material so that boundary pairs can be introduced also when the endpoints are singular. Section 6.9 on Dirichlet forms is inspired by Rellich [654] and Kalf [448]. The proof of Lemma 6.9.7 seems to be new. Section 6.10 is concerned with principal and nonprincipal solutions; cf. [198, 368, 369, 535] and, in particular, [616]. The Hardy type inequalities in Lemma 6.10.1 go back to [449]. The treatment is in some sense folklore, but Theorem 6.10.9 seems to be new. Our handling of the regular case in Section 6.8 serves as model for the semibounded singular cases in Section 6.11 and Section 6.12. These two sections are influenced by Kalf [448]; see also [671, 672] and a recent contribution [168]. The determination of the Friedrichs extension in our treatment goes hand in hand with the boundary pair; see also [310, 311, 508, 597, 600, 601, 615, 616, 654, 661, 671, 672, 779, 782].

The case of an integrable potential on a half-line is treated in Section 6.13. It appears already in [740], where the corresponding spectral measure is determined. The construction of solutions with a given asymptotic behavior can be found for instance in [542, 740]. The example with the Pöschl–Teller potential at the very end of Section 6.13 can be found in, e.g., [2].

There is a large amount of literature devoted to special topics. For λ-dependent boundary conditions we just refer to [117, 313, 314, 675, 687, 688, 689, 749] and [261, 263] for Pontryagin exit spaces. The determination of the Kreĭn–von Neumann extension and other nonnegative extensions can be found in [212, 350]. For singular perturbations associated with Sturm–Liouville operators, see for instance [331] and the later papers [9, 28, 29, 124, 171, 182, 282, 315, 316, 479, 480, 481, 482, 507, 531, 532, 549, 550]; for δ-point interactions we refer to [8, 293]. Special properties of the Titchmarsh–Weyl coefficient have been studied in many papers; we just mention [130, 292, 366, 367]. Already early in the 20th century there was an interest in boundary value problems with conditions at interior points and, more general, integral boundary conditions; for instance, see [135, 583] and for a later review [762]. It was pointed out by Coddington [203, 204] that such conditions can be described in tems of relation extensions of a restriction of the usual minimal operator; see for a brief selection also [188, 189, 190, 203, 204, 205, 206, 238, 264, 267, 270, 399, 403, 484, 485, 486, 697, 784].

The topic of semibounded self-adjoint extensions in Section 5.5 and Section 5.6 is closely related to problems that are called left-definite in the literature; we only mention [13, 131, 473, 474, 488, 545, 698, 699, 708, 748] and the references therein. Another case of interest is when the weight function in the Sturm–Liouville equation changes sign, in which case Kreĭn spaces come up naturally: the following is just a limited selection [116, 122, 123, 138, 223, 225, 304, 305]. Under certain circumstances the Weyl function in the limit-point case (of a usual Sturm–Liouville operator) belongs to the Kac class; see for instance [420]. For further results in this direction, see [384, 385, 386, 387]; in these cases there is a distinguished self-adjoint extension, namely the generalized Friedrichs extension; see the notes on Chapter 5. Sturm–Liouville equations with vector-valued coefficients are beyond the scope of this text; see [345, 346] and for a recent contribution [330].

Notes on Chapter 7

Canonical systems of differential equations are discussed in the monographs [72, 340, 653, 681]. The spectral theory for such systems has been developed in varying degrees of generality in an abundance of papers [62, 63, 64, 85, 161, 162, 163,

213, 263, 265, 279, 422, 423, 424, 425, 427, 428, 487, 575, 592, 612, 613, 614, 620, 667, 668, 669, 687, 688, 689, 692, 693, 694, 695, 696]. In [520] there is a general procedure to reduce systems to a canonical form. Many investigations concerning boundary problems can be written in terms of canonical systems; see for a general procedure [619, 693], so that for instance [129, 211] can be subsumed. As a particular example we mention the system of second-order differential equations studied in the dissertation of a student of Titchmarsh [191, 192, 193, 194]. Frequently conditions are imposed on the canonical system so that there is a full analogy with the usual operator treatment. However, in general one is confronted with relations.

The first treatment of canonical systems in terms of relations is due to Orcutt [619]; see also [97, 408, 471, 524, 538] and [11, 593, 594]. In our opinion the present case of 2×2 systems already serves as a good illustration of the various phenomena that may occur. The interest in 2×2 systems is justified by the work of de Branges [146, 147, 148, 149]; see also [278, 504, 528, 715, 716, 766, 767, 768, 769, 770, 771, 772]. Via these systems one may also approach Sturm–Liouville equations with distributional coefficients as, for instance, in [281, 283]. In Remling [658] and the forthcoming book [666] by Romanov our systems are treated with the de Branges results in mind. For a number of topics we rely on the lecture notes by Kaltenbäck and Woracek [460].

Sections 7.1, 7.2, and 7.3 contain preparatory material. The inequalites (7.1.1) and (7.1.3) are standard for Bochner integrals. The construction of the Hilbert space $\mathcal{L}^2_\Delta(\imath)$ follows the treatment in [460]; see also [276, 670]. For further information concerning these and more general spaces, see [440] and the expositions in [2, 276]. The treatment of the square-integrable solutions in Section 7.4 is based on the monotonicity principle in Theorem 5.2.11; cf. [97]. For a different treatment, see for instance [612]. Section 7.5 is devoted to definite systems. The present notion of definiteness can be found in [340, 619]; in the literature sometimes a more restrictive form of definiteness is used. Proposition 7.5.4 can be found in [614, Hilfsatz (3.1)] and [471]; for a more abstract treatment, see [98]. The modification of solutions is avoided in [619]; the present argument in Proposition 7.5.6 seems to be new. The minimal and maximal (multivalued) relations associated with canonical systems can be found in Section 7.6. They were originally introduced by Orcutt [619]; see also Kac [442, 443, 444, 445] and [408]. The extension theory for them naturally involves (multivalued) relations. In [97] it is indicated how all such systems fit in the boundary triplet scheme; see also [471, 538].

In Section 7.7 it is assumed that the endpoints are (quasi)regular, in which case a boundary triplet is constructed in Theorem 7.7.2. The resolvent of the self-adjoint extension $\ker \Gamma_0$ is an integral operator whose kernel belongs to the Hilbert–Schmidt class. In this way we can show that the operator part of the minimal relation is simple. In Section 7.8 it is assumed that one of the endpoints is in the limit-point case. We prove the simplicity of the operator part of the minimal relation along the lines of [337], cf. [263] (see also Chapter 6 for the Sturm–Liouville case). The treatment of the Fourier transform in the limit-point case uses

Appendix B and parallels the treatment in Chapter 6. Subordinate solutions for canonical systems are introduced in Section 7.9; cf. [388]. Here again we follow the results for the Sturm–Liouville case in Section 6.7; see also the corresponding notes in Chapter 5, where the appropriate references can be found.

The discussion of the special cases in Section 7.10 is just an indication of the possibilities; we could also pay attention to, for instance, the so-called Dirac systems. The connection with the Sturm–Liouville case and its generalization in [281] is only briefly indicated. The special case in Theorem 7.10.1 is modelled on [460]. For the connection of canonical systems with strings see, for instance, [453]. Finally, we would like to mention that the approach via boundary triplets allows also λ-dependent boundary conditions; for some related papers we refer to [221, 263, 265, 675, 687, 688, 689, 742].

Notes on Chapter 8

The notion of Gelfand triples or riggings of Hilbert spaces in Section 8.1 is often used in the treatment of partial differential equations and Sobolev spaces, and can be found in a similar form in Berezanskiĭ's monograph [133] (see also the textbook [774]); cf. [132] and the contributions [534, 537] by Lax and Leray. There are many well-known textbooks on Sobolev spaces, among which we mention here only the monographs [3, 5, 164, 136, 284, 291, 351, 569, 584, 744, 783]. In our opinion a very useful source for trace maps, the Green identities, and similar related results (also for nonsmooth domains) is the monograph [573] by McLean, see also the list of references therein. For the description of the spaces $H^s(\partial\Omega)$ in Corollary 8.2.2 as domains of powers of the Laplace–Beltrami operator on $\partial\Omega$ see [322, 544, 567]. In some cases it is also convenient to use powers of a Dirichlet-to-Neumann map, as in [114].

The discussion of the minimal and maximal operators in Section 8.3 is standard, e.g., Proposition 8.3.1 can be found in Triebel's textbook [745]. The H^2-regularity in Theorem 8.3.4 for smooth domains can be found in [167] and [84], see also [4, 308, 544, 517] or [1, Theorem 7.2] for a recent very general result on the H^2-regularity of the Neumann operator (and more general) on certain classes of nonsmooth bounded and unbounded domains. As explained in Section 8.7, for the class of Lipschitz domains the H^2-regularity of the Dirichlet and Neumann Laplacian up to the boundary fails and has to be replaced by the weaker $H^{3/2}$-regularity; cf. [431, 432] and [92, 115, 323, 324, 325, 326] for some recent closely related works dealing with Schrödinger operators on Lipschitz domains. In this context we also mention the papers [585, 586, 587, 588] by Mitrea and Taylor for general layer potential methods in Lipschitz domains on Riemannian manifolds. It is worth mentioning that the resolvent of the Neumann Laplacian in Proposition 8.3.3 is not necessarily compact if the boundary of the bounded domain Ω is not of class C^2 (or not Lipschitz), see [415] for a well-known counterexample using a rooms-and-passages domain. For similar *unusual* spectral properties we also mention Example 8.4.9 on the essential spectrum of self-adjoint realizations of the

Laplacian on a bounded domain, which however is very different from a technical point of view. In a related context the existence of self-adjoint extensions with prescribed point spectrum, absolutely continuous, and singular continuous spectrum in spectral gaps of a fixed underlying symmetric operator was also discussed in [6, 7, 154, 156, 158, 159, 160]. For our purposes Theorem 8.3.9 and Theorem 8.3.10 play an important role; for the case of C^∞-smooth domains such results can be found in [544], see also [352, 354]. The present versions of the extension theorems are inspired by slightly different considerations in [115]; the variant for Lipschitz domains in Theorem 8.7.5 can be proved by means of a similar technique (see also [92] for a comprehensive discussion). We remark that the definition and topologies on the spaces $\mathscr{G}_0$ and $\mathscr{G}_1$ in (8.7.4)–(8.7.5) from [92, 115] are partly inspired by abstract considerations in [246] for generalized boundary triplets. Concerning Section 8.3, as a final comment we mention that for more general second-order elliptic operators in bounded and unbounded domains Proposition 8.3.13 can be found in [118, 119, 120].

The boundary triplet in Theorem 8.4.1 can also be found in, e.g., [105, 169, 173, 359, 557, 643] and extends with some simple modifications to second-order and $2m$-th order elliptic operators with variable coefficients. In a different form this boundary triplet is already essentially contained in the well-known work of Grubb [352], where all closed extensions of a minimal operator elliptic partial differential operator were characterised by nonlocal boundary conditions; see also the early contribution [747] by Vishik and the fundamental paper [140] by Birman. In fact, it seems that the more recent paper [27] by Amrein and Pearson on a generalisation of Weyl–Titchmarsh theory for Schrödinger operators inspired many operator theorists to investigate partial differential operators from an extension theory point of view. Various papers based on boundary triplet techniques and related methods were published in the last decade; besides those papers listed before we mention as a selection here [18, 90, 91, 103, 107, 108, 251, 323, 324, 360, 429, 566, 568, 635, 647, 678] in which, e.g., Dirichlet-to-Neumann maps and Kreĭn type resolvent formulas are treated. For self-adjoint realizations of the Laplacians, Schrödinger operators, and more general second-order elliptic differential operators in nonsmooth domains we refer to, e.g., [1, 92, 115, 326, 358]. Particular attention has been paid to the spectral properties of realizations with local and nonlocal Robin boundary conditions in [41, 106, 179, 180, 226, 325, 361, 413, 543, 628, 629, 665]. Furthermore, the recent contributions [35, 36, 37, 38, 39, 40] by Arendt, ter Elst, and coauthors form an interesting series of papers on elliptic differential operators and Dirichlet-to-Neumann operators based mainly on form methods.

The present treatment of semibounded Schrödinger operator in Section 8.5 with the help of boundary pairs and boundary triplets seems to be new. However, the problem of lower boundedness was also discussed with different methods in [349, 355, 362]. The Kreĭn–von Neumann extension which is of special interest in this context was investigated in, e.g., [14, 68, 69, 70, 71, 93, 181, 356, 362, 578, 579]. For coupling methods for elliptic differential operators based on boundary triplet techniques in the spirit of Section 8.6 we refer to the recent paper [88], where also an abstract version of the third Green identity was proved. Finally, we mention that various classes of λ-dependent boundary value problems for elliptic operators can be treated in such a context, see, e.g., [87, 103, 137, 285, 286].

Notes on Appendices A–D

The appendices were included for the convenience of the reader. Here we collect some notes for each of the appendices A–D.

In Appendix A we present the basic properties of Nevanlinna functions that are needed in the text. The Stieltjes inversion formula for the Borel transform is folklore. The integral representation of scalar Nevanlinna functions is derived in a classical way involving the Helly and Helly–Bray theorems; see [72, 763]. The inversion formula in Lemma A.2.7 is standard; see [272, 446]. For the present purposes it suffices to approach the integration with respect to operator-valued measures in terms of improper Riemann integrals. For the operator measure version see for instance [691]. For Kac functions, Stieltjes functions, and inverse Stieltjes functions see [49, 52, 329, 407, 446, 447, 505]. As far as we know, the proof of Proposition A.5.4 is new; cf. [572]. For related work see [318, 319, 329, 602]. The case of generalized Nevanlinna functions was initiated in the works of Kreĭn and Langer [499, 500]; see [145, 225, 232, 239, 242, 255, 380, 411, 434, 435, 437, 546, 547, 548] for further developments.

For the general notion of Fourier transforms in Appendix B associated with Sturm–Liouville equations, see [539, 540, 542, 700, 701, 740, 780, 781]. Our basic idea here is inspired by the treatment in [208]. The arguments establishing the surjectivity of the Fourier transform are also inspired by [281, 715, 716]. Fourier transforms that are partially isometric go back at least to [204] in a treatment connected with multivalued operators. The discussion in this section has connections with the treatment of Kreĭn's directing functionals in [495, 518, 526, 527].

Necessary and sufficient conditions (as in Appendix C) for the sum of closed subspaces have a long history. The concept of opening of a pair of subspaces goes back to Friedrichs and Dixmier; see for instance [214, 249, 250]. Lemma C.4 goes back to [246].

The main reference for Appendix D is the paper by Douglas [273]; see also [30, 299]. There have been many generalizations of the Douglas paper. We only mention a particular direction of extension, namely [639, 684]; see also [402]. The application in Proposition 1.5.11 in Chapter 1 has the same flavor, but is obtained by reduction to the case in this appendix; see [390].

Bibliography

[1] H. Abels, G. Grubb, and I. Wood. Extension theory and Krein-type resolvent formulas for nonsmooth boundary value problems. J. Functional Analysis 266 (2014) 4037–4100.

[2] N.I. Achieser and I.M. Glasman. *Theorie der linearen Operatoren im Hilbertraum.* 8th edition. Akademie Verlag, 1981.

[3] R.A. Adams and J.J.F. Fournier. *Sobolev spaces.* 2nd edition. Academic Press, 2003.

[4] S. Agmon, A. Douglis, and L. Nirenberg. Estimates near the boundary for solutions of elliptic partial differential equations satisfying general boundary conditions I. Comm. Pure Appl. Math. 12 (1959) 623–727.

[5] M.S. Agranovich. *Sobolev spaces, their generalizations and elliptic problems in smooth and Lipschitz domains.* Monographs in Mathematics, Springer, 2015.

[6] S. Albeverio, J.F. Brasche, M.M. Malamud, and H. Neidhardt. Inverse spectral theory for symmetric operators with several gaps: scalar-type Weyl functions. J. Functional Analysis 228 (2005) 144–188.

[7] S. Albeverio, J.F. Brasche, and H. Neidhardt. On inverse spectral theory for self-adjoint extensions: mixed types of spectra. J. Functional Analysis 154 (1998) 130–173.

[8] S. Albeverio, F. Gesztesy, R. Hoegh-Krohn, and H. Holden. *Solvable models in quantum mechanics.* 2nd edition. AMS Chelsea Publishing, Amer. Math. Soc., 2005.

[9] S. Albeverio, A. Kostenko, and M.M. Malamud. Spectral theory of semibounded Sturm–Liouville operators with local interactions on a discrete set. J. Math. Physics 51 (2010) 1–24.

[10] S. Albeverio and P. Kurasov. *Singular perturbations of differential operators and Schrödinger type operators.* Cambridge University Press, 2000.

[11] S. Albeverio, M.M. Malamud, and V.I. Mogilevskiĭ. On Titchmarsh–Weyl functions and eigenfunction expansions of first-order symmetric systems. Integral Equations Operator Theory 77 (2013) 303–354.

[12] B.P. Allakhverdiev. On the theory of dilatation and on the spectral analysis of dissipative Schrödinger operators in the case of the Weyl limit circle (Russian). Izv. Akad. Nauk SSSR Ser. Mat. 54 (1990) 242–257 [Math. USSR-Izv. 36 (1991) 247–262].

[13] W. Allegretto and A.B. Mingarelli. Boundary problems of the second order with an indefinite weight function. J. Reine Angew. Math. 398 (1989) 1–24.

[14] A. Alonso and B. Simon. The Birman–Kreĭn–Vishik theory of self-adjoint extensions of semibounded operators. J. Operator Theory 4 (1980) 251–270; 6 (1981) 407.

[15] D. Alpay. *Reproducing kernel Krein spaces of analytic functions and inverse scattering*. Doctoral dissertation, The Weizmann Institute of Science, Rehovot, 1985.

[16] D. Alpay. Some remarks on reproducing kernel Krein spaces. Rocky Mountain J. Math. 21 (1991) 1189–1205.

[17] D. Alpay. A theorem on reproducing kernel Hilbert spaces of pairs. Rocky Mountain J. Math. 22 (1992) 1243–1258.

[18] D. Alpay and J. Behrndt. Generalized Q-functions and Dirichlet-to-Neumann maps for elliptic differential operators. J. Functional Analysis 257 (2009) 1666–1694.

[19] D. Alpay, P. Bruinsma, A. Dijksma, and H.S.V. de Snoo. A Hilbert space associated with a Nevanlinna function. In: *Signal processing, scattering and operator theory, and numerical methods*. Birkhäuser, 1990, pp. 115–122.

[20] D. Alpay, P. Bruinsma, A. Dijksma, and H.S.V. de Snoo. Interpolation problems, extensions of operators and reproducing kernel spaces I. Oper. Theory Adv. Appl. 50 (1990) 35–82.

[21] D. Alpay, P. Bruinsma, A. Dijksma, and H.S.V. de Snoo. Interpolation problems, extensions of symmetric operators and reproducing kernel spaces II. Integral Equations Operator Theory 14 (1991) 465–500.

[22] D. Alpay, P. Bruinsma, A. Dijksma, and H.S.V. de Snoo. Interpolation problems, extensions of symmetric operators and reproducing kernel spaces II (missing Section 3). Integral Equations Operator Theory 15 (1992) 378–388.

[23] D. Alpay, A. Dijksma, J. Rovnyak, and H.S.V. de Snoo. *Schur functions, operator colligations, and Pontryagin spaces*. Oper. Theory Adv. Appl., Vol. 96, Birkhäuser, 1997.

[24] D. Alpay and H. Dym. Hilbert spaces of analytic functions, inverse scattering and operator models I. Integral Equations Operator Theory 7 (1984) 589–641.

[25] D. Alpay and H. Dym. Hilbert spaces of analytic functions, inverse scattering and operator models II. Integral Equations Operator Theory 8 (1985) 145–180.

[26] W.O. Amrein, A.M. Hinz, and D.B. Pearson (editors). *Sturm–Liouville theory: past and present*. Birkhäuser, 2005.

[27] W.O. Amrein and D.B. Pearson. M-operators: a generalisation of Weyl–Titchmarsh theory. J. Comp. Appl. Math. 171 (2004) 1–26.

[28] A.Yu. Ananieva and V. Budyika. To the spectral theory of the Bessel operator on finite interval and half-line. J. Math. Sciences 211 (2015) 624–645.

[29] A.Yu. Ananieva and V.S. Budyika. On the spectral theory of the Bessel operator on a finite interval and the half-line (Russian). Differ. Uravn. 52 (2016) 1568–1572 [Differential Equations 52 (2016) 1517–1522].

[30] W.N. Anderson Jr. and G.E. Trapp. Shorted operators. II. SIAM J. Appl. Math. 28 (1975) 60–71.

[31] T. Ando. *Linear operators on Krein spaces*. Division of Applied Mathematics, Research Institute of Applied Electricity, Hokkaido University, Sapporo, 1979.

[32] T. Ando. *Reproducing kernel spaces and quadratic inequalities*. Division of Applied Mathematics, Research Institute of Applied Electricity, Hokkaido University, Sapporo, 1987.

[33] T. Ando. *De Branges spaces and analytic operator functions*. Division of Applied Mathematics, Research Institute of Applied Electricity, Hokkaido University, Sapporo, 1990.

[34] T. Ando and K. Nishio. Positive selfadjoint extensions of positive symmetric operators. Tohoku Math. J. 22 (1970) 65–75.

[35] W. Arendt and A.F.M. ter Elst. The Dirichlet-to-Neumann operator on rough domains. J. Differential Equations 251 (2011) 2100–2124.

[36] W. Arendt and A.F.M. ter Elst. Sectorial forms and degenerate differential operators. J. Operator Theory 67 (2012) 33–72.

[37] W. Arendt and A.F.M. ter Elst. From forms to semigroups. Oper. Theory Adv. Appl. 221 (2012) 47–69.

[38] W. Arendt and A.F.M. ter Elst. The Dirichlet-to-Neumann operator on exterior domains. Potential Analysis 43 (2015) 313–340.

[39] W. Arendt, A.F.M. ter Elst, J.B. Kennedy, and M. Sauter. The Dirichlet-to-Neumann operator via hidden compactness. J. Functional Analysis 266 (2014) 1757–1786.

[40] W. Arendt, A.F.M. ter Elst, and M. Warma. Fractional powers of sectorial operators via the Dirichlet-to-Neumann operator. Comm. Partial Differential Equations 43 (2018) 1–24.

[41] W. Arendt and M. Warma. The Laplacian with Robin boundary conditions on arbitrary domains. Potential Analysis 19 (2003) 341–363.

[42] R. Arens. Operational calculus of linear relations. Pacific J. Math. 11 (1961) 9–23.

[43] Yu.M. Arlinskiĭ. Positive spaces of boundary values and sectorial extensions of nonnegative symmetric operators. Ukrainian Math. J. 40 (1988) 8–15.

[44] Yu.M. Arlinskiĭ. Maximal sectorial extensions and closed forms associated with them. Ukrainian Math. J. 48 (1996) 723–739.

[45] Yu.M. Arlinskiĭ. Extremal extensions of sectorial linear relations. Mat. Studii 7 (1997) 81–96.

[46] Yu.M. Arlinskiĭ. On a class of nondensely defined contractions and their extensions. J. Math. Sciences 79 (1999) 4391–4419.

[47] Yu.M. Arlinskiĭ. Abstract boundary conditions for maximal sectorial extensions of sectorial operators. Math. Nachr. 209 (2000) 5–35.

[48] Yu.M. Arlinskiĭ. Boundary triplets and maximal accretive extensions of sectorial operators. In: *Operator methods for boundary value problems*. London Math. Soc. Lecture Note Series, Vol. 404, 2012, pp. 35–72.

[49] Yu.M. Arlinskiĭ, S. Belyi, and E.R. Tsekanovskiĭ. *Conservative realizations of Herglotz-Nevanlinna functions.* Oper. Theory Adv. Appl., Vol. 217, Birkhäuser, 2011.

[50] Yu.M. Arlinskiĭ and S. Hassi. Q-functions and boundary triplets of nonnegative operators. Oper. Theory Adv. Appl. 244 (2015) 89–130.

[51] Yu.M. Arlinskiĭ and S. Hassi. Compressed resolvents of selfadjoint contractive exit space extensions and holomorphic operator-valued functions associated with them. Methods Funct. Anal. Topology 21 (2015) 199–224.

[52] Yu.M. Arlinskiĭ and S. Hassi. Stieltjes and inverse Stieltjes holomorphic families of linear relations and their representations. Studia Math. To appear.

[53] Yu.M. Arlinskiĭ, S. Hassi, Z. Sebestyén, and H.S.V. de Snoo. On the class of extremal extensions of a nonnegative operator. Oper. Theory Adv. Appl. 127 (2001) 41–81.

[54] Yu.M. Arlinskiĭ, Yu. Kovalev, and E.R. Tsekanovskiĭ. Accretive and sectorial extensions of nonnegative symmetric operators. Complex Anal. Oper. Theory 6 (2012) 677–718.

[55] Yu. M. Arlinskiĭ and E.R. Tsekanovskiĭ. Quasiselfadjoint contractive extensions of a Hermitian contraction (Russian). Teor. Funktsii Funktsional. Anal. i Prilozhen (1988) 9–16 [J. Soviet Math. 49 (1990) 1241–1247].

[56] Yu.M. Arlinskiĭ and E.R. Tsekanovskiĭ. On von Neumann's problem in extension theory of nonnegative operators. Proc. Amer. Math. Soc. 131 (2003) 3143–3154.

[57] Yu.M. Arlinskiĭ and E.R. Tsekanovskiĭ. The von Neumann problem for nonnegative symmetric operators. Integral Equations Operator Theory 51 (2005) 319–356.

[58] Yu.M. Arlinskiĭ and E.R. Tsekanovskiĭ. M. Kreĭn's research on semi-bounded operators, its contemporary developments, and applications. Oper. Theory Adv. Appl. 190 (2009) 65–112.

[59] N. Aronszajn. Theory of reproducing kernels. Trans. Amer. Math. Soc. 68 (1950) 337–404.

[60] N. Aronszajn. On a problem of Weyl in the theory of singular Sturm–Liouville equations. Amer. J. Math. 79 (1957) 597–610.

[61] N. Aronszajn and W.F. Donoghue. On exponential representations of analytic functions. J. Anal. Math. 5 (1956/1957) 321–388.

[62] D.Z. Arov and H. Dym. *J*-inner matrix functions, interpolation and inverse problems for canonical systems, I. Foundations. Integral Equations Operator Theory 29 (1997) 373–454.

[63] D.Z. Arov and H. Dym. *J*-inner matrix functions, interpolation and inverse problems for canonical systems, II. The inverse monodromy problem. Integral Equations Operator Theory 36 (2000) 11–70.

[64] D.Z. Arov and H. Dym. *J*-inner matrix functions, interpolation and inverse problems for canonical systems, III. More on the inverse monodromy problem. Integral Equations Operator Theory 36 (2000) 127–181.

[65] D.Z. Arov, M. Kurula, and O.J. Staffans. Boundary control state/signal systems and boundary triplets. In: *Operator methods for boundary value problems*. London Math. Soc. Lecture Note Series, Vol. 404, 2012, pp. 73–85.

[66] D.Z. Arov, M. Kurula, and O.J. Staffans. Passive state/signal systems and conservative boundary relations. In: *Operator methods for boundary value problems*. London Math. Soc. Lecture Note Series, Vol. 404, 2012, pp. 87–119.

[67] D.Z. Arov and O.J. Staffans. State/signal linear time-invariant systems theory. Part I: discrete time systems. Oper. Theory Adv. Appl. 161 (2006) 115–177.

[68] M.S. Ashbaugh, F. Gesztesy, A. Laptev, M. Mitrea, and S. Sukhtaiev. A bound for the eigenvalue counting function for Krein–von Neumann and Friedrichs extensions. Adv. Math. 304 (2017) 1108–1155.

[69] M.S. Ashbaugh, F. Gesztesy, M. Mitrea, R. Shterenberg, and G. Teschl. The Krein–von Neumann extension and its connection to an abstract buckling problem. Math. Nachr. 210 (2010) 165–179.

[70] M.S. Ashbaugh, F. Gesztesy, M. Mitrea, R. Shterenberg, and G. Teschl. A survey on the Krein-von Neumann extension, the corresponding abstract buckling problem, and Weyl-type spectral asymptotics for perturbed Krein Laplacians in nonsmooth domains. Oper. Theory Adv. Appl. 232 (2013) 1–106.

[71] M.S. Ashbaugh, F. Gesztesy, M. Mitrea, and G. Teschl. Spectral theory for perturbed Krein Laplacians in nonsmooth domains. Adv. Math. 223 (2010) 1372–1467.

[72] F.V. Atkinson. *Discrete and continuous boundary problems*. Academic Press, 1964.

[73] T.Ya. Azizov. Extensions of J-isometric and J-symmetric operators (Russian). Funktsional. Anal. i Prilozhen 18 (1984) 57–58 [Funct. Anal. Appl. 18 (1984) 46–48].

[74] T.Ya. Azizov, J. Behrndt, P. Jonas, and C. Trunk. Compact and finite rank perturbations of An Overview of Linear Relations. Integral Equations Operator Theory 63 (2009) 151–163.

[75] T.Ya. Azizov, J. Behrndt, P. Jonas, and C. Trunk. Spectral points of definite type and type π for linear operators and relations in Krein spaces. J. London Math. Soc. 83 (2011) 768–788.

[76] T.Ya. Azizov, B. Ćurgus, and A. Dijksma. Standard symmetric operators in Pontryagin spaces: a generalized von Neumann formula and minimality of boundary coefficients. J. Functional Analysis 198 (2003) 361–412.

[77] T.Ya. Azizov and I.S. Iokhvidov. *Linear operators in spaces with indefinite metric*. John Wiley and Sons, 1989.

[78] T.Ya. Azizov, P. Jonas, and C. Trunk. Spectral points of type π_+ and π_- of selfadjoint operators in Krein spaces. J. Functional Analysis 226 (2005) 114–137.

[79] T.Ya. Azizov and L.I. Soukhotcheva. Linear operators in almost Krein spaces. Oper. Theory Adv. Appl. 175 (2007) 1–11.

[80] D. Baidiuk. Boundary triplets and generalized resolvents of isometric operators in a Pontryagin space (Russian). Ukr. Mat. Visn. 10 (2013), no. 2, 176–200, 293 [J. Math. Sciences 194 (2013) 513–531].

[81] D. Baidiuk and S. Hassi. Completion, extension, factorization, and lifting of operators. Math. Ann. 364 (2016) 1415–1450.

[82] A.G. Baskakov and K.I. Chernyshov. Spectral analysis of linear relations and degenerate operator semigroups. Mat. Sb. 193 (2002) 3–42 [Sb. Math. 193 (2002) 1573–1610].

[83] C.J.K. Batty and A.F.M. ter Elst. On series of sectorial forms. J. Evolution Equations 14 (2014) 29–47.

[84] R. Beals. Non-local boundary value problems for elliptic operators. Amer. J. Math. 87 (1965) 315–362.

[85] H. Behncke and D. Hinton. Two singular point linear Hamiltonian systems with an interface condition. Math. Nachr. 283 (2010) 365–378.

[86] J. Behrndt. Boundary value problems with eigenvalue depending boundary conditions. Math. Nachr. 282 (2009) 659–689.

[87] J. Behrndt. Elliptic boundary value problems with λ-dependent boundary conditions. J. Differential Equations 249 (2010) 2663–2687.

[88] J. Behrndt, V.A. Derkach, F. Gesztesy, and M. Mitrea. Coupling of symmetric operators and the third Green identity. Bull. Math. Sciences 8 (2018) 49–80.

[89] J. Behrndt, V.A. Derkach, S. Hassi, and H.S.V. de Snoo. A realization theorem for generalized Nevanlinna pairs. Operators Matrices 5 (2011) 679–706.

[90] J. Behrndt and A.F.M. ter Elst. Dirichlet-to-Neumann maps on bounded Lipschitz domains. J. Differential Equations 259 (2015) 5903–5926.

[91] J. Behrndt, F. Gesztesy, H. Holden, and R. Nichols. Dirichlet-to-Neumann maps, abstract Weyl–Titchmarsh M-functions, and a generalized index of unbounded meromorphic operator-valued functions. J. Differential Equations 261 (2016) 3551–3587.

[92] J. Behrndt, F. Gesztesy, and M. Mitrea. Sharp boundary trace theory and Schrödinger operators on bounded Lipschitz domains. In preparation

[93] J. Behrndt, F. Gesztesy, M. Mitrea, and T. Micheler. The Krein-von Neumann realization of perturbed Laplacians on bounded Lipschitz domains. Oper. Theory Adv. Appl. 255 (2014) 49–66.

[94] J. Behrndt, S. Hassi, and H.S.V. de Snoo. Functional models for Nevanlinna families. Opuscula Mathematica 28 (2008) 233–245.

[95] J. Behrndt, S. Hassi, and H.S.V. de Snoo. Boundary relations, unitary colligations, and functional models. Complex Anal. Oper. Theory 3 (2009) 57–98.

[96] J. Behrndt, S. Hassi, H.S.V. de Snoo, and H.L. Wietsma. Monotone convergence theorems for semibounded operators and forms with applications. Proc. Royal Soc. Edinburgh 140A (2010) 927–951.

[97] J. Behrndt, S. Hassi, H.S.V. de Snoo, and H.L. Wietsma. Square-integrable solutions and Weyl functions for singular canonical systems. Math. Nachr. 284 (2011) 1334–1384.

[98] J. Behrndt, S. Hassi, H.S.V. de Snoo, and H.L. Wietsma. Limit properties of monotone matrix functions. Lin. Alg. Appl. 436 (2012) 935–953.

[99] J. Behrndt and P. Jonas. On compact perturbations of locally definitizable self-adjoint relations in Krein spaces. Integral Equations Operator Theory 52 (2005) 17–44.

[100] J. Behrndt and P. Jonas. Boundary value problems with local generalized Nevanlinna functions in the boundary condition. Integral Equations Operator Theory 55 (2006) 453–475.

[101] J. Behrndt and H.-C. Kreusler. Boundary relations and generalized resolvents of symmetric relations in Krein spaces. Integral Equations Operator Theory 59 (2007) 309–327.

[102] J. Behrndt, M. Kurula, A.J. van der Schaft, and H. Zwart. Dirac structures and their composition on Hilbert spaces. J. Math. Anal. Appl. 372 (2010) 402–422.

[103] J. Behrndt and M. Langer. Boundary value problems for elliptic partial differential operators on bounded domains. J. Functional Analysis 243 (2007) 536–565.

[104] J. Behrndt and M. Langer. On the adjoint of a symmetric operator. J. London Math. Soc. 82 (2010) 563–580.

[105] J. Behrndt and M. Langer. Elliptic operators, Dirichlet-to-Neumann maps and quasi boundary triples. In: *Operator methods for boundary value problems*. London Math. Soc. Lecture Note Series, Vol. 404, 2012, pp. 121–160.

[106] J. Behrndt, M. Langer, I. Lobanov, V. Lotoreichik, and I.Yu. Popov. A remark on Schatten–von Neumann properties of resolvent differences of generalized Robin Laplacians on bounded domains. J. Math. Anal. Appl. 371 (2010) 750–758.

[107] J. Behrndt, M. Langer, and V. Lotoreichik. Spectral estimates for resolvent differences of self-adjoint elliptic operators. Integral Equations Operator Theory 77 (2013) 1–37.

[108] J. Behrndt, M. Langer, and V. Lotoreichik. Schrödinger operators with δ and δ'-potentials supported on hypersurfaces. Ann. Henri Poincaré 14 (2013) 385–423.

[109] J. Behrndt, M. Langer, V. Lotoreichik, and J. Rohleder. Quasi boundary triples and semibounded self-adjoint extensions. Proc. Roy. Soc. Edinburgh 147A (2017) 895–916.

[110] J. Behrndt, M. Langer, V. Lotoreichik, and J. Rohleder. Spectral enclosures for non-self-adjoint extensions of symmetric operators. J. Functional Analysis 275 (2018) 1808–1888.

[111] J. Behrndt and A. Luger. An analytic characterization of the eigenvalues of self-adjoint extensions. J. Functional Analysis 242 (2007) 607–640.

[112] J. Behrndt, M.M. Malamud, and H. Neidhardt. Scattering theory for open quantum systems with finite rank coupling. Math. Phys. Anal. Geom. 10 (2007) 313–358.

[113] J. Behrndt, M.M. Malamud, and H. Neidhardt. Scattering matrices and Weyl functions. Proc. London Math. Soc. 97 (2008) 568–598.

[114] J. Behrndt, M.M. Malamud, and H. Neidhardt. Scattering matrices and Dirichlet-to-Neumann maps. J. Functional Analysis 273 (2017) 1970–2025.

[115] J. Behrndt and T. Micheler. Elliptic differential operators on Lipschitz domains and abstract boundary value problems. J. Functional Analysis 267 (2014) 3657–3709.

[116] J. Behrndt and F. Philipp. Spectral analysis of singular ordinary differential operators with indefinite weights. J. Differential Equations 248 (2010) 2015–2037.

[117] J. Behrndt and F. Philipp. Finite rank perturbations in Pontryagin spaces and a Sturm–Liouville problem with λ-rational boundary conditions. Oper. Theory Adv. Appl. 263 (2018) 163–189.

[118] J. Behrndt and J. Rohleder. An inverse problem of Calderón type with partial data. Comm. Partial Differential Equations 37 (2012) 1141–1159.

[119] J. Behrndt and J. Rohleder. Spectral analysis of self-adjoint elliptic differential operators, Dirichlet-to-Neumann maps, and abstract Weyl functions. Adv. Math. 285 (2015) 1301–1338.

[120] J. Behrndt and J. Rohleder. Titchmarsh–Weyl theory for Schrödinger operators on unbounded domains. J. Spectral Theory 6 (2016) 67–87.

[121] J. Behrndt and H.S.V. de Snoo. On Kreĭn's formula. J. Math. Anal. Appl. 351 (2009) 567–578.

[122] J. Behrndt and C. Trunk. Sturm–Liouville operators with indefinite weight functions and eigenvalue depending boundary conditions. J. Differential Equations 222 (2006) 297–324.

[123] J. Behrndt and C. Trunk. On the negative squares of indefinite Sturm–Liouville operators. J. Differential Equations 238 (2007) 491–519.

[124] A. Beigl, J. Eckhardt, A. Kostenko, and G. Teschl. On spectral deformations and singular Weyl functions for one-dimensional Dirac operators. J. Math. Phys. 56 (2015), no. 1, 012102, 11 pp.

[125] S. Di Bella and C. Trapani. Some representation theorems for sesquilinear forms. J. Math. Anal. Appl. 451 (2017) 64–83.

[126] S. Belyi, G. Menon, and E.R. Tsekanovskiĭ. On Krein's formula in the case of non-densely defined symmetric operators. J. Math. Anal. Appl. 264 (2001) 598–616.

[127] A. Ben-Israel and T.N.E. Greville. Generalized inverses: theory and applications. CMS Books in Mathematics, Springer, 2003.

[128] C. Bennewitz. Symmetric relations on a Hilbert space. Springer Lecture Notes Math. 280 (1972) 212–218.

[129] C. Bennewitz. Spectral theory for pairs of differential operators. Ark. Mat. 15 (1977) 33–61.

[130] C. Bennewitz. Spectral asymptotics for Sturm–Liouville equations. Proc. London Math. Soc. (3) 59 (1989) 294–338.

[131] C. Bennewitz and W.N. Everitt. On second order left definite boundary value problems. Springer Lecture Notes Math. 1032 (1983) 31–67.

[132] Yu.M. Berezanskiĭ. Spaces with negative norm (Russian). Uspehi Mat. Nauk 18 no. 1 (1963) 63–96.

[133] Yu.M. Berezanskiĭ. Expansions in eigenfunctions of selfadjoint operators (Russian). Naukova Dumka, Kiev, 1965 [Transl. Math. Monographs, Vol. 17, Amer. Math. Soc., 1968].

[134] G. Berkolaiko and P. Kuchment. *Introduction to quantum graphs.* Mathematical Surveys and Monographs, Vol. 186, Amer. Math. Soc., 2013.

[135] F. Betschler. *Über Integraldarstellungen welche aus speziellen Randwertproblemen bei gewöhnlichen linearen inhomogenen Differentialgleichungen entspringen.* Doctoral dissertation, Julius-Maximilians-Universität, Würzburg, 1912.

[136] P.K. Bhattacharyya. *Distributions. Generalized functions with applications in Sobolev spaces.* de Gruyter Textbook, 2012.

[137] P. Binding, R. Hryniv, H. Langer, and B. Najman. Elliptic eigenvalue problems with eigenparameter dependent boundary conditions. J. Differential Equations 174 (2001) 30–54.

[138] P. Binding, H. Langer, and M. Möller. Oscillation results for Sturm–Liouville problems with an indefinite weight function. J. Comput. Appl. Math. 171 (2004) 93–101.

[139] M.S. Birman. On the self-adjoint extensions of positive definite operators (Russian). Mat. Sb. 38 (1956) 431–450.

[140] M.S. Birman. Perturbations of the continuous spectrum of a singular elliptic operator by varying the boundary and the boundary conditions (Russian). Vestnik Leningrad Univ. 17 (1962) 22–55 [Amer. Math. Soc. Transl. 225 (2008), 19–53].

[141] M.S. Birman and M.Z. Solomjak. *Spectral theory of self-adjoint operators in Hilbert space.* Kluwer, 1987.

[142] J. Bognar. *Indefinite inner product spaces.* Ergebnisse der Mathematik und ihrer Grenzgebiete, Vol. 78, Springer, 1974.

[143] A.A. Boitsev. Boundary triplets approach for Dirac operator. In: *Mathematical results in quantum mechanics.* World Scientific, 2015, pp. 213–219.

[144] A.A. Boitsev, J.F. Brasche, M.M. Malamud, H. Neidhardt, and I.Yu. Popov. Boundary triplets, tensor products and point contacts to reservoirs. Ann. Henri Poincaré 19 (2018) 2783–2837.

[145] M. Borogovac and A. Luger. Analytic characterizations of Jordan chains by pole cancellation functions of higher order. J. Functional Analysis 267 (2014) 4499–4518.

[146] L. de Branges. Some Hilbert spaces of entire functions. Trans. Amer. Math. Soc. 96 (1960) 259–295.

[147] L. de Branges. Some Hilbert spaces of entire functions II. Trans. Amer. Math. Soc. 99 (1961) 118–152.

[148] L. de Branges. Some Hilbert spaces of entire functions III. Trans. Amer. Math. Soc. 100 (1961) 73–115.

[149] L. de Branges. Some Hilbert spaces of entire functions IV. Trans. Amer. Math. Soc. 105 (1962) 43–83.

[150] L. de Branges. The expansion theorem for Hilbert spaces of entire functions. In: *Entire functions and related parts of analysis.* Proc. Symp. Pure Math., Vol. 11, Amer. Math. Soc., 1968, pp. 79-148.

[151] L. de Branges, *Hilbert spaces of entire functions.* Prentice-Hall, 1968.

[152] L. de Branges and J. Rovnyak. *Square summable power series.* Holt, Rinehart and Winston, 1966.

[153] L. de Branges and J. Rovnyak. Appendix on square summable power series, Canonical models in quantum scattering theory. In: *Perturbation theory and its applications in quantum mechanics*. John Wiley and Sons, 1966, pp. 295–392.

[154] J.F. Brasche. Spectral theory for self-adjoint extensions. In: *Spectral theory of Schrödinger operators*. Contemp. Math., Vol. 340, Amer. Math. Soc., 2004, pp. 51–96.

[155] J.F. Brasche, M.M. Malamud, and H. Neidhardt. Weyl function and spectral properties of selfadjoint extensions. Integral Equations Operator Theory 43 (2002) 264–289.

[156] J.F. Brasche, M.M. Malamud, and H. Neidhardt. Selfadjoint extensions with several gaps: finite deficiency indices. Oper. Theory Adv. Appl. 162 (2006) 85–101.

[157] J.F. Brasche and H. Neidhardt. Some remarks on Krein's extension theory. Math. Nachr. 165 (1994) 159–181.

[158] J.F. Brasche and H. Neidhardt. On the absolutely continuous spectrum of self-adjoint extensions. J. Functional Analysis 131 (1995) 364–385.

[159] J.F. Brasche and H. Neidhardt. On the singular continuous spectrum of self-adjoint extensions. Math. Z. 222 (1996) 533–542.

[160] J.F. Brasche, H. Neidhardt, and J. Weidmann. On the point spectrum of self-adjoint extensions. Math. Z. 214 (1993) 343–355.

[161] F. Brauer. Singular self-adjoint boundary value problems for the differential equation $Lx = \lambda Mx$. Trans. Amer. Math. Soc. 88 (1958) 331–345.

[162] F. Brauer. Spectral theory for the differential equation $Lu = \lambda Mu$. Can. J. Math. 10 (1958) 431–446.

[163] F. Brauer. Spectral theory for linear systems of differential equations. Pacific J. Math. 10 (1960) 17–34.

[164] H. Brezis. *Functional analysis, Sobolev spaces and partial differential equations*. Universitext, Springer, 2011.

[165] M.S. Brodskiĭ. *Triangular and Jordan representations of linear operators*. Transl. Math. Monographs, Vol. 32, Amer. Math. Soc., 2006.

[166] M.S. Brodskiĭ. Unitary operator colligations and their characteristic functions (Russian). Uspekhi Mat. Nauk 33 (1978) 141–168 [Russian Math. Surveys 33 (1978) 159–191].

[167] F.E. Browder, On the spectral theory of elliptic differential operators. I. Math. Ann. 142 (1961) 22–130.

[168] B.M. Brown and W.D. Evans. Selfadjoint and m-sectorial extensions of Sturm–Liouville operators. Integral Equations Operator Theory 85 (2016) 151–166.

[169] B.M. Brown, G. Grubb, and I. Wood. M-functions for closed extensions of adjoint pairs of operators with applications to elliptic boundary problems. Math. Nachr. 282 (2009) 314–347.

[170] B.M. Brown, J. Hinchcliffe, M. Marletta, S. Naboko, and I. Wood. The abstract Titchmarsh–Weyl M-function for adjoint operator pairs and its relation to the spectrum. Integral Equations Operator Theory 63 (2009) 297–320.

[171] B.M. Brown, H. Langer, and M. Langer. Bessel-type operators with an inner singularity. Integral Equations Operator Theory 75 (2013) 257–300.

[172] B.M. Brown, H. Langer, and C. Tretter. Compressed resolvents and reduction of spectral problems on star graphs. Complex Anal. Oper. Theory 13 (2019) 291–320.

[173] B.M. Brown, M. Marletta, S. Naboko, and I. Wood. Boundary triplets and M-functions for non-selfadjoint operators, with applications to elliptic PDEs and block operator matrices. J. London Math. Soc. (2) 77 (2008) 700–718.

[174] B.M. Brown, M. Marletta, S. Naboko, and I. Wood. Inverse problems for boundary triples with applications. Studia Mathematica 237 (2017) 241–275.

[175] B.M. Brown, D.K.R. McCormack, W.D. Evans, and M. Plum. On the spectrum of second-order differential operators with complex coefficients. Proc. Roy. Soc. London A 455 (1999) 1235–1257.

[176] V.M. Bruk. A certain class of boundary value problems with a spectral parameter in the boundary condition (Russian). Mat. Sb. 100 (1976) 210–216 [Sb. Math. 29 (1976) 186–192].

[177] V.M. Bruk. Extensions of symmetric relations (Russian). Mat. Zametki 22 (1977) 825–834 [Math. Notes 22 (1977) 953–958].

[178] V.M. Bruk. Linear relations in a space of vector functions (Russian). Mat. Zametki 24 (1978) 499–511 [Math. Notes 24 (1978) 767–773].

[179] V. Bruneau, K. Pankrashkin, and N. Popoff. Eigenvalue counting function for Robin Laplacians on conical domains. J. Geom. Anal. 28 (2018) 123–151.

[180] V. Bruneau and N. Popoff. On the negative spectrum of the Robin Laplacian in corner domains. Anal. PDE. 9 (2016) 1259–1283.

[181] V. Bruneau and G. Raikov. Spectral properties of harmonic Toeplitz operators and applications to the perturbed Krein Laplacian. Asymptotic Analysis 109 (2018) 53–74.

[182] R. Brunnhuber, J. Eckhardt, A. Kostenko, and G. Teschl. Singular Weyl–Titchmarsh–Kodaira theory for one-dimensional Dirac operators. Monatsh. Math. 174 (2014) 515–547.

[183] J. Brüning, V. Geyler, and K. Pankrashkin. Cantor and band spectra for periodic quantum graphs with magnetic fields. Commun. Math. Phys. 269 (2007) 87–105.

[184] J. Brüning, V. Geyler, and K. Pankrashkin. Spectra of self-adjoint extensions and applications to solvable Schrödinger operators. Rev. Math. Phys. 20 (2008) 1–70.

[185] C. Cacciapuoti, D. Fermi, and A. Posilicano. On inverses of Kreĭn's Q-functions. Rend. Mat. Appl. (7) 39 (2018) 229–240.

[186] J.W. Calkin. Abstract symmetric boundary conditions. Trans. Amer. Math. Soc. 45 (1939) 369–442.

[187] J.W. Calkin. General self-adjoint boundary conditions for certain partial differential operators. Proc. Nat. Acad. Sci. U.S.A. 25 (1939) 201–206.

[188] R. Carlson. Eigenfunction expansions for self-adjoint integro-differential operators. Pacific J. Math. 81 (1979) 327–347.

[189] R. Carlson. Expansions associated with non-self-adjoint boundary-value problems. Proc. Amer. Math. Soc. 73 (1979) 173–179.

[190] E.A. Catchpole. An integro-differential operator. J. London Math. Soc. (2) 6 (1973) 513–523.

[191] N.K. Chakravarty. Some problems in eigenfunction expansions I. Quart. J. Math. 16 (1965) 135–150.

[192] N.K. Chakravarty. Some problems in eigenfunction expansions II. Quart. J. Math. 19 (1965) 213–224.

[193] N.K. Chakravarty. Some problems in eigenfunction expansions III. Quart. J. Math. 19 (1968) 397–415.

[194] N.K. Chakravarty. Some problems in eigenfunction expansions IV. Indian J. Pure Appl. Math. 1 (1970) 347–353.

[195] J. Chaudhuri and W.N. Everitt. On the spectrum of ordinary second-order differential operators. Proc. Royal Soc. Edinburgh 68A (1969) 95–119.

[196] K.D. Cherednichenko, A.V. Kiselev, and L.O. Silva. Functional model for extensions of symmetric operators and applications to scattering theory. Netw. Heterog. Media 13 (2018) 191–215.

[197] S. Clark, F. Gesztesy, and M. Mitrea. Boundary data maps for Schrödinger operators on a compact interval. Math. Model. Nat. Phenom. 5 (2010) 73–121.

[198] S. Clark, F. Gesztesy, and R. Nichols. Principal solutions revisited. In: *Stochastic and infinite dimensional analysis*. Trends in Mathematics, Birkhäuser, 2016, pp. 85–117.

[199] S. Clark, F. Gesztesy, R. Nichols, and M. Zinchenko. Boundary data maps and Krein's resolvent formula for Sturm–Liouville operators on a finite interval. Operators Matrices 8 (2014) 1–71.

[200] E.A. Coddington. The spectral representation of ordinary self-adjoint differential operators. Ann. Math. (2) 60 (1954) 192–211.

[201] E.A. Coddington. Generalized resolutions of the identity for symmetric ordinary differential operators. Ann. Math. (2) 68 (1958) 378–392.

[202] E.A. Coddington. Extension theory of formally normal and symmetric subspaces. Mem. Amer. Math. Soc. 134 (1973).

[203] E.A Coddington. Self-adjoint subspace extensions of non-densely defined symmetric operators. Adv. Math. 14 (1974) 309–332.

[204] E.A. Coddington. Self-adjoint problems for nondensely defined ordinary differential operators and their eigenfunction expansions. Adv. Math. 15 (1975) 1–40.

[205] E.A. Coddington and A. Dijksma. Self-adjoint subspaces and eigenfunction expansions for ordinary differential subspaces. J. Differential Equations 20 (1976) 473–526.

[206] E.A. Coddington and A. Dijksma. Adjoint subspaces in Banach spaces, with applications to ordinary differential subspaces. Annali di Matematica Pura ed Applicata 118 (1978) 1–118.

[207] E.A. Coddington and R.C. Gilbert. Generalized resolvents of ordinary differential operators. Trans. Amer. Math. Soc. 93 (1959) 216–241.

[208] E.A. Coddington and N. Levinson. On the nature of the spectrum of singular second order linear differential equations. Canadian J. Math. 3 (1951) 335–338.

[209] E.A. Coddington and N. Levinson. *Theory of ordinary differential equations.* McGraw-Hill, 1955.

[210] E.A. Coddington and H.S.V. de Snoo. Positive selfadjoint extensions of positive subspaces. Math. Z. 159 (1978) 203–214.

[211] E.A. Coddington and H.S.V. de Snoo. Differential subspaces associated with pairs of ordinary differential expressions. J. Differential Equations 35 (1980) 129–182.

[212] E.A. Coddington and H.S.V. de Snoo. *Regular boundary value problems associated with pairs of ordinary differential expressions.* Lecture Notes in Mathematics, Vol. 858, Springer, 1981.

[213] S.D. Conte and W.C. Sangren. An expansion theorem for a pair of first order equations. Canadian J. Math. 6 (1954) 554–560.

[214] G. Corach and A. Maestripieri. Redundant decompositions, angles between subspaces and oblique projections. Publicacions Matemàtiques 54 (2010) 461–484.

[215] R. Corso. A Kato's second type representation theorem for solvable sesquilinear forms. J. Math. Anal. Appl. 462 (2018) 982–998.

[216] R. Corso and C. Trapani. Representation theorems for solvable sesquilinear forms. Integral Equations Operator Theory 89 (2017) 43–68.

[217] M.G. Crandall. Norm preserving extensions of linear transformations on Hilbert spaces. Proc. Amer. Math. Soc. 21 (1969) 335–340.

[218] R. Cross. *Multi-valued linear operators.* Marcel Dekker, 1998.

[219] M.E. Cumakin. Generalized resolvents of isometric operators (Russian). Sibirsk. Mat. Z. 8 (1967) 876–892.

[220] B. Ćurgus. Definitizable extensions of positive symmetric operators in a Kreĭn space. Integral Equations Operator Theory 12 (1989) 615–631.

[221] B. Ćurgus, A. Dijksma, and T. Read. The linearization of boundary eigenvalue problems and reproducing kernel Hilbert spaces. Lin. Alg. Appl. 329 (2001) 97–136.

[222] B. Ćurgus, A. Dijksma, H. Langer, and H.S.V. de Snoo. Characteristic functions of unitary colligations and of bounded operators in Krein spaces. Oper. Theory Adv. Appl. 41 (1989) 125–152.

[223] B. Ćurgus and H. Langer. A Krein-space approach to symmetric ordinary differential operators with an indefinite weight function. J. Differential equations 79 (1989) 31–61.

[224] P. Deift and R. Killip. On the absolutely continuous spectrum of one-dimensional Schrödinger operators with square summable potentials. Comm. Math. Phys. 203 (1999) 341–347.

[225] K. Daho and H. Langer. Sturm–Liouville operators with indefinite weight function. Proc. Royal Soc. Edinburgh 78A (1977) 161–191.

[226] D. Daners. Robin boundary value problems on arbitrary domains. Trans. Amer. Math. Soc. 352 (2000) 4207–4236.

[227] V.A. Derkach. Extensions of a Hermitian operator in a Kreĭn space (Russian). Dokl. Akad. Nauk Ukrain. SSR Ser. A 1988, no. 5, 5–9.

[228] V.A. Derkach. Extensions of a Hermitian operator that is not densely defined in a Kreĭn space (Russian). Dokl. Akad. Nauk Ukrain. SSR Ser. A 1990, no. 10, 15–19.

[229] V.A. Derkach. On Weyl function and generalized resolvents of a Hermitian operator in a Krein space. Integral Equations Operator Theory 23 (1995) 387–415.

[230] V.A. Derkach. On generalized resolvents of Hermitian relations in Krein spaces. J. Math. Sciences 97 (1999) 4420–4460.

[231] V.A. Derkach. Abstract interpolation problem in Nevanlinna classes. Oper. Theory Adv. Appl. 190 (2009) 197–236.

[232] V.A. Derkach and S. Hassi. A reproducing kernel space model for N_κ-functions. Proc. Amer. Math. Soc. 131 (2003) 3795–3806.

[233] V.A. Derkach, S. Hassi, and M.M. Malamud. Generalized boundary triples, Weyl functions, and inverse problems. arXiv:1706.0794824 (2017), 104 pp.

[234] V.A. Derkach, S. Hassi, M.M. Malamud, and H.S.V. de Snoo. Generalized resolvents of symmetric operators and admissibility. Methods Funct. Anal. Topology 6 (2000) 24–55.

[235] V.A. Derkach, S. Hassi, M.M. Malamud, and H.S.V. de Snoo. Boundary relations and their Weyl families. Dokl. Russian Akad. Nauk 39 (2004) 151–156.

[236] V.A. Derkach, S. Hassi, M.M. Malamud, and H.S.V. de Snoo. Boundary relations and their Weyl families. Trans. Amer. Math. Soc. 358 (2006) 5351–5400.

[237] V.A. Derkach, S. Hassi, M.M. Malamud, and H.S.V. de Snoo. Boundary relations and generalized resolvents of symmetric operators. Russian J. Math. Phys. 16 (2009) 17–60.

[238] V.A. Derkach, S. Hassi, M.M. Malamud, and H.S.V. de Snoo. Boundary triplets and Weyl functions. Recent developments. In: *Operator methods for boundary value problems*. London Math. Soc. Lecture Note Series, Vol. 404, 2012, pp. 161–220.

[239] V.A. Derkach, S. Hassi, and H.S.V. de Snoo. Operator models associated with singular perturbations. Methods Funct. Anal. Topology 7 (2001) 1–21.

[240] V.A. Derkach, S. Hassi, and H.S.V. de Snoo. Rank one perturbations in a Pontryagin space with one negative square. J. Functional Analysis 188 (2002) 317–349.

[241] V.A. Derkach, S. Hassi, and H.S.V. de Snoo. Singular perturbations of self-adjoint operators. Math. Phys. Anal. Geom. 6 (2003) 349–384.

[242] V.A. Derkach, S. Hassi, and H.S.V. de Snoo. Asymptotic expansions of generalized Nevanlinna functions and their spectral properties. Oper. Theory Adv. Appl. 175 (2007) 51–88.

[243] V.A. Derkach and M.M. Malamud. Weyl function of Hermitian operator and its connection with characteristic function (Russian). Preprint 85-9 (104) Donetsk Fiz-Techn. Institute AN Ukrain. SSR, 1985.

[244] V.A. Derkach and M.M Malamud. On the Weyl function and Hermitian operators with gaps (Russian). Soviet Math. Dokl. 35 (1987) 393–398 [Dokl. Akad. Nauk SSSR 293 (1987) 1041–1046].

[245] V.A. Derkach and M.M. Malamud, Generalized resolvents and the boundary value problems for Hermitian operators with gaps. J. Functional Analysis 95 (1991) 1–95.

[246] V.A. Derkach and M.M. Malamud. The extension theory of Hermitian operators and the moment problem. J. Math. Sciences 73 (1995) 141–242.

[247] V.A. Derkach and M.M. Malamud. *Extension theory of symmetric operators and boundary value problems.* Proceedings of Institute of Mathematics of NAS of Ukraine, Vol. 104, 2017.

[248] V.A. Derkach, M.M. Malamud, and E.R. Tsekanovskiĭ. Sectorial extensions of positive operator (Russian). Ukr. Mat. Zh. 41 (1989) 151–158 [Ukrainian Math. J. 41 (1989) 136–142].

[249] F. Deutsch. The angle between subspaces of a Hilbert space. In: *Approximation theory, wavelets and applications.* Kluwer, 1995, pp. 107–130.

[250] F. Deutsch. *Best approximation in inner product spaces.* CMS Books in Mathematics, Springer, 2001.

[251] N.C. Dias, A. Posilicano, and J.N. Prata. Self-adjoint, globally defined Hamiltonian operators for systems with boundaries. Commun. Pure Appl. Anal. 10 (2011), no. 6, 1687–1.

[252] J.F. van Diejen and A. Tip. Scattering from generalized point interaction using self-adjoint extensions in Pontryagin spaces. J. Math. Phys. 32 (1991) 631–641.

[253] A. Dijksma and H. Langer. Operator theory and ordinary differential operators. In: *Lectures on operator theory and its applications.* Fields Institute Monographs, Vol. 3, Amer. Math. Soc., 1996, pp. 73–139.

[254] A. Dijksma and H. Langer. Compressions of self-adjoint extensions of a symmetric operator and M.G. Krein's resolvent formula. Integral Equations Operator Theory 90 (2018) 41 30 pp.

[255] A. Dijksma, H. Langer, A. Luger, and Yu. Shondin. A factorization result for generalized Nevanlinna functions of the class N_κ. Integral Equations Operator Theory 36 (2000) 121–125.

[256] A. Dijksma, H. Langer, Yu. Shondin, and C. Zeinstra. Self-adjoint operators with inner singularities and Pontryagin spaces. Oper. Theory Adv. Appl. 118 (2000) 105–175.

[257] A. Dijksma, H. Langer, and H.S.V. de Snoo. Selfadjoint Π_κ-extensions of symmetric subspaces: an abstract approach to boundary problems with spectral parameter in the boundary condition. Integral Equations Operator Theory 90 (1984) 459–515.

[258] A. Dijksma, H. Langer, and H.S.V. de Snoo. Unitary colligations in Π_κ-spaces, characteristic functions and Štraus extensions. Pacific J. Math. 125 (1986) 347–362.

[259] A. Dijksma, H. Langer, and H.S.V. de Snoo. Characteristic functions of unitary operator colligations in Π_κ-spaces. Oper. Theory Adv. Appl. 19 (1986) 125–194.

[260] A. Dijksma, H. Langer, and H.S.V. de Snoo. Unitary colligations in Krein spaces and their role in the extension theory of isometries and symmetric An Overview of Linear Relations. Springer Lecture Notes Math. 1242 (1987) 1–42.

[261] A. Dijksma, H. Langer, and H.S.V. de Snoo. Symmetric Sturm–Liouville operator with eigenvalue depending boundary conditions. Canadian Math. Soc. Conf. Proc. 8 (1987) 87–116.

[262] A. Dijksma, H. Langer, and H.S.V. de Snoo. Representations of holomorphic operator functions by means of resolvents of unitary or selfadjoint operators in Krein spaces. Oper. Theory Adv. Appl. 24 (1987) 123–143.

[263] A. Dijksma, H. Langer, and H.S.V. de Snoo. Hamiltonian systems with eigenvalue depending boundary conditions. Oper. Theory Adv. Appl. 35 (1988) 37–83.

[264] A. Dijksma, H. Langer, and H.S.V. de Snoo. Generalized coresolvents of standard isometric operators and generalized resolvents of standard symmetric relations in Kreĭn spaces. Oper. Theory Adv. Appl. 48 (1990) 261–274.

[265] A. Dijksma, H. Langer, and H.S.V. de Snoo. Eigenvalues and pole functions of Hamiltonian systems with eigenvalue depending boundary conditions. Math. Nachr. 161 (1993) 107–154.

[266] A. Dijksma and H.S.V. de Snoo. Self-adjoint extensions of symmetric subspaces. Pacific J. Math. 54 (1974) 71–100.

[267] A. Dijksma and H.S.V. de Snoo. Eigenfunction expansions for non-densely defined differential operators. J. Differential Equations 60 (1975) 21–56.

[268] A. Dijksma and H.S.V. de Snoo. Symmetric and selfadjoint relations in Kreĭn spaces I. Oper. Theory Adv. Appl. 24 (1987) 145–166.

[269] A. Dijksma and H.S.V. de Snoo. Symmetric and selfadjoint relations in Kreĭn spaces II. Anal. Acad. Sci. Fenn. Ser. A I Math. 12 (1987) 199–216.

[270] A. Dijksma, H.S.V. de Snoo, and A.A. El Sabbagh. Selfadjoint extensions of regular canonical systems with Stieltjes boundary conditions. J. Math. Anal. Appl. 152 (1990) 546–583.

[271] W.F. Donoghue. On the perturbation of spectra. Comm. Pure Appl. Math. 18 (1965) 559–579.

[272] W.F. Donoghue. *Monotone matrix functions and analytic continuation.* Springer, 1974.

[273] R.G. Douglas. On majorization, factorization, and range inclusion of operators on Hilbert space. Proc. Amer. Math. Soc. 17 (1966) 413–416.

[274] M.A. Dritschel and J. Rovnyak. Extension theorems for contraction operators on Krein spaces. Oper. Theory Adv. Appl. 47 (1990) 221–305.

[275] M.A. Dritschel and J. Rovnyak. Operators on indefinite inner product spaces. In: *Lectures on operator theory and its applications.* Fields Institute Monographs, Vol. 3, Amer. Math. Soc., 1996, pp. 141–232.

[276] N. Dunford and J.T. Schwarz. *Linear operators, Part II: Spectral theory.* John Wiley and Sons, 1963.

[277] H. Dym. *J-contractive matrix functions, reproducing kernel Hilbert spaces and interpolation.* Regional conference series in mathematics, Vol. 71, Amer. Math. Soc., 1989.

[278] H. Dym and McKean. *Gaussian processes, function theory, and the inverse spectral problem.* Academic Press, 1976.

[279] H. Dym and A. Iacob. Positive definite extensions, canonical equations and inverse problems. Oper. Theory Adv. Appl. 12 (1984) 141–240.

[280] W.F. Eberlein. Closure, convexity and linearity in Banach spaces. Ann. Math. 47 (1946) 688–703.

[281] J. Eckhardt, F. Gesztesy, R. Nichols, and G. Teschl. Weyl–Titchmarsh theory for Sturm–Liouville operators with distributional potentials. Opuscula Math. 33 (2013) 467–563.

[282] J. Eckhardt, F. Gesztesy, R. Nichols, and G. Teschl. Supersymmetry and Schrödinger-type operators with distributional matrix-valued potentials. J. Spectral Theory 4 (2014) 715–768.

[283] J. Eckhardt, A. Kostenko, and G. Teschl. Spectral asymptotics for canonical systems. J. Reine Angew. Math. 736 (2018) 285–315.

[284] D.E. Edmunds and W.D. Evans. *Spectral theory and differential operators.* Oxford University Press, 1987.

[285] J. Ercolano and M. Schechter. Spectral theory for operators generated by elliptic boundary problems with eigenvalue parameter in boundary conditions. I. Comm. Pure Appl. Math. 18 (1965) 83–105.

[286] J. Ercolano and M. Schechter. Spectral theory for operators generated by elliptic boundary problems with eigenvalue parameter in boundary conditions. II. Comm. Pure Appl. Math. 18 (1965) 397–414.

[287] Y. Ershova, I. Karpenko, and A. Kiselev. Isospectrality for graph Laplacians under the change of coupling at graph vertices. J. Spectral Theory 6 (2016) 43–66.

[288] Y. Ershova and A. Kiselev. Trace formulae for graph Laplacians with applications to recovering matching conditions. Methods Funct. Anal. Topology 18 (2012) 343–359.

[289] Y. Ershova and A. Kiselev. Trace formulae for Schrödinger operators on metric graphs with applications to recovering matching conditions. Methods Funct. Anal. Topology 20 (2014) 134–148.

[290] A. Etkin. On an abstract boundary value problem with the eigenvalue parameter in the boundary condition. Fields Inst. Commun. 25 (2000) 257–266.

[291] L.C. Evans. *Partial differential equations.* Graduate Series in Mathematics, Vol. 19, Amer. Math. Soc., 2010.

[292] W.N. Everitt. On a property of the m-coefficient of a second-order linear differential equation. J. London Math. Soc. (2) 4 (1972) 443–457.

[293] P. Exner. Seize ans après. In: Appendix K to *Solvable models in quantum mechanics.* 2nd edition. AMS Chelsea Publishing, Amer. Math. Soc., 2005.

[294] P. Exner, A. Kostenko, M.M. Malamud, and H. Neidhardt. Spectral theory of infinite quantum graphs. Ann. Henri Poincaré 19 (2018) 3457–3510.

[295] W. Faris. *Self-adjoint operators*. Lecture Notes in Mathematics, Vol. 437, Springer, 1975.

[296] B. Farkas and M. Matolcsi. Commutation properties of the form sum of positive, symmetric operators. Acta Sci. Math. (Szeged) 67 (2001) 777–790.

[297] A. Favini and A. Yagi. *Degenerate differential equations in Banach spaces*. CRC Press, 1998.

[298] M. Fernandez-Miranda and J.Ph. Labrousse. The Cayley transform of linear relations. Proc. Amer. Math. Soc. 133 (2005) 493–498.

[299] P. Fillmore and J. Williams. On operator ranges. Adv. Math. 7 (1971) 254–281.

[300] C. Fischbacher. The nonproper dissipative extensions of a dual pair. Trans. Amer. Math. Soc. 370 (2018) 8895–8920.

[301] C. Fischbacher. A Birman–Krein–Vishik–Grubb theory for sectorial operators. Complex Anal. Oper. Theory. To appear.

[302] C. Fischbacher, S. Naboko, and I. Wood. The proper dissipative extensions of a dual pair. Integral Equations Operator Theory 85 (2016) 573–599.

[303] A. Fleige, S. Hassi, and H.S.V. de Snoo. A Kreĭn space approach to representation theorems and generalized Friedrichs extensions. Acta Sci. Math. 66 (2000) 633–650.

[304] A. Fleige, S. Hassi, H.S.V. de Snoo, and H. Winkler. Generalized Friedrichs extensions associated with interface conditions for Sturm–Liouville operators. Oper. Theory Adv. Appl. 163 (2005) 135–145.

[305] A. Fleige, S. Hassi, H.S.V. de Snoo, and H. Winkler. Sesquilinear forms corresponding to a non-semibounded Sturm–Liouville operator. Proc. Royal Soc. Edinburgh 140A (2010) 291–318.

[306] A. Fleige, S. Hassi, H.S.V. de Snoo, and H. Winkler. Non-semibounded closed symmetric forms associated with a generalized Friedrichs extension. Proc. Roy. Soc. Edinburgh 144A (2014) 731–745.

[307] K.-H. Förster and M.M. Nafalska. A factorization of extremal extensions with applications to block operator matrices. Acta Math. Hungar. 129 (2010) 112–141.

[308] R. Freeman. Closed operators and their adjoints associated with elliptic differential operators. Pacific J. Math. 22 (1967) 71–97.

[309] H. Freudenthal. Über die Friedrichssche Fortsetzung halb-beschränkter hermite-scher Operatoren. Proc. Akad. Amsterdam 39 (1936) 832–833.

[310] K.O. Friedrichs. Spektraltheorie halbbeschränkter Operatoren und Anwendung auf die Spektralzerlegung von Differentialoperatoren. Math. Ann. 109 (1933/1934) 465–487, 685–713; 110 (1935) 777–779.

[311] K.O. Friedrichs. Über die ausgezeichnete Randbedingung in der Spektraltheorie der halbbeschränkten gewöhnlichen Differentialoperatoren zweiter Ordnung. Math. Ann. 112 (1935/6) 1–23.

[312] C.T. Fulton. Parametrizations of Titchmarsh's $m(\lambda)$-functions in the limit circle case. Trans. Amer. Math. Soc. 229 (1977) 51–63.

[313] C.T. Fulton. Two-point boundary value problems with eigenvalue parameter contained in the boundary condition. Proc. Roy. Soc. Edinburgh 77A (1977) 293–308.

[314] C.T. Fulton. Singular eigenvalue problems with eigenvalue parameter contained in the boundary conditions. Proc. Roy. Soc. Edinburgh 87A (1980/81) 1–34.

[315] C.T. Fulton and H. Langer. Sturm–Liouville operators with singularities and generalized Nevanlinna functions. Complex Anal. Oper. Theory 4 (2010) 179–243.

[316] C.T. Fulton, H. Langer, and A. Luger. Mark Krein's method of directing functionals and singular potentials. Math. Nachr. 285 (2012) 1791–1798.

[317] F. Gesztesy, N.J. Kalton, K.A. Makarov, and E.R. Tsekanovskiĭ. Some applications of operator-valued Herglotz functions. Oper. Theory Adv. Appl. 123 (2001) 271–321.

[318] F. Gesztesy and K.A. Makarov. $SL_2(\mathbb{R})$, exponential Herglotz representations, and spectral averaging. St. Petersburg Math. J. 15 (2004) 393–418.

[319] F. Gesztesy, K.A. Makarov, and S.N. Naboko. The spectral shift operator. Oper. Theory Adv. Appl. 108 (1999) 59–90.

[320] F. Gesztesy, K.A. Makarov, and E.R. Tsekanovskiĭ. An addendum to Krein's formula. J. Math. Anal. Appl. 222 (1998) 594–606.

[321] F. Gesztesy, K.A. Makarov, and M. Zinchenko. Essential closures and ac spectra for reflectionless CMV, Jacobi, and Schrödinger operators revisited. Acta Applicandae Math. 103 (2008) 315–339.

[322] F. Gesztesy, D. Mitrea, I. Mitrea, and M. Mitrea. On the nature of the Laplace–Beltrami operator on Lipschitz manifolds. J. Math. Sci. 172 (2011) 279–346.

[323] F. Gesztesy and M. Mitrea. Generalized Robin boundary conditions, Robin-to-Dirichlet maps, and Krein-type resolvent formulas for Schrödinger operators on bounded Lipschitz domains. In: *Perspectives in partial differential equations, harmonic analysis and applications*. Proc. Sympos. Pure Math., Vol. 79, Amer. Math. Soc., 2008, pp. 105–173.

[324] F. Gesztesy and M. Mitrea, Robin-to-Robin maps and Krein-type resolvent formulas for Schrödinger operators on bounded Lipschitz domains. Oper. Theory Adv. Appl. 191 (2009) 81–113.

[325] F. Gesztesy and M. Mitrea. Nonlocal Robin Laplacians and some remarks on a paper by Filonov on eigenvalue inequalities. J. Differential Equations 247 (2009) 2871–2896.

[326] F. Gesztesy and M. Mitrea. A description of all self-adjoint extensions of the Laplacian and Krein-type resolvent formulas on non-smooth domains. J. Math. Anal. Appl. 113 (2011) 53–172.

[327] F. Gesztesy, S.N. Naboko, R. Weikard, and M. Zinchenko. Donoghue-type m-functions for Schrödinger operators with operator-valued potentials. J. Anal. Math. 137 (2019) 373–427.

[328] F. Gesztesy, R. Nowell, and W. Pötz. One-dimensional scattering theory for quantum systems with nontrivial spatial asymptotics. Differential Integral Equations 10 (1997) 521–546.

[329] F. Gesztesy and E.R. Tsekanovskiĭ. On matrix-valued Herglotz functions. Math. Nachr. 218 (2000) 61–138.

[330] F. Gesztesy, R. Weikard, and M. Zinchenko. On spectral theory for Schrödinger operators with operator-valued potentials. J. Differential Equations 255 (2013) 1784–1827.

[331] F. Gesztesy and M. Zinchenko. On spectral theory for Schrödinger operators with strongly singular potentials. Math. Nachr. 279 (2006) 1041–1082.

[332] F. Gesztesy and M. Zinchenko. Symmetrized perturbation determinants and applications to boundary data maps and Krein-type resolvent formulas. Proc. Lond. Math. Soc. (3) 104 (2012) 577–612.

[333] D.J. Gilbert. On subordinacy and analysis of the spectrum of Schrödinger operators with two singular endpoints. Proc. Roy. Soc. Edinburgh 112A (1989) 213–229.

[334] D.J. Gilbert. On subordinacy and spectral multiplicity for a class singular differential operators. Proc. Roy. Soc. Edinburgh 128A (1998) 549–584.

[335] D.J. Gilbert and D.B. Pearson. On subordinacy and analysis of the spectrum of one-dimensional Schrödinger operators. J. Math. Anal. Appl. 128 (1987) 30–56.

[336] R.C. Gilbert. Spectral multiplicity of selfadjoint dilations. Proc. Amer. Math. Soc. 19 (1968) 477–482.

[337] R.C. Gilbert. Simplicity of linear ordinary differential operators. J. Differential Equations 11 (1972) 672–681.

[338] R.C. Gilbert. Spectral representation of selfadjoint extensions of a symmetric operator. Rocky Mountain J. Math. 2 (1972) 75–96.

[339] I.M. Glazman. On the theory of singular differential operators. Uspekhi Mat. Nauk 5 (1950) 102–135.

[340] I. Gohberg and M.G. Kreĭn. The basic propositions on defect numbers, root numbers and indices of linear operators (Russian). Uspekhi Mat. Nauk. 12 (1957) 43–118 [Transl. Amer. Math. Soc. (2) 13 (1960) 185–264].

[341] I. Gohberg and M.G. Kreĭn. *Theory and applications of Volterra operators in Hilbert space.* Transl. Math. Monographs, Vol. 24, Amer. Math. Soc., 1970.

[342] N. Goloshchapova. On the multi-dimensional Schrödinger operators with point interactions. Methods Funct. Anal. Topology 17 (2011) 126–143.

[343] N. Goloshchapova, M.M. Malamud, and V. Zastavnyi. Radial positive definite functions and spectral theory of the Schrödinger operators with point interactions. Math. Nachr. 285 (2012) 1839–1859.

[344] N. Goloshchapova and L. Oridoroga. On the negative spectrum of one-dimensional Schrödinger operators with point interactions. Integral Equations Operator Theory 67 (2010) 1–14.

[345] M.L. Gorbachuk. Self-adjoint boundary value problems for differential equation of the second order with unbounded operator coefficient (Russian). Funktsional. Anal. i Prilozhen 5 (1971) 10–21 [Funct. Anal. Appl. 5 (1971) 9–18].

[346] V.I. Gorbachuk and M.L. Gorbachuk. *Boundary value problems for operator differential equations* (Russian). Naukova Dumka, Kiev, 1984 [Kluwer, 1991].

[347] V.I. Gorbachuk, M.L. Gorbachuk, and A.N. Kochubei. Extension theory for symmetric operators and boundary value problems for differential equations (Russian). Ukr. Mat. Zh. 41 (1989) 1299–1313 [Ukrainian Math. J. 41 (1989) 1117–1129].

[348] V.I. Gorbachuk and M.L. Gorbachuk. *M.G. Krein's lectures on entire operators*. Oper. Theory Adv. Appl., Vol. 97, Birkhäuser, 1991.

[349] M.L. Gorbachuk and V.A. Mikhailets. Semibounded selfadjoint extensions of symmetric operators (Russian). Dokl. Akad. Nauk SSSR 226 (1976) 765–768 [Sov. Math. Dokl. 17 (1976) 185–186].

[350] Y.I. Granovskyi and L.L. Oridoroga. Krein–von Neumann extension of an even order differential operator on a finite interval. Opuscula Math. 38 (2018) 681–698.

[351] P. Grisvard. *Elliptic problems in nonsmooth domains*. Pitman, 1985.

[352] G. Grubb. A characterization of the non-local boundary value problems associated with an elliptic operator. Ann. Scuola Norm. Sup. Pisa (3) 22 (1968) 425–513.

[353] G. Grubb. Les problems aux limites généraux d'un opérateur elliptique, provenant de la théorie variationnelle. Bull. Sci. Math. 94 (1970) 113–157.

[354] G. Grubb. On coerciveness and semiboundedness of general boundary problems. Israel J. Math. 10 (1971) 32–95.

[355] G. Grubb. Properties of normal boundary problems for elliptic even-order systems. Ann. Scuola Norm. Sup. Pisa (4) 1 (1974) 1–61.

[356] G. Grubb. Spectral asymptotics for the "soft" selfadjoint extension of a symmetric elliptic differential operator. J. Operator Theory 10 (1983) 9–20.

[357] G. Grubb. Known and unknown results on elliptic boundary problems. Bull. Amer. Math. Soc. 43 (2006) 227–230.

[358] G. Grubb. Krein resolvent formulas for elliptic boundary problems in nonsmooth domains. Rend. Semin. Mat. Univ. Politec. Torino 66 (2008) 13–39.

[359] G. Grubb. *Distributions and operators*. Graduate Texts in Mathematics, Vol. 252, Springer, 2009.

[360] G. Grubb. The mixed boundary value problem, Krein resolvent formulas and spectral asymptotic estimates. J. Math. Anal. Appl. 382 (2011) 339–363.

[361] G. Grubb. Spectral asymptotics for Robin problems with a discontinuous coefficient. J. Spectral Theory 1 (2011) 155–177.

[362] G. Grubb. Krein-like extensions and the lower boundedness problem for elliptic operators. J. Differential Equations 252 (2012) 852–885.

[363] G. Grubb. Extension theory for elliptic partial differential operators with pseudodifferential methods. In: *Operator methods for boundary value problems*. London Math. Soc. Lecture Note Series, Vol. 404, 2012, pp. 221–258.

[364] L. Grubišić, V. Kostrykin, K.A. Makarov, and K. Veselić. Representation theorems for indefinite quadratic forms revisited. Mathematika 59 (2013) 169–189.

[365] M. Haase. *The functional calculus for sectorial operators*. Oper. Theory Adv. Appl., Vol. 169, Birkhäuser, 2006.

[366] B.J. Harris. The asymptotic form of the Titchmarsh–Weyl m-function. J. London Math. Soc. (2) 30 (1984) 110–118.

[367] B.J. Harris. The asymptotic form of the Titchmarsh–Weyl m-function associated with a second order differential equation with locally integrable coefficient. Proc. Roy. Soc. Edinburgh 102A (1986) 243–251.

[368] P. Hartman. Self-adjoint, non-oscillatory systems of ordinary, second order, linear differential equations. Duke Math. J. 24 (1957) 25–35.

[369] P. Hartman. Ordinary differential equations. 2nd edition. SIAM Classics in Mathematics, 1982.

[370] S. Hassi. On the Friedrichs and the Kreĭn–von Neumann extension of nonnegative relations. Acta Wasaensia 122 (2004) 37–54.

[371] S. Hassi, M. Kaltenbäck, and H.S.V. de Snoo. Triplets of Hilbert spaces and Friedrichs extensions associated with the subclass N_1 of Nevanlinna functions. J. Operator Theory 37 (1997) 155–181.

[372] S. Hassi, M. Kaltenbäck, and H.S.V. de Snoo. A characterization of semibounded selfadjoint operators. Proc. Amer. Math. Soc. 124 (1997) 2681–2692.

[373] S. Hassi, M. Kaltenbäck, and H.S.V. de Snoo. Selfadjoint extensions of the orthogonal sum of symmetric relations. I. 16th Oper. Theory Conf. Proc. (1997) 163–178.

[374] S. Hassi, M. Kaltenbäck, and H.S.V. de Snoo. Generalized Kreĭn–von Neumann extensions and associated operator models. Acta Sci. Math. (Szeged) 64 (1998) 627–655.

[375] S. Hassi, M. Kaltenbäck, and H.S.V. de Snoo. Selfadjoint extensions of the orthogonal sum of symmetric relations. II. Oper. Theory Adv. Appl. 106 (1998) 187–200.

[376] S. Hassi, M. Kaltenbäck, and H.S.V. de Snoo. The sum of matrix Nevanlinna functions and selfadjoint exit space extensions. Oper. Theory Adv. Appl. 103 (1998) 137–154.

[377] S. Hassi and S. Kuzhel. On J-self-adjoint operators with stable C-symmetries. Proc. Roy. Soc. Edinburgh 143A (2013) 141–167.

[378] S. Hassi, J.Ph. Labrousse, and H.S.V. de Snoo. Selfadjoint extensions of relations whose domain and range are orthogonal. Methods Funct. Anal. Topology. To appear.

[379] S. Hassi, H. Langer, and H.S.V. de Snoo. Selfadjoint extensions for a class of symmetric operators with defect numbers (1,1). 15th Oper. Theory Conf. Proc. (1995) 115–145.

[380] S. Hassi and A. Luger. Generalized zeros and poles of N_κ-functions: on the underlying spectral structure. Methods Funct. Anal. Topology 12 (2006) 131–150.

[381] S. Hassi, M.M. Malamud, and V.I. Mogilevskiĭ. Generalized resolvents and boundary triplets for dual pairs of linear relations. Methods Funct. Anal. Topology 11 (2005) 170–187.

[382] S. Hassi, M.M. Malamud, and V.I. Mogilevskiĭ. Unitary equivalence of proper extensions of a symmetric operator and the Weyl function. Integral Equations Operator Theory 77 (2013) 449–487.

[383] S. Hassi, M.M. Malamud, and H.S.V. de Snoo. On Kreĭn's extension theory of nonnegative operators. Math. Nachr. 274/275 (2004) 40–73.

[384] S. Hassi, M. Möller, and H.S.V. de Snoo. Sturm–Liouville operators and their spectral functions. J. Math. Anal. Appl. 282 (2003) 584–602.

[385] S. Hassi, M. Möller, and H.S.V. de Snoo. Singular Sturm–Liouville equations whose coefficients depend rationally on the eigenvalue parameter. J. Math. Anal. Appl. 295 (2004) 258–275.

[386] S. Hassi, M. Möller, and H.S.V. de Snoo. A class of Nevanlinna functions associated with Sturm–Liouville operators. Proc. Amer. Math. Soc. 134 (2006) 2885–2893.

[387] S. Hassi, M. Möller, and H.S.V. de Snoo. Limit-point/limit-circle classification for Hain-Lüst type equations. Math. Nachr. 291 (2018) 652–668.

[388] S. Hassi, C. Remling, and H.S.V. de Snoo. Subordinate solutions and spectral measures of canonical systems. Integral Equations Operator Theory 37 (2000) 48–63.

[389] S. Hassi, A. Sandovici, and H.S.V. de Snoo. Factorized sectorial relations, their maximal sectorial extensions, and form sums. Banach J. Math. Anal. 13 (2019) 538–564.

[390] S. Hassi, A. Sandovici, H.S.V. de Snoo, and H. Winkler. Form sums of nonnegative selfadjoint operators. Acta Math. Hungar. 111 (2006) 81–105.

[391] S. Hassi, A. Sandovici, H.S.V. de Snoo, and H. Winkler. A general factorization approach to the extension theory of nonnegative operators and relations. J. Operator Theory 58 (2007) 351–386.

[392] S. Hassi, A. Sandovici, H.S.V. de Snoo, and H. Winkler. Extremal extensions for the sum of nonnegative selfadjoint relations. Proc. Amer. Math. Soc. 135 (2007) 3193–3204.

[393] S. Hassi, A. Sandovici, H.S.V. de Snoo, and H. Winkler. Extremal maximal sectorial extensions of sectorial relations. Indag. Math. 28 (2017) 1019–1055.

[394] S. Hassi, A.J. van der Schaft, H.S.V. de Snoo, and H.J. Zwart. Dirac structures and boundary relations. In: *Operator methods for boundary value problems*. London Math. Soc. Lecture Note Series, Vol. 404, 2012, pp. 259–274.

[395] S. Hassi, Z. Sebestyén, and H.S.V. de Snoo. Lebesgue type decompositions for nonnegative forms. J. Functional Analysis 257 (2009) 3858–3894.

[396] S. Hassi, Z. Sebestyén, and H.S.V. de Snoo. Domain and range descriptions for adjoint relations, and parallel sums and differences of forms. Oper. Theory Adv. Appl. 198 (2009) 211–227.

[397] S. Hassi, Z. Sebestyén, and H.S.V. de Snoo. Lebesgue type decompositions for linear relations and Ando's uniqueness criterion. Acta Sci. Math. (Szeged) 84 (2018) 465–507.

[398] S. Hassi, Z. Sebestyén, H.S.V. de Snoo, and F.H. Szafraniec. A canonical decomposition for linear operators and linear relations. Acta Math. Hungar. 115 (2007) 281–307.

[399] S. Hassi and H.S.V. de Snoo. One-dimensional graph perturbations of selfadjoint relations. Ann. Acad. Sci. Fenn. Ser. A I Math. 22 (1997) 123–164.

[400] S. Hassi and H.S.V. de Snoo. On rank one perturbations of selfadjoint operators. Integral Equations Operator Theory 29 (1997) 288–300.

[401] S. Hassi and H.S.V. de Snoo. Nevanlinna functions, perturbation formulas and triplets of Hilbert spaces. Math. Nachr. 195 (1998) 115–138.

[402] S. Hassi and H.S.V. de Snoo. Factorization, majorization, and domination for linear relations. Annales Univ. Sci. Budapest 58 (2015) 53–70.

[403] S. Hassi, H.S.V. de Snoo, A.E. Sterk, and H.Winkler. Finite-dimensional graph perturbations of selfadjoint Sturm–Liouville operators. In: *Operator theory, structured matrices, and dilations*. Theta Foundation, 2007, pp. 205–226.

[404] S. Hassi, H.S.V. de Snoo, and F.H. Szafraniec. Componentwise and Cartesian decompositions of linear relations. Dissertationes Mathematicae 465 (2009) 59 pp.

[405] S. Hassi, H.S.V. de Snoo, and F.H. Szafraniec. Infinite-dimensional perturbations, maximally nondensely defined symmetric operators, and some matrix representations. Indag. Math. 23 (2012) 1087–1117.

[406] S. Hassi, H.S.V. de Snoo, and F.H. Szafraniec. Normal intermediate extensions of symmetric relations. Acta Math. Szeged 80 (2014) 195–232.

[407] S. Hassi, H.S.V. de Snoo, and A.D.I. Willemsma. Smooth rank one perturbations of selfadjoint operators. Proc. Amer. Math. Soc. 126 (1998) 2663–2675.

[408] S. Hassi, H.S.V. de Snoo, and H. Winkler. Boundary-value problems for two-dimensional canonical systems. Integral Equations Operator Theory 36 (2000) 445–479.

[409] S. Hassi, H.S.V. de Snoo, and H. Woracek. Some interpolation problems of Nevanlinna-Pick type. The Kreĭn–Langer method. Operator Theory Adv. Appl. 106 (1998) 201–216.

[410] S. Hassi and H.L. Wietsma, On Calkin's abstract symmetric boundary conditions. In: *Operator methods for boundary value problems*. London Math. Soc. Lecture Note Series, Vol. 404, 2012, pp. 3–34.

[411] S. Hassi and H.L. Wietsma. Products of generalized Nevanlinna functions with symmetric rational functions. J. Functional Analysis 266 (2014) 3321–3376.

[412] E. Heinz. Halbbeschränktheit gewöhnlicher Differentialoperatoren höherer Ordnung. Math. Ann. 135 (1958) 1–49.

[413] B. Helffer and A. Kachmar. Eigenvalues for the Robin Laplacian in domains with variable curvature. Trans. Amer. Math. Soc. 369 (2017) 3253–3287.

[414] G. Hellwig. *Differential operators of mathematical physics*. Addison-Wesley, 1967.

[415] R. Hempel, L. Seco, and B. Simon. The essential spectrum of Neumann Laplacians on some bounded, singular domains. J. Functional Analysis 102 (1991) 448–483.

[416] E. Hewitt and K. Stromberg. *Real and abstract analysis*. Springer, 1965.

[417] E. Hilb. Über Integraldarstellungen willkürlicher Funktionen. Math. Ann. 66 (1909) 1–66.

[418] E. Hilb. Über Reihentwicklungen nach den Eigenfunktionen linearer Differentialgleichungen 2ter Ordnung. Math. Ann. 71 (1911) 76–87.

[419] E. Hilb. Über Reihentwicklungen, welche aus speziellen Randwertproblemen bei gewöhnlichen linearen inhomogenen Differentialgleichungen entspringen. J. Reine Angew. Math. 140 (1911) 205–229.

[420] E. Hille. *Lectures on ordinary differential equations.* Addison-Wesley, 1969.

[421] E. Hille and R.S. Phillips. *Functional analysis and semi-groups.* Amer. Math. Soc. Colloquium Publications, Volume 31, 1996.

[422] D.B. Hinton and A. Schneider. On the Titchmarsh–Weyl coefficients for singular S-Hermitian systems I. Math. Nachr. 163 (1993) 323–342.

[423] D.B. Hinton and A. Schneider. On the Titchmarsh–Weyl coefficients for singular S-Hermitian systems II. Math. Nachr. 185 (1997) 67–84.

[424] D.B. Hinton and A. Schneider. Titchmarsh–Weyl coefficients for odd-order linear Hamiltonian systems. J. Spectral Math. Appl. 1 (2006) 1–36.

[425] D.B. Hinton and J.K. Shaw. On Titchmarsh–Weyl $m(\lambda)$-functions for linear Hamiltonian systems. J. Differential Equations 40 (1981) 316–342.

[426] D.B. Hinton and J.K. Shaw. Titchmarsh–Weyl theory for Hamiltonian systems. In: *Spectral theory and differential operators.* North-Holland Math. Studies, Vol. 55, 1981, pp. 219–231.

[427] D.B. Hinton and J.K. Shaw. Hamiltonian systems of limit point or limit circle type with both endpoints singular. J. Differential Equations 50 (1983) 444–464.

[428] D.B. Hinton and J.K. Shaw. On boundary value problems for Hamiltonian systems with two singular endpoints. SIAM J. Math. Anal. 15 (1984) 272–286.

[429] A. Ibort, F. Lledó, and J.M. Pérez-Pardo. Self-adjoint extensions of the Laplace–Beltrami operator and unitaries at the boundary. J. Functional Analysis 268 (2015) 634–670.

[430] I.S. Iohvidov, M.G. Krein, and H. Langer. *Introduction to the spectral theory of operators in spaces with an indefinite metric.* Mathematische Forschung, Band 9, Akademie Verlag, 1982.

[431] D. Jerison and C. Kenig. The Neumann problem in Lipschitz domains. Bull. Amer. Math. Soc. 4 (1981) 203–207.

[432] D. Jerison and C. Kenig. The inhomogeneous Dirichlet problem in Lipschitz domains. J. Functional Analysis 113 (1995) 161–219.

[433] S.Y. Jitomirskaya, Y. Last, R. del Rio, and B. Simon. Operators with singular continuous spectrum. IV. Hausdorff dimensions, rank one perturbations, and localization. J. Anal. Math. 69 (1996) 153–200.

[434] P. Jonas. A class of operator-valued meromorphic functions on the unit disc. Ann. Acad. Sci. Fenn. Math. 17 (1992) 257–284.

[435] P. Jonas. Operator representations of definitizable functions. Ann. Acad. Sci. Fenn. Math. 25 (2000) 41–72.

[436] P. Jonas. On locally definite operators in Krein spaces. In: *Spectral analysis and its applications.* Theta Foundation, 2003, pp. 95–127.

[437] P. Jonas. On operator representations of locally definitizable functions. Oper. Theory Adv. Appl. 162 (2005) 165–190.

[438] K. Jörgens and F. Rellich. *Eigenwerttheorie gewöhnlicher Differentialgleichungen.* Springer, 1976.

[439] P.E.T. Jorgensen. Unbounded operators; perturbations and commutativity problems. J. Functional Analysis 39 (1980) 281–307.

[440] I.S. Kac. On the Hilbert spaces generated by monotone Henrmitian matrix functions. Zap. Mat. Otd. Fiz.-Mat. Fak. i Har'kov. Mat. Obšč 22 (1950) 95–113 (1951).

[441] I.S. Kac. Spectral multiplicity of a second-order differential operator and expansion in eigenfunctions. Izv. Akad Nauk. SSSR Ser. Mat 27 (1963) 1081–1112.

[442] I.S. Kac. Linear relations generated by canonical differential equations (Russian). Funktsional. Anal. i Prilozhen 17 (1983) 86-87.

[443] I.S. Kac. Linear relations, generated by a canonical differential equation on an interval with regular endpoints, and the expansibility in eigenfunctions (Russian). Deposited Paper, Odessa, 1984.

[444] I.S. Kac. Expansibility in eigenfunctions of a canonical differential equation on an interval with singular endpoints and associated linear relations (Russian). Deposited in Ukr NIINTI, No. 2111, 1986. (VINITI Deponirovannye Nauchnye Raboty, No. 12 (282), b.o. 1536, 1986).

[445] I.S. Kac. Linear relations, generated by a canonical differential equation and expansibility in eigenfunctions (Russian). Algebra i Analiz 14 (2002) 86–120 [St. Petersburg Math. J. 14 (2003) 429–452].

[446] I.S. Kac and M.G. Kreĭn. R-functions – analytic functions mapping the upper halfplane into itself. Supplement I to the Russian edition of F.V. Atkinson, *Discrete and continuous boundary problems* (Russian). Mir, 1968 [Amer. Math. Soc. Transl. (2) 103 (1974) 1–18].

[447] I.S. Kac and M.G. Kreĭn. On the spectral functions of the string. Supplement II to the Russian edition of F.V. Atkinson, *Discrete and continuous boundary problems* (Russian). Mir, 1968 [Amer. Math. Soc. Transl. (2) 103 (1974), 19–102].

[448] H. Kalf. A characterization of the Friedrichs extension of Sturm–Liouville operators. J. London Math. Soc. (2) 17 (1978) 511–521.

[449] H. Kalf and J. Walter. Strongly singular potentials and essential self-adjointness of singular elliptic operators in $C_0^\infty(\mathbb{R}^n \setminus \{0\})$. J. Functional Analysis 10 (1972) 114–130.

[450] M. Kaltenbäck. *Ausgewählte Kapitel Operatortheorie*. Lecture Notes, TU Wien.

[451] M. Kaltenbäck. *Funktionalanalysis 2*. Lecture Notes, TU Wien.

[452] M. Kaltenbäck, H. Winkler, and H. Woracek. Almost Pontryagin spaces. Oper. Theory Adv. Appl. 160 (2005) 253–271.

[453] M. Kaltenbäck, H. Winkler, and H. Woracek. Strings, dual strings, and related canonical systems. Math. Nachr. 280 (2007) 1518–1536.

[454] M. Kaltenbäck and H. Woracek. Pontryagin spaces of entire functions I. Integral Equations Operator Theory 33 (1999) 34–97.

[455] M. Kaltenbäck and H. Woracek. Pontryagin spaces of entire functions II. Integral Equations Operator Theory 33 (1999) 305–380.

[456] M. Kaltenbäck and H. Woracek. Pontryagin spaces of entire functions III. Acta Sci. Math. (Szeged) 69 (2003) 241–310.

[457] M. Kaltenbäck and H. Woracek. Pontryagin spaces of entire functions IV. Acta Sci. Math. (Szeged) 72 (2006) 709–835.

[458] M. Kaltenbäck and H. Woracek. Pontryagin spaces of entire functions V. Acta Sci. Math. (Szeged) 77 (2011) 223–336.

[459] M. Kaltenbäck and H. Woracek. Pontryagin spaces of entire functions VI. Acta Sci. Math. (Szeged) 76 (2010) 511–560.

[460] M. Kaltenbäck and H. Woracek. *Kanonische Systeme: Direkte Spektraltheorie.* Winter School, Reichenau, 2015.

[461] I.M. Karabash, A. Kostenko, and M.M. Malamud. The similarity problem for *J*-nonnegative Sturm–Liouville operators. J. Differential Equations 246 (2009) 964–997.

[462] T. Kato. *Perturbation theory for linear operators.* 1980 edition. Classics in Mathematics, Springer, 1995.

[463] S. Khan and D.B. Pearson. Subordinacy and spectral theory for infinite matrices. Helv. Phys. Acta 65 (1992) 505–527.

[464] Y. Kilpi. Über selbstadjungierte Fortsetzungen symmetrischer Transformationen im Hilbert-Raum. Ann. Acad. Sci. Fenn. Ser. A I 264 (1959) 23 pp.

[465] T. Kimura and M. Takahashi. Sur les opérateurs différentiels ordinaires linéaires formellement autoadjoints. I. Funkcial. Ekvac. 7 (1965) 35–90.

[466] A.N. Kochubeĭ. On extensions of symmetric operators and symmetric binary relations (Russian). Mat. Zametki 17 (1975) 41–48 [Math. Notes 17 (1975), 25–28].

[467] A.N. Kochubeĭ. Extensions of a positive definite symmetric operator (Russian). Dokl. Akad. Nauk Ukrain. SSR Ser. A 3 (1979) 168–171, 237 [Amer. Math. Soc. Transl. (2) 124 (1984) 139–142].

[468] A.N. Kochubeĭ. Characteristic functions of symmetric operators and their extensions (Russian). Izv. Akad. Nauk Armyan. SSR Ser. Mat. 15 (1980) 219–232 [Soviet J. Contemporary Math. Anal. 15 (1980)].

[469] K. Kodaira. The eigenvalue problem for ordinary differential equations of the second order and Heisenberg's theory of *S*-matrices. Amer. J. Math. 71 (1949) 921–945.

[470] K. Kodaira. On ordinary differential equations of any even order and the corresponding eigenfunction expansions. Amer. J. Math. 72 (1950) 501–544.

[471] V.I. Kogan and F.S. Rofe-Beketov. On square-integrable solutions of symmetric systems of differential equations of arbitrary order. Proc. Roy. Soc. Edinburgh 74A (1976) 5–40.

[472] V.Yu. Kolmanovich and M.M. Malamud. Extensions of sectorial operators and dual pairs of contractions (Russian). Deposited at VINITI No. 4428–85, Donetsk, 1985.

[473] Q. Kong, M. Möller, H. Wu, and A. Zettl. Indefinite Sturm–Liouville problems. Proc. Roy. Soc. Edinburgh 133A (2003) 639–652.

[474] Q. Kong, H. Wu, and A. Zettl. Singular left-definite Sturm–Liouville problems. J. Differential Equations 206 (2004) 1–29.

[475] N.D. Kopachevskiĭ and S.G. Kreĭn. Abstract Green formula for a triple of Hilbert spaces, abstract boundary-value and spectral problems (Russian). Ukr. Mat. Visn. 1 (2004) 69–97 [Ukr. Math. Bull. 1 (2004), 77–105].

[476] K. Koshmanenko. *Singular quadratic forms in perturbation theory.* Kluwer, 1999.

[477] A. Kostenko and M.M. Malamud. 1-D Schrödinger operators with local point interactions on a discrete set. J. Differential Equations 249 (2010) 253–304.

[478] A. Kostenko, M.M. Malamud, and D.D. Natyagajlo. The matrix Schrödinger operator with δ-interactions (Russian). Mat. Zametki 100 (2016) 59–77 [Math. Notes 100 (2016) 49–65].

[479] A. Kostenko, A. Sakhnovich, and G. Teschl. Weyl–Titchmarsh theory for Schrödinger operators with strongly singular potentials. Int. Math. Res. Not. IMRN 2012 (8) 1699–1747.

[480] A. Kostenko, A. Sakhnovich, and G. Teschl. Commutation methods for Schrödinger operators with strongly singular potentials. Math. Nachr. 285 (2012) 392–410.

[481] A. Kostenko and G. Teschl. On the singular Weyl–Titchmarsh function of perturbed spherical Schrödinger operators. J. Differential Equations 250 (2011) 3701–3739.

[482] A. Kostenko and G. Teschl. Spectral asymptotics for perturbed spherical Schrödinger operators and applications to quantum scattering. Comm. Math. Phys. 322 (2013) 255–275.

[483] Yu.G. Kovalev. 1-D nonnegative Schrödinger operators with point interactions. Mat. Stud. 39 (2013) 150–163.

[484] A.M. Krall. Differential-boundary operators. Trans. Amer. Math. Soc. 154 (1971) 429–458.

[485] A.M. Krall. The development of general differential and general differential boundary systems. Rocky Mountain J. Math. 5 (1975) 493–542.

[486] A.M. Krall. *n*th Order Stieltjes differential boundary operators and Stieltjes differential boundary systems. J. Differential Equations 24 (1977) 253–267.

[487] A.M. Krall. $M(\lambda)$-theory for singular Hamiltonian systems with one singular endpoint. SIAM J. Math. Anal. 20 (1989) 664–700.

[488] A.M. Krall. Left-definite regular Hamiltonian systems. Math. Nachr. 174 (1995) 203–217.

[489] A.M. Krall. *Hilbert space, boundary value problems and orthogonal polynomials.* Oper. Theory Adv. Appl., Vol. 133, Birkhäuser, 2002.

[490] M.A. Krasnoselskiĭ. On selfadjoint extensions of Hermitian operators. Ukr. Mat. Zh. 1 no. 1 (1949) 21–38.

[491] M.G. Kreĭn. On Hermitian operators whose deficiency indices are 1. Dokl. Akad. Nauk URSS 43 (1944) 323–326.

[492] M.G. Kreĭn. Concerning the resolvents of an Hermitian operator with the deficiency-index (m, m). Dokl. Akad. Nauk URSS 52 (1946) 651–654.

[493] M.G. Kreĭn. The theory of selfadjoint extensions of semibounded Hermitian operators and its applications I. Mat. Sb. 20 (62) (1947) 431–495.

[494] M.G. Kreĭn. The theory of selfadjoint extensions of semibounded Hermitian operators and its applications II. Mat. Sb. 21 (63) (1947) 366–404.

[495] M.G. Kreĭn. On Hermitian operators with directing functionals. Sbirnik Praz Inst. Math. AN URSS 10 (1948) 83–106.

[496] M.G. Kreĭn. The fundamental propositions of the theory of representations of Hermitian operators with deficiency index (m, m) (Russian). Ukr. Mat. Zh. 1 no. 2 (1949) 3–66.

[497] M.G. Kreĭn and H. Langer. On defect subspaces and generalized resolvents of Hermitian operator in Pontryagin space. Funct. Anal. Appl. 5 (1971) 136–146; 217–228.

[498] M.G. Kreĭn and H. Langer. Über die verallgemeinerten Resolventen und die charakteristische Funktion eines isometrischen Operators in Raume Π_κ. In: *Colloquia Math. Soc. Janos Bolyai* 5. North Holland, 1972, pp. 353–399.

[499] M.G. Kreĭn and H. Langer. Über die Q-Funktion eines π-hermiteschen Operators im Raume Π_κ. Acta Sci. Math. (Szeged) 34 (1973) 191–230.

[500] M.G. Kreĭn and H. Langer. Über einige Fortsetzungsprobleme, die eng mit der Theorie hermitescher Operatoren im Raume Π_κ zusammenhängen. I. Einige Funktionenklassen und ihre Darstellungen. Math. Nachr. 77 (1977) 187–236.

[501] M.G. Kreĭn and H. Langer. Über einige Fortsetzungsprobleme, die eng mit der Theorie hermitescher Operatoren im Raume Π_κ zusammenhängen. II. Verallgemeinerte Resolventen, u-Resolventen und ganze Operatoren. J. Functional Analysis 30 (1978) 390–447.

[502] M.G. Kreĭn and H. Langer. On some extension problems which are closely connected with the theory of Hermitian operators in a space Π_κ. III. Indefinite analogues of the Hamburger and Stieltjes moment problems. Part (I). Beiträge Analysis 14 (1979) 25–40. Part (II). Beiträge Analysis 15 (1980) 27–45.

[503] M.G. Kreĭn and H. Langer. On some continuation problems which are closely related to the theory of operators in spaces Π_κ. IV. Continuous analogues of orthogonal polynomials on the unit circle with respect to an indefinite weight and related continuation problems for some classes of functions. J. Operator Theory 13 (1985) 299–417.

[504] M.G. Kreĭn and H. Langer. Continuation of Hermitian positive definite functions and related questions. Integral Equations Operator Theory 78 (2014) 1–69.

[505] M.G. Kreĭn and A.A. Nudelman. *Markov moment problem and extremal problems.* Transl. Mathematical Monographs, Vol. 50, Amer. Math. Soc., 1977.

[506] M.G. Kreĭn and Yu.L. Shmuljan. On linear fractional transformations with operator coefficients (Russian). Mat. Issled 2 (1967) 64–96 [Amer. Math. Soc. Transl. (2) 103 (1974) 125–152].

[507] P. Kurasov and A. Luger. An operator theoretic interpretation of the generalized Titchmarsh–Weyl coefficient for a singular Sturm–Liouville problem. Math. Phys. Anal. Geom. 14 (2011) 115–151.

[508] T. Kusche, R. Mennicken, and M. Möller. Friedrichs extension and essential spectrum of systems of differential operators of mixed order. Math. Nachr. 278 (2005) 1591–1606.

[509] A. Kuzhel and S. Kuzhel. *Regular extensions of Hermitian operators.* Kluwer, 1998.

[510] S. Kuzhel. On some properties of unperturbed operators. Methods Funct. Anal. Topology 2 (1996) 78–84.

[511] S. Kuzhel and M. Znojil. Non-self-adjoint Schrödinger operators with nonlocal one-point interactions. Banach J. Math. Anal. 11 (2017) 923–944.

[512] J.Ph. Labrousse. Geodesics in the space of linear relations in a Hilbert space. 18th Oper. Theory Conf. Proc. (2000) 213–234.

[513] J.Ph. Labrousse. Idempotent linear relations. In: *Spectral theory and its applications.* The Theta Foundation, 2003, pp. 121–141.

[514] J.Ph. Labrousse. Bisectors, isometries and connected components in Hilbert spaces. Oper. Theory Adv. Appl. 198 (2009) 239–258.

[515] J.Ph. Labrousse, A. Sandovici, H.S.V. de Snoo, and H. Winkler. The Kato decomposition of quasi-Fredholm relations. Operators Matrices 4 (2010) 1–51.

[516] J.Ph. Labrousse, A. Sandovici, H.S.V. de Snoo, and H. Winkler. Closed linear relations and their regular points. Operators Matrices 6 (2012) 681–714.

[517] O.A. Ladyzhenskaya and N.N. Uraltseva. *Linear and quasilinear elliptic equations.* Mathematics in Science and Engineering, Vol. 46, Academic Press, 1968.

[518] H. Langer. Über die Methode der richtenden Funktionale von M.G. Kreĭn. Acta Math. Acad. Sci. Hung. 21 (1970) 207–224.

[519] H. Langer. Spectral functions of definitizable operators in Krein spaces. Springer Lecture Notes Math. 948 (1982) 1–46.

[520] H. Langer and R. Mennicken. A transformation of right-definite S-hermitian systems to canonical systems. Differential Integral Equations 3 (1990) 901–908.

[521] H. Langer and M. Möller. Linearization of boundary eigenvalue problems. Integral Equations Operator Theory 14 (1991) 105–119.

[522] H. Langer and P. Sorjonen. Verallgemeinerte Resolventen hermitischer und isometrischer Operatoren im Pontryaginraum. Ann. Acad. Sci. Fenn. Ser. A I561 (1974) 45 pp.

[523] H. Langer and B. Textorius. On generalized resolvents and Q-functions of symmetric linear relations (subspaces) in Hilbert space. Pacific J. Math. 72 (1977) 135–165.

[524] H. Langer and B. Textorius. L-resolvent matrices of symmetric linear relations with equal defect numbers; applications to canonical differential relations. Integral Equations Operator Theory 5 (1982) 208–243.

[525] H. Langer and B. Textorius. Generalized resolvents of dual pairs of contractions. Oper. Theory Adv. Appl. 6 (1982) 103–118.

[526] H. Langer and B. Textorius. Spectral functions of a symmetric linear relation with a directing mapping, I. Proc. Roy. Soc. Edinburgh 97A (1984) 165–176.

[527] H. Langer and B. Textorius. Spectral functions of a symmetric linear relation with a directing mapping, II. Proc. Roy. Soc. Edinburgh 101A (1985) 111–124.

[528] H. Langer and H. Winkler. Direct and inverse spectral problems for generalized strings. Integral Equations Operator Theory 30 (1998) 409–431.

[529] M. Langer. A general HELP inequality connected with symmetric operators. Proc. Roy. Soc. London 462A (2006) 587–606.

[530] M. Langer and A. Luger. Scalar generalized Nevanlinna functions: realizations with block operator matrices. Oper. Theory Adv. Appl. 162 (2006) 253–276.

[531] M. Langer and H. Woracek. Dependence of the Weyl coefficient on singular interface conditions. Proc. Edinburgh Math. Soc. 52 (2009) 445–487.

[532] M. Langer and H. Woracek. A local inverse spectral theorem for Hamiltonian systems. Inverse Problems 27 (2011), no. 5, 055002, 17 pp.

[533] Y. Last. Quantum dynamics and decompositions of singular continuous spectra. J. Functional Analysis 142 (1996) 406–445.

[534] P.D. Lax. On Cauchy's problem for hyperbolic equations and the differentiability of solutions of elliptic equations. Comm. Pure Appl. Math. 8 (1955) 615–633.

[535] W. Leighton and M. Morse. Singular quadratic functions. Trans. Amer. Math. Soc. 40 (1936) 252–288.

[536] D. Lenz, C. Schubert, and I. Veselić. Unbounded quantum graphs with unbounded boundary conditions. Math. Nachr. 287 (2014) 962–979.

[537] J. Leray. *Lectures on hyperbolic equations with variable coefficients.* Institute for Advanced Studies, Princeton, 1952.

[538] M. Lesch and M.M. Malamud. On the deficiency indices and self-adjointness of symmetric Hamiltonian systems. J. Differential Equations 189 (2003) 556–615.

[539] N. Levinson. A simplified proof of the expansion theorem for singular second order linear differential equations. Duke Math. J. 18 (1951) 57–71.

[540] N. Levinson. The expansion theorem for singular self-adjoint differential operators. Ann. Math. (2) 59 (1954) 300–315.

[541] B.M. Levitan. Proof of the theorem on the expansion in the eigenfunctions of selfadjoint differential equations. Dokl. Akad. Nauk SSSR 73 (1950) 651–654.

[542] B.M. Levitan and I.S. Sargsjan. *Sturm–Liouville and Dirac operators.* Kluwer, 1991.

[543] M. Levitin and L. Parnovski. On the principal eigenvalue of a Robin problem with a large parameter. Math. Nachr. 281 (2008) 272–281.

[544] J.L. Lions and E. Magenes. *Non-homogeneous boundary value problems and applications, Vol. 1.* Springer, 1972.

[545] L. Littlejohn and R. Wellman. On the spectra of left-definite operators. Complex Anal. Oper. Theory 7 (2011) 437–455.

[546] A. Luger. A factorization of regular generalized Nevanlinna functions. Integral Equations Operator Theory 43 (2002) 326–345.

[547] A. Luger. Generalized poles and zeros of generalized Nevanlinna functions, a survey. In: *Proc. algorithmic information theory conference.* Vaasan Yliop. Julk. Selvityksiä Rap., Vol. 124, Vaasan Yliopisto, 2005, pp. 85–94.

[548] A. Luger. A characterization of generalized poles of generalized Nevanlinna functions. Math. Nachr. 279 (2006) 891–910.

[549] A. Luger and C. Neuner. An operator theoretic interpretation of the generalized Titchmarsh–Weyl function for perturbed spherical Schrödinger operators. Complex Anal. Oper. Theory 9 (2015) 1391–1410.

[550] A. Luger and C. Neuner. On the Weyl solution of the 1-dim Schrödinger operator with inverse fourth power potential. Monatsh. Math. 180 (2016) 295–303.

[551] A. Lunyov. Spectral functions of the simplest even order ordinary differential operator. Methods Funct. Anal. Topology 19 (2013) 319–326.

[552] V.E. Lyantse and O.G. Storozh. *Methods of the theory of unbounded operators.* Naukova Dumka, Kiev, 1983.

[553] M.M. Malamud. Some classes of extensions of a Hermitian operator with gaps (Russian). Ukr. Mat. Zh. 44 (1992) 215–233 [Ukrainian Math. J. 44 (1992) 190–204].

[554] M.M. Malamud. On a formula for the generalized resolvents of a non-densely defined Hermitian operator (Russian). Ukr. Mat. Zh. 44 (1992) 1658–1688 [Ukrainian Math. J. 44 (1992) 1522–1547].

[555] M.M. Malamud. On some classes of extensions of sectorial operators and dual pairs of contractions. Oper. Theory Adv. Appl. 124 (2001) 401–449.

[556] M.M. Malamud. Operator holes and extensions of sectorial operators and dual pairs of contractions. Math. Nachr. 279 (2006) 625–655.

[557] M.M. Malamud. Spectral theory of elliptic operators in exterior domains. Russ. J. Math. Phys. 17 (2010) 96–125.

[558] M.M. Malamud and S.M. Malamud. Spectral theory of operator measures in a Hilbert space (Russian). Algebra i Analiz 15 (2003) 1–77 [St. Petersburg Math. J. 15 (2004) 323–373].

[559] M.M. Malamud and V.I. Mogilevskiĭ. On Weyl functions and Q-functions of dual pairs of linear relations. Dopov. Nats. Akad. Nauk Ukr. Mat. Prirodozn. Tekh. Nauki 1999, no. 4, 32–37.

[560] M.M. Malamud and V.I. Mogilevskiĭ. Krein type formula for canonical resolvents of dual pairs of linear relations. Methods Funct. Anal. Topology 8 (2002) 72–100.

[561] M.M. Malamud and V.I. Mogilevskiĭ. Generalized resolvents of an isometric operator (Russian). Mat. Zametki 73 (2003) 460–465 [Math. Notes 73 (2003) 429–435].

[562] M.M. Malamud and H. Neidhardt. On the unitary equivalence of absolutely continuous parts of self-adjoint extensions. J. Functional Analysis 260 (2011) 613–638.

[563] M.M. Malamud and H. Neidhardt. Perturbation determinants for singular perturbations. Russ. J. Math. Phys. 21 (2014) 55–98.

[564] M.M. Malamud and H. Neidhardt. Trace formulas for additive and non-additive perturbations. Adv. Math. 274 (2015) 736–832.

[565] M.M. Malamud and K. Schmüdgen. Spectral theory of Schrödinger operators with infinitely many point interactions and radial positive definite functions. J. Functional Analysis 263 (2012) 3144–3194.

[566] J. Malinen and O. J. Staffans. Impedance passive and conservative boundary control systems. Complex Anal. Oper. Theory 1 (2007) 279–300.

[567] A. Mantile, A. Posilicano, and M. Sini. Self-adjoint elliptic operators with boundary conditions on not closed hypersurfaces. J. Differential Equations 261 (2016) 1–55.

[568] M. Marletta. Eigenvalue problems on exterior domains and Dirichlet to Neumann maps. J. Comput. Appl. Math. 171 (2004) 367–391.

[569] V.G. Mazya. *Sobolev spaces with applications to elliptic partial differential equations.* Second, revised and augmented edition. Grundlehren der Mathematischen Wissenschaften, Vol. 342, Springer, 2011.

[570] A.G.R. McIntosh. Hermitian bilinear forms which are not semibounded. Bull. Amer. Math. Soc. 76 (1970) 732–737.

[571] A.G.R. McIntosh. Bilinear forms in Hilbert space. J. Math. Mech. 19 (1970) 1027–1045.

[572] R. McKelvey. Spectral measures, generalized resolvents, and functions of positive type. J. Math. Anal. Appl. 11 (1965) 447–477.

[573] W. McLean. *Strongly elliptic systems and boundary integral equations.* Cambridge University Press, 2000.

[574] R. Mennicken and M. Möller. *Non-self-adjoint boundary eigenvalue problems.* North-Holland Mathematics Studies, Vol. 192, North-Holland Publishing Co., 2003.

[575] R. Mennicken and H.-D. Niessen. $(S, \tilde{S})$-adjungierte Randwertoperatoren. Math. Nachr. 44 (1970) 259–284.

[576] Y. Mezroui. Projection orthogonale sur le graphe d'une relation linéaire fermée. Trans. Amer. Math. Soc. 352 (2000) 2789–2800.

[577] V.A. Mikhailets. Spectra of operators and boundary value problems (Russian). In: *Spectral analysis of differential operators.* Akad. Nauk Ukrain. SSR, Inst. Mat., Kiev, 1980, pp. 106–131, 134.

[578] V.A. Mikhailets. Distribution of the eigenvalues of finite multiplicity of Neumann extensions of an elliptic operator (Russian). Differentsial'nye Uravneniya 30 (1994) 178–179 [Differential Equations 30 (1994) 167–168].

[579] V.A. Mikhailets. Discrete spectrum of the extreme nonnegative extension of the positive elliptic differential operator. In: *Proceedings of the Ukrainian Mathematical Congress 2001, Section 7, Nonlinear Analysis.* Kyiv, 2006, pp. 80–94.

[580] O.Ya. Milyo and O.G. Storozh. On general form of maximal accretive extension of positive definite operator (Russian). Dokl. Akad. Nauk Ukr. SSR 6 (1991) 19–22.

[581] O.Ya. Milyo and O.G. Storozh. Maximal accretive extensions of positive definite operator with finite defect number (Russian). Lviv University, 31 pages, Deposited in GNTB of Ukraine 28.10.93, no 2139 Uk93.

[582] O.Ya. Milyo and O.G. Storozh. Maximal accretiveness and maximal nonnegativeness conditions for a class of finite-dimensional perturbations of a positive definite operator. Mat. Studii 12 (1999) 90–100.

[583] R. von Mises. Beitrag zum Oszillationsproblem. *Festschrift Heinrich Weber zu seinem siebzigsten Geburtstag am 5. März 1912 gewidmet von Freunden und Schülern.* B.G. Teubner, 1912, pp. 252-282.

[584] D. Mitrea. *Distributions, partial differential equations, and harmonic analysis.* Universitext, Springer, 2013.

[585] M. Mitrea and M. Taylor. Boundary layer methods for Lipschitz domains in Riemannian manifolds. J. Functional Analysis 163 (1999) 181–251.

[586] M. Mitrea and M. Taylor. Potential theory on Lipschitz domains in Riemannian manifolds: Sobolev–Besov space results and the Poisson problem. J. Functional Analysis 176 (2000) 1–79.

[587] M. Mitrea and M. Taylor. Potential theory on Lipschitz domains in Riemannian manifolds: Hölder continuous metric tensors. Comm. Partial Differential Equations 25 (2000) 1487–1536.

[588] M. Mitrea and M. Taylor. Potential theory on Lipschitz domains in Riemannian manifolds: L^p, Hardy and Hölder type estimates. Commun. Anal. Geometry 9 (2001) 369–421.

[589] V.I. Mogilevskiĭ. Boundary triplets and Krein type resolvent formula for symmetric operators with unequal defect numbers. Methods Funct. Anal. Topology 12 (2006) 258–280.

[590] V.I. Mogilevskiĭ. Boundary triplets and Titchmarsh–Weyl functions of differential operators with arbitrary deficiency indices. Methods Funct. Anal. Topology 15 (2009) 280–300.

[591] V.I. Mogilevskiĭ. Minimal spectral functions of an ordinary differential operator. Proc. Edinburgh Math. Soc. 55 (2012) 731–769.

[592] V.I. Mogilevskiĭ. Boundary pairs and boundary conditions for general (not necessarily definite) first-order symmetric systems with arbitrary deficiency indices. Math. Nachr. 285 (2012) 1895–1931.

[593] V.I. Mogilevskiĭ. On eigenfunction expansions of first-order symmetric systems and ordinary differential operators of an odd order. Integral Equations Operator Theory 82 (2015) 301–337.

[594] V.I. Mogilevskiĭ. Spectral and pseudospectral functions of Hamiltonian systems: development of the results by Arov–Dym and Sakhnovich. Methods Funct. Anal. Topology 21 (2015) 370–402.

[595] V.I. Mogilevskiĭ. Symmetric extensions of symmetric linear relations (operators) preserving the multivalued part. Methods Funct. Anal. Topology 24 (2018) 152–177.

[596] V.I. Mogilevskiĭ. On compressions of self-adjoint extensions of a symmetric linear relation. Integral Equations Operator Theory 91 (2019) 9 30 pp.

[597] M. Möller. On the unboundedness below of the Sturm–Liouville operator. Proc. Roy. Soc. Edinburgh 129A (1999) 1011–1015.

[598] M. Möller and V. Pivovarchik. *Spectral theory of operator pencils, Hermite–Biehler functions, and their applications.* Oper. Theory Adv. Appl., Vol. 246, Birkhäuser, 2015.

[599] M. Möller and F.H. Szafraniec. Adjoints and formal adjoints of matrices of unbounded operators. Proc. Amer. Math. Soc. 136 (2008) 2165–2176.

[600] M. Möller and A. Zettl. Semi-boundedness of ordinary differential operators. J. Differential Equations 115 (1995) 24–49.

[601] M. Möller and A. Zettl. Symmetric differential operators and their Friedrichs extension. J. Differential Equations 115 (1995) 50–69.

[602] S.N. Naboko. Nontangential boundary values of operator R-functions in a half-plane (Russian). Algebra i Analiz 1 (1989) 197–222 [Leningrad Math. J. 1 (1990) 1255–1278].

[603] B. Sz.-Nagy, C. Foias, H. Bercovici, and L. Kérchy. *Harmonic analysis of operators on Hilbert space*. 2nd edition. Universitext, Springer, 2010 [enlarged and revised edition of 1st edition by B. Sz.-Nagy and C. Foias].

[604] B. Sz.-Nagy and A. Koranyi. Operatortheoretische Behandlung und Verallgemeinerung eines Problemkreises in der komplexen Funktionentheorie. Acta Math. 100 (1958) 171–202.

[605] M.A. Naĭmark. Self-adjoint extensions of the second kind of a symmetric operator. Bull. Acad. Sci. URSS Ser. Math. 4 (1940) 53–104.

[606] M.A. Naĭmark. Spectral functions of a symmetric operator. Izv. Akad. Nauk SSSR. Ser. Mat. 4 (1940) 277–318.

[607] M.A. Naĭmark. On spectral functions of a symmetric operator. Izv. Akad. Nauk SSSR. Ser. Mat. 7 (1943) 285–296.

[608] M.A. Naĭmark. *Linear differential operators II*. Ungar Pub. Co., 1967.

[609] M.Z. Nashed (editor). *General inverses and applications*. Academic Press, 1976.

[610] J. von Neumann. Allgemeine Eigenwerttheorie Hermitescher Funktionaloperatoren. Math. Ann. 102 (1930) 49–131.

[611] J. von Neumann. Über adjungierte Operatoren. Ann. Math. 33 (1932) 294–310.

[612] H.-D. Niessen. Singuläre S-hermitesche Rand-Eigenwertprobleme. Manuscripta Math. 3 (1970) 35–68.

[613] H.-D. Niessen. Zum verallgemeinerten zweiten Weylschen Satz. Arch. Math. 22 (1971) 648–656.

[614] H.-D. Niessen. Greensche Matrix und die Formel von Titchmarsh–Kodaira für singuläre S-hermitesche Eigenwertprobleme. J. Reine Angew. Math. 261 (1973) 164–193.

[615] H.-D. Niessen and A. Zettl. The Friedrichs extension of regular ordinary differential operators. Proc. Roy. Soc. Edinburgh 114A (1990) 229–236.

[616] H.-D. Niessen and A. Zettl. Singular Sturm–Liouville problems: The Friedrichs extension and comparison of eigenvalues. Proc. London Math. Soc. 64 (1992) 545–578.

[617] A.A. Nudelman. On a new problem of moment problem type (Russian) Dokl. Akad. Nauk SSSR 233 (1977) 792–795 [Soviet Math. Dokl. 18 (1977) 507–510].

[618] E.-M. Ouhabaz. Second order elliptic operators with essential spectrum $[0, \infty)$ on L^p. Comm. Partial Differential Equations 20 (1995) 763–773.

[619] B.C. Orcutt. *Canonical differential equations*. Dissertation, University of Virginia, 1969.

[620] S.A. Orlov. Nested matrix disks depending analytically on a parameter, and the theorem on invariance of the ranks of the radii of limiting matrix disks. Izv. Akad. Nauk SSSR Ser. Mat 40 (1967) 593–644.

[621] S. Ôta. Decomposition of unbounded derivations in operator algebras. Tôhoku Math. J. 33 (1981) 215–225.

[622] S. Ôta. Closed linear operators with domain containing their range. Proc. Edinburgh Math. Soc. 27 (1984) 229–233.

[623] S. Ôta. On a singular part of an unbounded operator. Zeitschrift für Analysis und ihre Anwendungen 7 (1987) 15–18.

[624] K. Pankrashkin. Resolvents of self-adjoint extensions with mixed boundary conditions. Rep. Math. Phys. 58 (2006) 207–221.

[625] K. Pankrashkin. Spectra of Schrödinger operators on equilateral quantum graphs. Lett. Math. Phys. 77 (2006) 139–154.

[626] K. Pankrashkin. Unitary dimension reduction for a class of self-adjoint extensions with applications to graph-like structures. J. Math. Anal. Appl. 396 (2012) 640–655.

[627] K. Pankrashkin. An example of unitary equivalence between self-adjoint extensions and their parameters. J. Functional Analysis 265 (2013) 2910–2936.

[628] K. Pankrashkin and N. Popoff. Mean curvature bounds and eigenvalues of Robin Laplacians. Calculus Variations Partial Differential Equations 54 (2015) 1947–1961.

[629] K. Pankrashkin and N. Popoff. An effective Hamiltonian for the eigenvalue asymptotics of the Robin Laplacian with a large parameter. J. Math. Pures Appl. 106 (2016) 615–650.

[630] B.S. Pavlov. The theory of extensions and explicitly-solvable models. Russian Math. Surveys 42 (1987) 127–168.

[631] R.S. Phillips. Dissipative operators and hyperbolic systems of partial differential equations. Trans. Amer. Math. Soc. (1959) 193–254.

[632] R.S. Phillips. Dissipative operators and parabolic partial diferential equations. Commun. Pure Appl. Math. 12 (1959) 249–276.

[633] R.S. Phillips. The extension of dual subspaces invariant under an algebra. In: Proceedings international symposium linear algebra, Israel, 1960. Academic Press, 1961, pp. 366–398.

[634] R.S. Phillips. On dissipative operators. In: Lecture series in differential equations, Vol. 2. Van Nostrand, 1969, pp. 65–113.

[635] R. Picard, S. Seidler, S. Trostorff, and M. Waurick. On abstract grad-div systems. J. Differential Equations 260 (2016) 4888–4917.

[636] H.M. Pipa and O.G. Storozh. Accretive perturbations of some proper extensions of a positive definite operator. Mat. Studii 25 (2006) 181–190.

[637] V. Pivovarchik and H. Woracek. Sums of Nevanlinna functions and differential equations on star-shaped graphs. Operators Matrices 3 (2009) 451–501.

[638] A. Pleijel. Spectral theory for pairs of formally self-adjoint ordinary differential operators. J. Indian Math. Soc. 34 (1970) 259–268.

[639] D. Popovici and Z. Sebestyén. Factorizations of linear relations. Adv. Math. 233 (2013) 40–55.

[640] A. Posilicano. A Krein-like formula for singular perturbations of selfadjoint operators and applications. J. Functional Analysis 183 (2001) 109–147.

[641] A. Posilicano. Boundary triples and Weyl functions for singular perturbations of self-adjoint operators. Methods Funct. Anal. Topology 10 (2004) 57–63.

[642] A. Posilicano. Self-adjoint extensions of restrictions. Operators Matrices 2 (2008) 483–506.

[643] A. Posilicano and L. Raimondi. Krein's resolvent formula for self-adjoint extensions of symmetric second-order elliptic differential operators. J. Phys. A 42 (2009) 015204, 11 pp.

[644] O. Post. First-order operators and boundary triples. Russ. J. Math. Phys. 14 (2007), 482–492.

[645] O. Post. Equilateral quantum graphs and boundary triples. In: *Analysis on graphs and its applications*. Proc. Sympos. Pure Math., Vol. 77, Amer. Math. Soc., 2008, pp. 469–490.

[646] O. Post. Spectral analysis on graph-like spaces. Lecture Notes in Mathematics, Vol. 2039, Springer, 2012.

[647] O. Post. Boundary pairs associated with quadratic forms. Math. Nachr. 289 (2016) 1052–1099.

[648] V. Prokaj and Z. Sebestyén. On Friedrichs extensions of operators. Acta Sci. Math. 62 (1996) 243–246.

[649] M.C. Reed and B. Simon. *Methods of modern mathematical physics I. Functional analysis*. Academic Press, 1972.

[650] M.C. Reed and B. Simon. *Methods of modern mathematical physics II. Fourier analysis, self-adjointness*. Academic Press, 1975.

[651] M.C. Reed and B. Simon. *Methods of modern mathematical physics III. Scattering theory*. Academic Press, 1978.

[652] M.C. Reed and B. Simon. *Methods of modern mathematical physics IV. Analysis of operators*. Academic Press, 1977.

[653] W.T. Reid. *Ordinary differential equations*. John Wiley and Sons, 1971.

[654] F. Rellich. Halbbeschränkte gewöhnliche Differentialoperatoren zweiter Ordnung. Math. Ann. 122 (1950) 343–368.

[655] C. Remling. Relationships between the m-function and subordinate solutions of second order differential operators. J. Math. Anal. Appl. 206 (1997) 352–363.

[656] C. Remling. Spectral analysis of higher order differential operators I. General properties of the m-function. J. London Math. Soc. 58 (1998) 367–380.

[657] C. Remling. Schrödinger operators and de Branges spaces. J. Functional Analysis 196 (2002) 323–394.

[658] C. Remling. *Spectral theory of canonical systems*. Studies in Mathematics, Vol. 70, de Gruyter, 2018.

[659] F. Riesz and B. Sz.-Nagy. *Leçons d'analyse fonctionelle.* 6th edition. Gauthier-Villars, 1972.

[660] F.S. Rofe-Beketov. On selfadjoint extensions of differential operators in a space of vector-functions (Russian). Teor. Funktsii Funktsional. Anal. i Prilozhen 8 (1969) 3–24.

[661] F.S. Rofe-Beketov. Perturbations and Friedrichs extensions of semibounded operators on variable domains. Soviet Math. Dokl. 22 (1980) 795–799.

[662] F.S. Rofe-Beketov. The numerical range of a linear relation and maximum relations (Russian). Teor. Funktsii Funktsional. Anal. i Prilozhen 44 (1985) 103–112 [J. Math. Sciences 48 (1990) 329–336].

[663] F.S. Rofe-Beketov and A.M. Kholkin. *Spectral analysis of differential operators.* World Scientific Monograph Series in Math., Vol. 7, World Scientific, 2005.

[664] H. Röh. Self-adjoint subspace extensions satisfying λ-linear boundary conditions. Proc. Roy. Soc. Edinburgh 90A (1981) 107–124.

[665] J. Rohleder. Strict inequality of Robin eigenvalues for elliptic differential operators on Lipschitz domains. J. Math. Anal. Appl. 418 (2014) 978–984.

[666] R. Romanov. *Canonical systems and de Branges spaces.* arXiv:1408.6022.

[667] B.W. Roos and W.C. Sangren. Spectra for a pair of singular first order differential equations. Proc. Amer. Math. Soc. 12 (1961) 468–476.

[668] B.W. Roos and W.C. Sangren. Three spectral theorems for a pair of singular first order differential equations. Pacific J. Math. 12 (1962) 1047–1055.

[669] B.W. Roos and W.C. Sangren. Spectral theory of Dirac's radial relativistic wave equation. J. Math. Phys. 3 (1962) 882–890.

[670] M. Rosenberg. The square-integrability of matrix-valued functions with respect to a non-negative Hermitian measure. Duke Math. J. 31 (1964) 291–298.

[671] R. Rosenberger. *Charakterisierungen der Friedrichsfortsetzung von halbbeschränkten Sturm–Liouville Operatoren.* Doctoral dissertation, Technische Hochschule Darmstadt, 1984.

[672] R. Rosenberger. A new characterization of the Friedrichs extension of semibounded Sturm–Liouville operators. J. London Math. Soc. 31 (1985) 501–510.

[673] M. Rosenblum and J. Rovnyak. *Hardy classes and operator theory.* Oxford University Press, 1985.

[674] J. Rovnyak. Methods of Kreĭn space operator theory. Oper. Theory Adv. Appl. 134 (2002) 31–66.

[675] E.M. Russakovskiĭ. The matrix Sturm–Liouville problem with spectral parameter in the boundary condition. Algebraic and operator aspects. Trans. Moscow Math. Soc. 57 (1996) 159–184.

[676] W. Rudin. *Real and complex analysis.* 3rd edition. McGraw-Hill, 1986.

[677] V. Ryzhov. A general boundary value problem and its Weyl function. Opuscula Math. 27 (2007) 305–331.

[678] V. Ryzhov. Weyl–Titchmarsh function of an abstract boundary value problem, operator colligations, and linear systems with boundary control. Complex Anal. Oper. Theory 3 (2009) 289–322.

[679] Sh.N. Saakyan. Theory of resolvents of a symmetric operator with infinite defect numbers (Russian). Akad. Nauk Armjan. SSR Dokl. 41 (1965) 193–198.

[680] S. Saitoh. *Theory of reproducing kernels and its applications*. Longman Scientific and Technical, 1988.

[681] L.A. Sakhnovich. *Spectral theory of canonical differential systems. Method of operator identities*. Oper. Theory Adv. Appl., Vol. 107, Birkhäuser, 1999.

[682] S. Saks. *Theory of the integral*. 2nd revised edition. Dover Publications, 1964.

[683] A. Sandovici. Von Neumann's theorem for linear relations. Linear Multilinear Algebra 66 (2018) 1750–1756.

[684] A. Sandovici and Z. Sebestyén. On operator factorization of linear relations. Positivity 17 (2013) 1115–1122.

[685] A. Sandovici, H.S.V. de Snoo, and H. Winkler. The structure of linear relations in Euclidean spaces. Lin. Alg. Appl. 397 (2005) 141–169.

[686] A. Sandovici, H.S.V. de Snoo, and H. Winkler. Ascent, descent, nullity, and defect for linear relations in linear spaces. Lin. Alg. Appl. 423 (2007) 456–497.

[687] F.W. Schäfke and A. Schneider. S-hermitesche Rand-Eigenwertprobleme I. Math. Ann. 162 (1965) 9–26.

[688] F.W. Schäfke and A. Schneider. S-hermitesche Rand-Eigenwertprobleme II. Math. Ann. 165 (1966) 236–260.

[689] F.W. Schäfke and A. Schneider. S-hermitesche Rand-Eigenwertprobleme III. Math. Ann. 177 (1968) 67–94.

[690] M. Schechter. *Principles of functional analysis*. 2nd edition. Graduate Studies in Mathematics, Vol. 36, Amer. Math. Soc., 2001.

[691] K. Schmüdgen. *Unbounded self-adjoint operators on Hilbert space*. Graduate Texts in Mathematics, Vol. 265, Springer, 2012.

[692] A. Schneider. Untersuchungen über singuläre reelle S-hermitesche Differentialgleichungssysteme im Normalfall. Math. Z. 107 (1968) 271–296.

[693] A. Schneider. Zur Einordnung selbstadjungierter Rand-Eigenwertprobleme bei gewöhnlichen Differentialgleichungen in die Theorie S-hermitescher Rand-Eigenwertprobleme. Math. Ann. 178 (1968) 277–294.

[694] A. Schneider. Weitere Untersuchungen über singuläre reelle S-hermitesche Differentialgleichungssysteme im Normalfall. Math. Z. 109 (1969) 153–168.

[695] A. Schneider. Geometrische Bedeutung eines Satzes vom Weylschen Typ für S-hermitesche Differentialgleichungssysteme im Normalfall. Arch. Math. 20 (1969) 147–154.

[696] A. Schneider. Die Greensche Matrix S-hermitescher Rand-Eigenwertprobleme im Normalfall. Math. Ann. 180 (1969) 307–312.

[697] A. Schneider. On spectral theory for the linear selfadjoint equation $Fy = \lambda Gy$. Springer Lecture Notes Math. 846 (1981) 306–322.

[698] A. Schneider and H.-D. Niessen. Linksdefinite singuläre kanonische Eigenwertprobleme I. J. Reine Angew. Math. 281 (1976) 13–52.

[699] A. Schneider and H.-D. Niessen. Linksdefinite singuläre kanonische Eigenwertprobleme II. J. Reine Angew. Math. 289 (1977) 62–84.

[700] D.B. Sears. Integral transforms and eigenfunction theory. Quart. J. Math. (Oxford) 5 (1954) 47–58.

[701] D.B. Sears. Integral transforms and eigenfunction theory II. Quart. J. Math. (Oxford) 6 (1955) 213–217.

[702] Z. Sebestyén and E. Sikolya. On Kreĭn–von Neumann and Friedrichs extensions. Acta Sci. Math. (Szeged) 69 (2003) 323–336.

[703] Z. Sebestyén and J. Stochel. Restrictions of positive selfadjoint operators. Acta Sci. Math. Szeged 55 (1991) 149–154.

[704] Yu.L. Shmuljan. A Hellinger operator integral (Russian). Mat. Sb. (N.S.) 49 (91) (1959) 381–430 [Amer. Math. Soc. Transl. (2) 22 (1962) 289–337].

[705] Yu.L. Shmuljan. Theory of linear relations, and spaces with indefinite metric (Russian). Funktsional. Anal. i Prilozhen 10 (1976) 67–72.

[706] Yu.L. Shmuljan. Transformers of linear relations in J-spaces (Russian). Funktsional. Anal. i Prilozhen 14 (1980) 39–44 [Functional Anal. Appl. 14 (1980) 110–113].

[707] Yu.G. Shondin. Quantum-mechanical models in $\mathbb{R}^n$ associated with extensions of the energy operator in Pontryagin space (Russian). Teor. Mat. Fiz. 74 (1988) 331–344 [Theor. Math. Phys. 74 (1988) 220–230].

[708] D.A. Shotwell. *Boundary problems for the differential equation $Lu = \lambda \sigma u$ and associated eigenfunction-expansions*. Doctoral dissertation, University of Colorado, 1965.

[709] B. Simon. Lower semicontinuity of positive quadratic forms. Proc. Roy. Soc. Edinburgh 79A (1977) 267–273.

[710] B. Simon. A canonical decomposition for quadratic forms with applications to monotone convergence theorems. J. Functional Analysis 28 (1978) 377–385.

[711] B. Simon. m-functions and the absolutely continuous spectrum of one-dimensional almost periodic Schrödinger operators. In: *Differential equations*. North-Holland Math. Studies, Vol. 92, North-Holland, 1984, pp. 519.

[712] B. Simon. Bounded eigenfunctions and absolutely continuous spectra for one-dimensional Schrödinger operators. Proc. Amer. Math. Soc. 124 (1996) 3361–3369.

[713] S. Simonov and H. Woracek. Spectral multiplicity of selfadjoint Schrödinger operators on star-graphs with standard interface conditions. Integral Equations Operator Theory 78 (2014) 523–575.

[714] C.F. Skau. Positive self-adjoint extensions of operators affiliated with a von Neumann algebra. Math. Scand. 44 (1979) 171–195.

[715] H.S.V. de Snoo and H. Winkler. Canonical systems of differential equations with selfadjoint interface conditions on graphs. Proc. Roy. Soc. Edinburgh 135A (2005) 297–315.

[716] H.S.V. de Snoo and H. Winkler. Two-dimensional trace-normed canonical systems of differential equations and selfadjoint interface conditions. Integral Equations Operator Theory 51 (2005) 73–108.

[717] H.S.V. de Snoo and H. Woracek. Sums, couplings, and completions of almost Pontryagin spaces. Lin. Alg. Appl. 437 (2012) 559–580.

[718] H.S.V. de Snoo and H. Woracek. Restriction and factorization for isometric and symmetric operators in almost Pontryagin spaces. Oper. Theory Adv. Appl. 252 (2016) 123–170.

[719] H.S.V. de Snoo and H. Woracek. Compressed resolvents, Q-functions and h_0-resolvents in almost Pontryagin spaces. Oper. Theory Adv. Appl. 263 (2018) 425–483.

[720] H.S.V. de Snoo and H. Woracek. The Krein formula in almost Pontryagin spaces. A proof via orthogonal coupling. Indag. Math. 29 (2018) 714–729.

[721] P. Sorjonen. On linear relations in an indefinite inner product space. Ann. Acad. Sci. Fenn. Ser. A I Math. 4 (1978/1979) 169–192.

[722] P. Sorjonen. Extensions of isometric and symmetric linear relations in a Krein space. Ann. Acad. Sci. Fenn. Ser. A I Math. 5 (1980) 355–375.

[723] J. Stochel and F.H. Szafraniec. *Unbounded operators and subnormality.* To appear.

[724] M.H. Stone. *Linear transformations in Hilbert space and their applications to analysis.* Amer. Math. Soc. Colloquium Publications, Vol. 15, 1932.

[725] O.G. Storozh. Extremal extensions of a nonnegative operator, and accretive boundary value problems (Russian). Ukr. Mat. Zh. 42 (1990) 858–860 [Ukrainian Math. J. 42 (1990) 758–760 (1991)].

[726] A.V. Štraus. Generalized resolvents of symmetric operators (Russian). Izv. Akad. Nauk. SSSR Ser. Mat. 18 (1954) 51–86 [Math. USSR-Izv. 4 (1970) 179–208].

[727] A.V. Štraus. On spectral functions of differential operators (Russian). Izv. Akad. Nauk SSSR Ser. Mat. 19 (1955) 201–220.

[728] A.V. Štraus. Characteristic functions of linear operators (Russian). Izv. Akad. Nauk SSSR Ser. Mat. 24 (1960) 43–74 [Amer. Math. Soc. Transl. (2) 40 (1964) 1–37].

[729] A.V. Štraus. On selfadjoint operators in the orthogonal sum of Hilbert spaces (Russian). Dokl. Akad. Nauk SSSR 144 (1962) 512–515.

[730] A.V. Štraus. On the extensions of symmetric operators depending on a parameter (Russian). Izv. Akad. Nauk SSSR Ser. Mat. 29 (1965) 1389–1416 [Amer. Math. Soc. Transl. (2) 61 (1967) 113–141].

[731] A.V. Štraus. On one-parameter families of extensions of a symmetric operator (Russian). Izv. Akad. Nauk SSSR Ser. Mat. 30 (1966) 1325–1352 [Amer. Math. Soc. Transl. (2) 90 (1970) 135–164].

[732] A.V. Štraus. Extensions and characteristic function of a symmetric operator (Russian). Izv. Akad. Nauk SSSR Ser. Mat. 32 (1968) 186–207 [Math. USSR-Izv. 2 (1968) 181–204].

[733] A.V. Štraus. Extensions and generalized resolvents of a symmetric operator which is not densely defined (Russian). Izv. Akad. Nauk SSSR Ser. Mat. 34 (1970) 175–202 [Math. USSR-Izv. 4 (1970) 179–208].

[734] A.V. Štraus. On extensions of semibounded operators (Russian). Dokl. Akad. Nauk SSSR 211 (1973) 543–546 [Soviet Math. Dokl. 14 (1973) 1075–1079].

[735] A.V. Štraus. On the theory of extremal extensions of a bounded positive operator. Funkts. Anal. 18 (1982) 115–126.

[736] A.V. Štraus. Generalized resolvents of nondensely defined bounded symmetric operators. Amer. Math. Soc. Transl. (2) 154 (1992) 189–195.

[737] F.H. Szafraniec. The reproducing property: spaces and operators. To appear.

[738] G. Teschl. Mathematical methods in quantum mechanics with applications to Schrödinger operators. Graduate studies in mathematics, Vol. 99, Amer. Math. Soc., 2009.

[739] E.C. Titchmarsh. On expansions in eigenfunctions IV. Quart. J. Math. Oxford Ser. 12 (1941) 33–50.

[740] E.C. Titchmarsh. Eigenfunction expansions associated with second order differential equations I. 2nd ed. Oxford University Press, Oxford, 1962.

[741] E.C. Titchmarsh. Eigenfunction expansions associated with second order differential equations II. Oxford University Press, Oxford, 1958.

[742] C. Tretter. Boundary eigenvalue problems for differential equations $N\eta = \lambda P\eta$ and λ-polynomial boundary conditions. J. Differential Equations 170 (2001) 408–471.

[743] C. Tretter. Spectral theory of block operator matrices and applications. Imperial College Press, 2008.

[744] H. Triebel. Interpolation theory, function spaces, differential operators. 2nd revised and enlarged edition. Johann Ambrosius Barth Verlag, 1995.

[745] H. Triebel. Höhere Analysis. 2nd edition. Harri Deutsch, 1980.

[746] E.R. Tsekanovskiĭ and Yu.L. Shmuljan. The theory of biextensions of operators in rigged Hilbert spaces. Unbounded operator colligations and characteristic functions. Russian Math. Surveys 32 (1977) 73–131.

[747] M.I. Vishik. On general boundary conditions for elliptic differential equations. Trudy Moskov. Mat. Obsc. 1 (1952) 187–246 [Amer. Math. Soc. Transl. Ser. (2) 24 (1963) 107–172].

[748] R. Vonhoff. Spektraltheoretische Untersuchung linksdefiniter Differentialgleichungen im singulären Fall. Doctoral dissertation, Universität Dormund, 1995.

[749] J. Walter. Regular eigenvalue problems with eigenvalue parameter in the boundary condition. Math. Z. 133 (1973) 301–312.

[750] M. Waurick and S.A. Wegner. Some remarks on the notions of boundary systems and boundary triple(t)s. Math. Nachr. 291 (2018) 2489–2497.

[751] S.A. Wegner. Boundary triplets for skew-symmetric operators and the generation of strongly continuous semigroups. Anal. Math. 43 (2017) 657–686.

[752] J. Weidmann. Linear operators in Hilbert spaces. Graduate Texts in Mathematics, Vol. 68, Springer, 1980.

[753] J. Weidmann. Stetige Abhängigkeit der Eigenwerte und Eigenfunktionen elliptischer Differentialoperatoren vom Gebiet. Math. Scand. 54 (1984) 51–69.

[754] J. Weidmann. *Spectral theory of ordinary differential operators.* Lecture Notes in Mathematics, Vol. 1258, Springer, 1987.

[755] J. Weidmann. Strong operator convergence and spectral theory of ordinary differential operators. Universitatis Iagellonicae Acta Math. 34 (1997) 153–163.

[756] J. Weidmann. *Lineare Operatoren im Hilberträumen. Teil I: Grundlagen.* B.G. Teubner, 2000.

[757] J. Weidmann. *Lineare Operatoren im Hilberträumen. Teil II: Anwendungen.* B.G. Teubner, 2003.

[758] H. Weyl. Über gewöhnliche lineare Differentialgleichungen mit singulären Stellen und ihre Eigenfunktionen. Göttinger Nachr. (1909) 37–64.

[759] H. Weyl. Über gewöhnliche Differentialgleichungen mit Singularitäten und die zugehörigen Entwicklungen willkürlicher Funktionen. Math. Ann. 68 (1910) 220–269.

[760] H. Weyl. Über gewöhnliche lineare Differentialgleichungen mit singulären Stellen und ihre Eigenfunktionen. Göttinger Nachr. (1910) 442–467.

[761] H. Weyl. Ramifications, old and new, of the eigenvalue problem. Bull. Amer. Math. Soc. 56 (1950) 115–139.

[762] W.M. Whyburn. Differential equations with general boundary conditions. Bull. Amer. Math. Soc. 48 (1942) 692–704.

[763] D.V. Widder. *The Laplace transform.* Princeton Mathematical Series, Vol. 6, Princeton University Press, 1941.

[764] H.L. Wietsma. *On unitary relations between Kreĭn spaces.* Doctoral dissertation, University of Vaasa, 2012.

[765] H.L. Wietsma. Representations of unitary relations between Kreĭn spaces. Integral Equations Operator Theory 72 (2012) 309–344.

[766] H. Winkler. The inverse spectral problem for canonical systems. Integral Equations Operator Theory 22 (1995) 360–374.

[767] H. Winkler. On transformations of canonical systems. Oper. Theory Adv. Appl. 80 (1995) 276–288.

[768] H. Winkler. Canonical systems with a semibounded spectrum. Oper. Theory Adv. Appl. 106 (1998) 397–417.

[769] H. Winkler. Spectral estimations for canonical systems. Math. Nachr. 220 (2000) 115–141.

[770] H. Winkler. On generalized Friedrichs and Krein–von Neumann extensions and canonical systems. Math. Nachr. 236 (2002) 175–191.

[771] H. Winkler and H. Woracek. On semibounded canonical systems. Lin. Alg. Appl. 429 (2008) 1082–1092.

[772] H. Winkler and H. Woracek. Reparametrizations of non trace-normed Hamiltonians. Oper. Theory Adv. Appl. 221 (2012) 667–690.

[773] A. Wintner. *Spektraltheorie unendlicher Matrizen.* Hirzel, 1929.

[774] J. Wloka. *Partial differential equations.* Cambridge University Press, 1987.

[775] H. Woracek. *Spektraltheorie im Pontryaginraum*. Lecture Notes, TU Wien.

[776] H. Woracek. *Operatortheorie im Krein Raum 1+2*. Lecture Notes, TU Wien.

[777] H. Woracek. Reproducing kernel almost Pontryagin spaces. Lin. Alg. Appl. 461 (2014) 271–317.

[778] A. Yang. *A construction of Krein spaces of analytic functions*. Doctoral dissertation, Purdue University, 1990.

[779] S. Yao, J. Sun, and A. Zettl. The Sturm–Liouville Friedrichs extension. Appl. Math. 60 (2015) 299–320.

[780] K. Yosida. On Titchmarsh–Kodaira's formula concerning Weyl–Stone's eigenfunction expansion. Nagoya Math. J. 1 (1950) 49–58.

[781] K. Yosida. Correction. Nagoya Math. J. 6 (1953) 187–188.

[782] A. Zettl. *Sturm–Liouville theory*. Mathematical Surveys and Monographs, Vol. 121, Amer. Math. Soc., 2005.

[783] W.P. Ziemer. *Weakly differentiable functions*. Graduate Texts in Mathematics, Vol. 120, Springer, 1989.

[784] H.J. Zimmerberg. Linear integro-differential-boundary-parameter problems. Ann. Mat. Pura Appl. 105 (1975) 241–256.

Permissions

The contributors of this book come from diverse backgrounds, making this book a truly international effort. We would like to thank all the contributing authors for lending their expertise to make the book truly unique. They have played a crucial role in the development of this book. Without their invaluable contributions this book wouldn't have been possible. They have made vital efforts to compile up to date information on the varied aspects of this subject to make this book a valuable addition to the collection of many professionals and students.

This book was conceptualized with the vision of imparting up-to-date and integrated information in this field. To ensure the same, a matchless editorial board was set up. Every individual on the board went through rigorous rounds of assessment to prove their worth. After which they invested a large part of their time researching and compiling the most relevant data for our readers.

The editorial board has been involved in producing this book since its inception. They have spent rigorous hours researching and exploring the diverse topics which have resulted in the successful publishing of this book. They have passed on their knowledge of decades through this book. To expedite this challenging task, the publisher supported the team at every step. A small team of assistant editors was also appointed to further simplify the editing procedure and attain best results for the readers.

Apart from the editorial board, the designing team has also invested a significant amount of their time in understanding the subject and creating the most relevant covers. They scrutinized every image to scout for the most suitable representation of the subject and create an appropriate cover for the book.

The publishing team has been an ardent support to the editorial, designing and production team. Their endless efforts to recruit the best for this project, has resulted in the accomplishment of this book. They are a veteran in the field of academics and their pool of knowledge is as vast as their experience in printing. Their expertise and guidance has proved useful at every step. Their uncompromising quality standards have made this book an exceptional effort. Their encouragement from time to time has been an inspiration for everyone.

The publisher and the editorial board hope that this book will prove to be a valuable piece of knowledge for students, practitioners and scholars across the globe.

Index

Printed in the USA
CPSIA information can be obtained
at www.ICGtesting.com
JSHW011447221024
72173JS00004B/984